MW00454414

Engineering Design with Polymers and Composites

Second Edition

Engineering Design with Polymers and Composites

Second Edition

James C. Gerdeen, PhD, PE
Ronald A. L. Rorrer, PhD, PE

CRC Press is an imprint of the
Taylor & Francis Group, an **informa** business

CRC Press
Taylor & Francis Group
6000 Broken Sound Parkway NW, Suite 300
Boca Raton, FL 33487-2742

Printed in the United States of America on acid-free paper
Version Date: 2011901

International Standard Book Number: 978-1-4398-6052-6 (Hardback)

Library of Congress Cataloging-in-Publication Data

Gerdeen, James C.
Engineering design with polymers and composites. -- 2nd ed. / James C. Gerdeen and Ronald A.L. Rorrer.
p. cm.
Includes bibliographical references and index.
ISBN 978-1-4398-6052-6 (alk. paper)
1. Polymers. 2. Composite materials. 3. Engineering design. I. Rorrer, Ronald A. L. II. Title.

TA455.P58G47 2012
620.1'923--dc23 2011031501

Visit the Taylor & Francis Web site at
http://www.taylorandfrancis.com

and the CRC Press Web site at
http://www.crcpress.com

Contents

Preface xiii
Acknowledgment xv
Authors xvii

Chapter 1 Introduction 1

1.1 Introduction 1
1.2 History of Polymers 2
1.3 History of Composites 4
1.4 Examples of Polymers and Composites in Use 5
1.5 Definitions and Classifications 7
1.6 Identification of Plastics 9
1.7 Raw Materials and Production of Polymers 11
1.8 Chemical Structures 14
1.9 Glass Transition and Melting Temperatures 21
Homework Problems 25
References 26

Chapter 2 Mechanical Properties of Polymers 27

2.1 Introduction 27
2.2 Tensile Properties 27
2.2.1 Elongation 29
2.2.2 Elastic Modulus 30
2.2.3 Ultimate Tensile Strength 30
2.2.4 Yield Strength 30
2.3 Static Failure Theories 33
2.4 Creep Properties 38
2.5 Relaxation Properties 40
2.6 Dynamic Properties 40
2.6.1 Dynamic Tests 40
2.6.2 Dynamic Modulus and Damping 42
2.6.3 Dynamic Property Data 44
2.7 Large Strain Definitions 47
2.8 Analysis of Damping 47
2.9 Time Hardening Creep 50
2.10 Isochronous Creep Curves 51
Homework Problems 51
References 54

Chapter 3 Viscoelastic Behavior of Polymers ... 55

3.1 Mechanical Models ... 55
3.2 Mathematical Models ... 57
3.3 The Maxwell Fluid ... 57
3.4 The Kelvin Solid ... 60
3.5 The Four-Parameter Model ... 63
3.6 The Boltzmann Superposition Principle ... 64
3.7 Advanced Viscoelastic Models ... 67
3.8 The Viscoelastic Correspondence Principle ... 69
3.9 The Time–Temperature Equivalence Principle ... 71
Homework Problems ... 74
References ... 79

Chapter 4 Creep and Fatigue Failure ... 81

4.1 Creep Failure under Tension ... 81
4.2 Creep Failure under Compression ... 83
4.3 Fatigue of Polymers ... 85
4.4 Notch Sensitivity under Fatigue ... 90
4.5 Creep Buckling of Shells ... 91
Homework Problems ... 92
References ... 93

Chapter 5 Impact Strength and Fracture Toughness ... 95

5.1 Impact Strength ... 95
5.1.1 Thickness Effects ... 100
5.1.2 Rate Effects ... 101
5.1.3 Combined Stiffness and Impact Properties ... 102
5.2 Fracture Toughness ... 105
5.2.1 Brittle Fracture ... 106
5.2.2 Ductile Fracture ... 109
5.2.3 General Theory of Fracture Instability ... 110
5.3 Analysis of the Charpy and Izod Impact Tests Using Fracture Mechanics ... 115
5.4 Analysis of Impact Specimens at the Nanoscale ... 116
Homework Problems ... 117
References ... 119

Chapter 6 Selection of Polymers for Design Applications ... 121

6.1 Introduction ... 121
6.2 Basic Material Properties ... 121
6.3 Performance Parameters ... 122
6.4 Loading Conditions and Geometrical Configurations ... 122

6.5 Availability of Materials ... 122
6.6 A Rectangular Beam in Bending ... 123
6.7 Weighting-Factor Analysis ... 125
6.8 Thermal Gradient through a Beam ... 126
6.9 Rating Factors for Various Loading Requirements ... 128
6.10 Design Optimization ... 129
6.10.1 Graphical Solution ... 130
6.10.2 Computer Solution ... 135
6.10.3 Microsoft Excel Solver Routine ... 137
6.11 Computer Database Design Selection Procedure ... 140
6.11.1 Example Problem of Impact of a Beam ... 142
Homework Problems ... 142
References ... 143

Chapter 7 Design Applications of Some Polymers ... 145

7.1 Phenolic Resins with Fillers ... 145
7.2 Polycarbonate ... 147
7.3 Example Design with PC: Fan Impeller Blade ... 147
7.3.1 Creep Strain ... 149
7.3.2 Impact Failure ... 150
7.4 Example Design with PC: Snap/Fit Design ... 151
7.5 Example Design of PVC Pipe ... 152
7.6 Design with Fluorocarbon Resins ... 156
Homework Problems ... 158
References ... 159

Chapter 8 Composite Material Mechanics ... 161

8.1 Introduction ... 161
8.2 Composite Material Nomenclature and Definitions ... 161
8.3 Analysis of Composite Structures ... 165
8.3.1 Micromechanics of a Unidirectional Fiber-Reinforced Composite Layer (Lamina) ... 167
8.3.1.1 Determination of Apparent Longitudinal Young's Modulus ... 168
8.3.1.2 Determination of Major Poisson's Ratio of Unidirectional Lamina ... 171
8.3.1.3 Apparent Transverse Young's Modulus ... 172
8.3.1.4 Apparent Shear Modulus ... 174
8.3.1.5 Summary of Results from Micromechanics Analysis of Lamina Elastic Moduli ... 176
8.3.1.6 Prediction of Tensile Strength in Fiber Direction ... 177

8.3.2 Macromechanics of a Unidirectional Fiber-Reinforced Composite Layer of Lamina 178
8.3.2.1 Stress–Strain Relationships for Isotropic Materials 179
8.3.2.2 Anisotropic Materials: Contracted Notation 180
8.3.2.3 Orthotropic Lamina: Hooke's Law in Principal Material Coordinates 181
8.3.2.4 Stress (Strain) Relationships for Off-Axis Orientation 184
8.4 Experimental Determination of Engineering Elastic Constants 190
Homework Problems 194
Bibliography 195
References 195

Chapter 9 Composite Laminate Failure 197

9.1 Strength Properties and Failure Theories 197
9.1.1 A Review of Failure Theories for Isotropic Materials 197
9.1.2 Strength and Failure Theories for an Orthotropic Lamina 199
9.1.3 Failure by Fiber Pullout 205
9.2 Stiffness of Laminated Composites 206
9.2.1 Sandwich Beam 206
9.2.2 Orthotropic Plate 208
9.2.3 Laminated Plates 211
9.3 Thermal Stresses 216
9.4 Summary 216
Homework Problems 217
Bibliography 219
References 219

Chapter 10 Polymer Processing 221

10.1 Extrusion 221
10.2 Manufacture of PVC Pipe by Extrusion 223
10.3 Injection Molding 226
10.4 Thermoforming 228
10.5 Blow Molding 229
10.5.1 Inflation 232
10.5.2 Cooling Phase 236
Homework Problems 241
References 242

Chapter 11 Adhesion of Polymers and Composites 243

11.1 Introduction 243
11.2 Fundamentals of Adhesion 243
11.2.1 Wetting and Work of Adhesion 243
11.2.2 Measurement of Adhesion 246
11.2.3 Viscoelasticity of Adhesion 249
11.3 Adhesives 250
11.3.1 Common Polymeric Adhesives 250
11.3.2 Polymers as Matrix Materials (*In Situ* Adhesives) in Polymeric Composites 252
11.4 Enhancement of Adhesion in Composites 253
11.5 Curing of Adhesives 254
11.6 Summary 257
Homework Problems 257
References 258

Chapter 12 Polymer Fusing and Other Assembly Techniques 259

12.1 Introduction 259
12.2 Heated Tool Welding 260
12.3 Ultrasonic Welding 261
12.3.1 Joint Design 262
12.3.2 Staking 266
12.4 Friction Welding 267
12.4.1 Linear Vibration and Orbital Welding 267
12.4.2 Spin Welding 269
12.5 Laser Welding 269
12.6 Hot Gas 272
12.7 Resistance Welding 273
12.8 Induction Welding 274
12.9 Mechanical Fastener Connections 276
12.9.1 Screws 276
12.9.2 Inserts 278
Homework Problems 279
References 279

Chapter 13 Tribology of Polymers and Composites 281

13.1 Introduction 281
13.2 Contact Mechanics 282
13.3 Surface Topography 284
13.4 Friction 285
13.4.1 Static and Dynamic Coefficients of Friction 288
13.4.2 Adhesive and Abrasive Friction 289
13.5 Wear 289
13.5.1 Archard Wear Law 290

13.6 PV Limit 291
13.7 Rolling and Sliding 292
13.8 Modification of Polymers for Friction and Wear Performance 293
13.8.1 Internal Lubricants 293
13.8.2 Reinforcements 295
13.9 Composites 295
13.10 Wear of Composites 299
13.11 Heat Generation in Sliding Polymer Systems 300
13.11.1 Bulk Surface–Temperature Calculations 301
13.11.2 Flash Temperature 301
13.12 Special Considerations 302
13.12.1 Polymer-on-Polymer Sliding 302
13.12.2 Coatings 302
13.12.3 Effect of Surface Topography on Friction and Wear 302
13.12.4 Effect of Environment (Temperature, Humidity, Gases, and Liquids, etc.) on Friction and Wear 302
13.12.5 Friction-Induced Vibration 303
13.13 Simulative Laboratory Testing 304
Homework Problems 305
References 305

Chapter 14 Damping and Isolation with Polymers and Composites 307

14.1 Introduction 307
14.2 Relevance of the Thermomechanical Spectrum 308
14.3 Damping Methods of Material Modification (Cross-linking, *M*w, Structure) Polymers, and Composites Used in Damping and Isolation 309
14.3.1 Reduced Frequency Nomograph 310
14.4 Materials for Damping and Isolation 311
14.5 Fundamentals of Vibration Damping and Isolation 313
14.5.1 Dynamics of Vibrating Structures (Continuous and Discrete or Point) 313
14.6 Role of Dampers 319
14.7 Damping Layers 320
14.7.1 Application of Dampers and Isolators: Discrete Design of Dampers and Isolators for Equipment 320
Homework Problems 323
References 323

Chapter 15 Rapid Prototyping with Polymers 325

15.1 Introduction 325
15.2 Rapid Product Development, Tooling, and Manufacture 325

15.3 RP Techniques ... 326
15.4 RP Materials ... 332
15.4.1 Materials Used in FDM by Stratasys ... 332
15.4.2 Materials Used in SLA ... 334
15.4.3 Materials Used in LOM ... 334
15.4.4 Materials Used in SLS ... 334
15.5 Applications ... 334
Homework Problems ... 339
References ... 340

Chapter 16 Piezoelectric Polymers ... 341

16.1 Introduction ... 341
16.2 Piezoelectric Strain Behavior ... 342
16.3 Piezoelectric Material Properties ... 345
16.4 Hysteresis ... 346
16.5 Composites ... 348
Homework Problems ... 356
Further Reading ... 357
References ... 357

Appendix A: Conversion Factors ... 359

Appendix B: Area Moments of Inertia ... 361

Appendix C: Beam Reactions and Displacements ... 363

Appendix D: Laminate MATLAB® or Octave Code ... 367

Appendix E: Sample Input/Output for Laminate Program ... 377

Appendix F: Composite Materials Properties ... 383

Appendix G: Thermal and Electrical Properties ... 385

Index ... 389

Preface

As stated in the first edition in 2006, this book is written by engineers for the education of engineers. One can find many books on the chemistry of polymers, on the processing of polymers, and on the properties of polymers. However, there has been no adequate textbook, to our knowledge, on the analysis and design of mechanical components. That is the reason this book has been written. The first chapter is a history of polymers and composites. It is always interesting to place a technical topic in its historical context. This is useful for two reasons. An in-depth understanding of a topic is always enhanced by the historical understanding. Additionally, the opportunity to bore and annoy your colleagues cannot be overappreciated! The history of development makes the topic more interesting and less dry. Modern concepts in design such as weight-to-strength ratio and cost-to-strength ratio are used in selecting polymers and composites for design applications. Computer methods for selection of polymer materials from a database, for optimal design, and for laminated plate design are introduced.

This book is targeted toward two audiences. First and foremost, the intended audience is mechanical and civil engineering students. Second, it is also intended to be useful to the practicing engineer. However, unlike a reference handbook that contains a plethora of data, the purpose of this book is to provide a fundamental understanding of the topics for the reader to be able to understand phenomena that they encounter in applications or through subsequent study and research. As a text this book is intended for use in a one quarter or semester course at the senior level. It has been used for first-year graduate credit by assigning advanced topics. As a prerequisite, the student should have had a course in the mechanics of materials.

In this second edition, previous chapters have been rearranged and material has been added. The previous Chapter 4 on composites has been divided into two, Chapters 8 and 9. New chapters include Chapter 12 on polymer fusing and other assembly techniques, Chapter 15 on rapid prototyping, and Chapter 16 on piezoelectric polymers.

Acknowledgment

This book is dedicated to our wives Wanda Gerdeen and Velvet Rorrer, respectively (and respectfully), who have sacrificed their quality time to allow us to finish the work on this new edition.

Authors

James C. Gerdeen, PhD, PE, is professor emeritus, UCDHSC, University of Colorado at Denver and Health Science Center, distinguished professor, Michigan Technological University.

Professor Gerdeen has a BSME, 1959, from Michigan Technological University (MTU), an MS in 1962 from The Ohio State University, a PhD in 1965 from Stanford University, and a Master of Divinity degree. From 1959 to 1968, he was with Battelle Columbus Laboratories as a senior research engineer. He was director of manufacturing programs at both MTU and UCDHSC. He also has been an adjunct faculty at MAE, Missouri Science & Technology University, Rolla, Missouri. From 2005 to 2011, he has been teaching internet graduate classes in project management for Missouri State University, Springfield, Missouri.

Dr. Gerdeen is known internationally for his research work in pressure vessel design, structural analysis, metal working manufacturing, and mechanical design. He received the MTU Faculty Research Award in 1974, and the SME (Society of Manufacturing and Engineering) Educator of the Year Award for the Western Region in 1998. He has published over 65 papers, and more than 100 research reports.

Professor Gerdeen began teaching a new course on mechanical design with polymers and composites at MTU in 1982, and then compiled his own notes on the subject. These notes formed the basis for many of the chapters in the present text on the subject.

Ronald A. L. Rorrer, PhD, PE received his BS in mechanical engineering in 1984 from Virginia Polytechnic Institute and State University (VPI & SU, known as Virginia Tech), his MS in mechanical engineering from VPI & SU and his PhD in 1991 in mechanical engineering from VPI & SU. From 1986 until 1987 he was a precision mechanical design engineer at Martin Marietta in Orland, Florida. He worked as an advanced technology project leader at The Gates Rubber Company in Denver, Colorada from 1993 to 1997, first in the Adhesives and Lubricants Group, then in the Advanced Materials Research Group.

From 1994 to 1997, Dr. Rorrer was an adjunct professor in both the Division of Engineering at the Colorado School of Mines and the Department of Mechanical Engineering at the University of Colorado at Denver. In 1997, he became an assistant professor of mechanical engineering. In 2004 he was promoted to associate professor of the newly merged University of Colorado at Denver and Health Sciences Center.

Dr. Rorrer's research is in the fields of tribology, polymers, composites, and bioengineering. He has published over 30 papers and holds one patent.

1 Introduction

1.1 INTRODUCTION

The growth in the use of polymers and composites mirrored the incredible changes that occurred in the twentieth century and in this growth also contributed to those changes. For example, many polymers and composites were developed during the space race.

The more technical details of the introduction to polymers will occur later in this chapter. Here we introduce polymers in more general terms. What is a polymer? Historically, people have referred to polymers as plastics. Well, what is plastic? Technically plastic refers to the state when a material is deformed plastically and will not return to its original undeformed state. Historically, the polymers that most people are familiar with easily go into the plastic state. In general, the term plastic (from the Greek plastikos, meaning moldable) is more of a layperson term and the term polymer (from the Greek poly meros, meaning many unit) is used more in technical discussions. Thus, we will predominately use the term polymer in the text. Polymers are either hydrocarbon based (comprised largely of hydrogen and carbon atoms) or silicone based (comprised of silicone and other atoms). The polymers that we are most familiar with are biopolymers which are made naturally. The two most common examples are our skin and wood. Silicone-based polymers that we are familiar with are the silicone sealants used at home.

A plate of spaghetti noodles is the simplest analogy that can be made in relation to a polymeric material. Polymer chains are like long spaghetti noodles. Their properties are determined by the atoms that create the chain and the side chains that branch off the main chain. These chains are often very thin 1–3 atoms across their width or effective diameter and 100–10,000 of atoms long. Of course the spacing of the atoms is determined by their bonding and all of that other stuff you have forgotten from your chemistry classes. There are two types of polymers that we will discuss in much greater detail in the rest of the text. The first is thermoplastic, which means that the polymer chains are intertwined with each other and can eventually be pulled apart. Additionally, because of the noncovalent bonding between the chains these materials will melt. An example of a thermoplastic is nylon. Thermosets on the other hand are polymers that have covalently bonded chains (cross-linking) between the longer chains. These materials can carry loads indefinitely that do not rupture the cross-links. Due to the covalent bonding of the cross-links thermosets will degrade, not melt at high temperature. Examples of thermosets are virtually all rubbers or elastomers (some exceptions are some polyurethanes and obviously thermoplastic elastomers) and most epoxies.

Technical people have speculated for years that a national interest in oil should be based on the manufacture of polymers, not as fuel. This was stated during the 2003 U.S./Iraq war.

Design engineers have a responsibility to society to create designs that are not only functional and cost effective, but most importantly safe. Unlike an industrial product designer, who may just be concerned with kinematic function, ergonomics, or aesthetics, a design engineer must be ultimately concerned with whether or not the product will fail due to stress overload or excessive displacement. When working with the historically common engineering materials such as steel, aluminum, or titanium components typically fail due to static or fatigue stress failure. Polymers and composites have a myriad of failures that are rarely seen in the application of metals. For example, a static load on a polymer can eventually fail the polymer either due to stress or excessive deflection over time. In addition, polymers are very sensitive to solvents and chemicals. Granted stress corrosion cracking was evident in brass shells stored in barns during the U.S. Civil War, this is not a typical failure problem for most metals applications. Polymers on the other hand can evidence either inferior or superior chemical resistance relative to mechanical properties and thus performance. Creep which is the continual deformation under stress is a major design concern for polymers. In mechanical design, metals do not evidence significant creep, except under high temperature or extremely long time. One example is the leaf springs from cars manufactured in the 1950s and 1960s. The suspension sag of cars of this era and before is due to creep of the metal in the leaf springs.

There are many ways to learn about a subject. One can just study the technical fundamentals as is typically done by engineers and have a competent working knowledge and capability. However, the more depth that is, obtained in a topic leads to a better understanding. One of the ways to better understanding is to know the history.

1.2 HISTORY OF POLYMERS

If we exclude most of the biopolymers such as wood, cotton, silk, and others, one of the first polymers in use was natural rubber (NR). The Indians of South America were playing with rubber balls made from latex when the Spanish arrived in the Americas in the 1400s. The first description of the rubber ball dates back to 1496.

Polymers have replaced metals in components for a myriad of reasons. One of the major issues is cost. The lower cost of polymer parts is due to two major considerations. First, the cost of polymers per unit weight is cheaper than metals. Coupled with this is the typically net shape manufacture of polymer parts. While some polymeric materials are machined from solid material, most polymer components are made by processes such as injection molding. Often there is no additional machining, beyond removal of flash or molding sprues, to add to the cost of a polymer component. This is in contrast to metal components, which often require significant machining time and costs. Machining of a metallic component can often be the dominant cost of the part.

The development of most polymers have occurred at large chemical companies. One of the leaders in polymer development has been DuPont. In the United States many polymers are known generally by their DuPont trade names. Perhaps the best example of this is polytetrafluoroethylene (PTFE) known to the public as Teflon. A historical timeline of polymer development is shown in Table 1.1. The Nobel Prizes related to polymers are shown in Table 1.2.[1]

TABLE 1.1
Polymer Development

Pre-1400s	South-American Indians create rubber balls from NR
1496	Christopher Columbus brings balls back to Europe
1823	Charles MacIntosh coats cloth with rubber and benzene creating the waterproof MacIntosh
1839	Charles Goodyear vulcanizes NR with sulfur
1845	Scotsman R. W. Thompson applies for pneumatic tire patent
1870	Cellulose nitrate (CN), USA
1880s	Hilaire de Chardonnet creates rayon (artificial silk) from cellulose
1888	John Dunlop invents pneumatic tires for bicycle
1898	German chemist Hans von Pechmann accidently creates polyethylene
1905	Cellulose acetate (CA), Germany
1905	Bisphenol A precursor to polycarbonate and epoxy reported by Thomas Zwicke of Germany
1907	Chemist Leo Hendrick Baekeland creates bakelite from phenol and formaldehyde; this is the first truly synthetic polymer not found in nature. Material is now known as phenol formaldehyde or phenolic
1930	April, Dr. Wallace Carothers' team discovered neoprene (synthetic rubber) and polyester pre-cursor to nylon
	Bakelite invented
1933	Eric Fawcett and Reginald Gibson accidently create an industrial viable process for polyethylene
1933	PVC patented by Fritz Klatte
1934	Polymethyl methalcrylate or acrylic (PMMA)
1937	Otto Bayer and colleagues at Bayer discover basic polyurethane chemistry
1938	Heinrich Rinke creates first polyurethane polymer
1938	Dr. Roy J. Plunkett finds that frozen sample of PTFE has polymerized; DuPont will later trademark their brand of this material as Teflon
1938	Polyamide or nylon becomes the first fiber made wholly out of chemicals
1939	Low density polyethylene (LDPE)
1940s	Oil shortage of World War II leads to polymers based upon plants such as soy
	Silicone
1946	Polyester
1947	Epoxy
1951	J. Paul Hogan and Robert L. Banks discover polypropylene while attempting to convert ethylene and propylene into gasoline; in addition, they developed process for HDPE
1953	Polycarbonate discovered simultaneously by Dr. H. Schnell and D. W. Fox
1964	Stephanie Kwolek at the Dupont synthesizes Kevlar (generic name aramid)
1966	Kevlar patented
1971	Kevlar marketed
	Spandex invented
1983	Worldwide polymer passes steel consumption based on volume

continued

TABLE 1.1 (continued)
Polymer Development

1990s	Materials that change color with heat
	Sheet molding compound for body panels
~2000	Worldwide polymer passes steel consumption based on weight
2007	Brazilian company, Braskem produces first bio-based polyethylene derived from sugar cane

TABLE 1.2
Nobel Prizes Related to Polymers

1953	Hermann Staudinger of Freiburg receives Nobel Prize in Chemistry for his work on macromolecules
1963	Karl Ziegler of Mulheim and Giullo Natta of Milan share prize for catalysts used in polymerization
1974	Paul J. Flory of Stanford receives prize for theoretical and experimental investigations into the physical chemistry of macromolecules
1991	Physics Nobel Prize went to Pierre-Gilles de Gennes for applying order phenomena of simple systems to more complex systems such as liquid crystals and polymers
2000	Alan J. Heeger of Santa Barbara, Alan G. MacDiarmid of Pennsylvania, and Hideki Shirakawa of Tsukuba share the Nobel Prize in Chemistry for the discovery and development of conductive polymers

Source: Data from http://www.nobelprize.org

1.3 HISTORY OF COMPOSITES

Many individuals in the twentieth century would have associated composites primarily with aerospace applications such as planes or space craft. From the 1970s onward, this view changed as composites began to be increasingly utilized in sporting equipment, such as skis and tennis racquets.

Composites have historically not had the same advantage in material or manufacturing costs in the same way polymers have had over metals. In fact, prior to the 1990s, many of the composite materials that now exist in many people's homes were far costlier per unit weight than metals. In addition, many composite components required hand layup. With the exception of fiberglass, most fiber-reinforced composites were reserved for plane and aerospace applications as well as sports equipment, where the savings in weight and increase in performance over metals justified the exorbitant cost. However, as the price of composites has dropped and advanced computer-controlled manufacturing techniques have become more readily available, composite components are becoming increasingly more common.

The demarcation between polymers and composites is often a subject of debate. As a general rule, we will define composites to be materials that have distinct

TABLE 1.3
Composite Materials Development Timeline

Prehistory	Straw reinforced mud bricks used in construction of the pyramids
	Reinforced concrete
1880	Reinforced tires
1910	Reinforced NR belts
1940	First bamboo-reinforced concrete
	Fiberglass
1942	Fiberglass/polyester composite parts for airplanes produced by Owens-Corning
1950	Polymer dashboards
	Corvette with fiberglass body
1958	Boron fiber developed
1960	Boats hulls
1963	Carbon fiber developed
1970	Carbon fiber/epoxy
	Airplane wings
	Bicycle body tubes
1976	Composite tennis racquet patented
1980	Complete bicycle
1990	Complete fighter airplane autoclaved at Boeing

components that provide distinct functions. This definition is most obvious with a fiber-reinforced composite such as fiberglass, where the fiber provides the stiffness and strength, and the polyester matrix provides the transfer of stress and consolidation of the composite. The definition becomes murky is when there are reinforcements such as short fibers or particulates. In general, one does not tend to think of polymers that contains small-sized reinforcement as composite materials. In the strictest sense these materials are composites but they can be considered not to be, especially if the materials are virtually isotropic (i.e., same material properties regardless of direction). A timeline of development of composite materials is shown in Table 1.3.

1.4 EXAMPLES OF POLYMERS AND COMPOSITES IN USE

Perhaps the most notable examples of polymers and composites are the ones that we experience in our everyday lives, such as beverage containers, toys, dishes, and so on. However, let us examine some of the categories of polymers and composites. First beyond household goods are sporting goods. Tennis racquets and skis were some of the first consumer goods made from composites. The reason for this was that even though in the 1970s composite materials were costly and hard to manufacture, racquets and skis do not require a tremendous amount of material and people were willing to pay a premium for sporting goods. Examples of sporting goods are the cedar stripper canoe that is, covered with fiberglass and epoxy that was built by one of our

FIGURE 1.1 Fiberglass/epoxy-covered cedar stripper canoe.

alumni, Paul Rice, shown in Figure 1.1 and the polycarbonate kayak by Clear Blue Hawaii shown in Figure 1.2.

The largest static composite structures are bridges. Most bridge composites are decking materials comprised of fiberglass. Relative to structures that move are the B-2 stealth bomber which is made of graphite/epoxy. The B-2 is one of the most

FIGURE 1.2 Clear polycarbonate kayak.

costly composite structures. This material is chosen for its strength-to-weight ratio as well as its radar absorbing properties.

Polymers have found application in medical devices due to a combination of mechanical properties, ability to withstand the environment of the body, and also inertness relative to the body. Polymers and composites have been used as heart stents and valves, artificial joints, replacement vertebrae, bones, and so on.

Automotive applications include body panels, belts, tires, hoses, composite drive shafts. One of the authors knows local race car drivers in Denver, who will thermoform plastic door panels for their race cars with an ordinary shop vacuum. Recently, there has been a carbon fiber composite automobile frame proposed by Colorado Company. Aerospace applications are airplane wings, filament-wound rocket bodies. There are also homebuilt experimental aircraft which are rolled into the sun to cure the epoxy matrix. One of the largest composite structures in space is the composite arm made by Canada for the Space Station.

Other civil engineering examples of polymers and composites are geotechnical fabric for soils, polymer natural gas pipe, fiberglass reinforcing-bar (re-bar), composite I-beams, as well as the aforementioned decking for bridges.

Virtually everything that one could think of as made out of plastics is made out of plastics. To a metals person, there are metals and nonmetals; now, to a polymers person, there are polymers and nonpolymers!

One thing that many people do not realize is that there is an incredible amount of creating components from polymers and composites that can be performed in your own garage. You can cast components from polyurethane, you can make body panels for your car, or even lay up your own experimental aircraft. There are even heated injection molds that utilize a drill press as the injector.

In the past, the choice of engineering materials was limited to materials such as wood, steel, concrete, and other metals. Today an engineering design is not constrained by only the materials available, but the engineer can now design a material to meet certain specifications by blending different copolymers and creating special composites.

This text is intended primarily for design engineers and engineering students. However, the first chapter contains some chemistry, but only what is necessary to acquaint the student with the background necessary to understand the significant influence of chemical structure on mechanical properties of polymers.

1.5 DEFINITIONS AND CLASSIFICATIONS

The term "polymer" means "many mers." A "mer" is a monomer or one molecule such as the ethylene CH_2 molecule usually shown as

```
     H
     |
   — C —          CH2
     |
     H
```

A polymer is a chain of such monomers, such as the polyethylene molecule, usually shown as

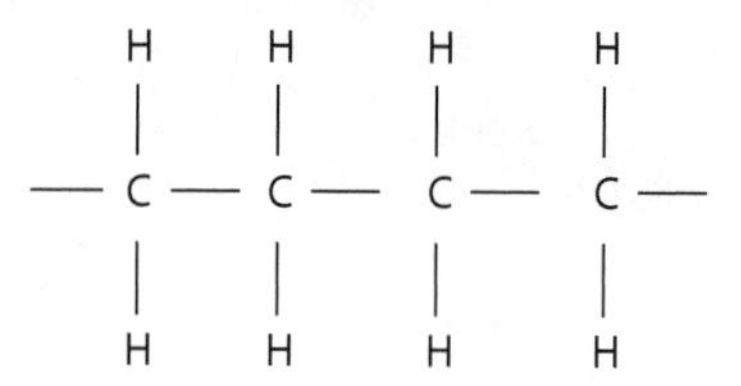

In general, the chemical names are the monomer name (X) and the polymer name poly (X). However, there are also copolymers that are formed from two monomers (XY) that are called poly (XY) or sometimes referred to as the X – Y copolymer.

Table 1.4 gives a list of major polymers and common abbreviated symbols, subdivided into the four types: thermoplastics, thermosets, elastomers, and thermoplastic

TABLE 1.4
Major Polymers

Group	Name	Acronym
Thermoplastic	Polyethylene[a]	PE
	Polyester	PET
	Polypropylene	PP
	Polystyrene	PS
	Polyvinyl chloride[a]	PVC
	Polyacetal	POM
	Acrylic	PMMA
	Polyamide (Nylon)	PA
	Polycarbonate	PC
	Polytetrafluorethylene	PTFE
	Polyurethane[a]	PUR
Thermosets	Epoxy	EP
	Melamine-formaldehyde	MF
	Urea-formaldehyde	UF
	Unsaturated polyester	UP
	Phenolic	PF
	Alkyd	
	Polyurethane[a]	PUR
	Polyethylene[a]	PE
	Polyvinyl chloride[a]	PVC
Elastomers	Natural rubber–polyisoprene	NR
	Styrene–butadiene rubber	SBR
	Polybutadiene	BR
	Butyl rubber	BUTYL

TABLE 1.4 (continued)
Major Polymers

Group	Name	Acronym
	Polychloroprene (Neoprene)	CR
	Synthetic polyisoprene	IR
	Nitrile (nitrile butadiene)	NBR
	Silicone rubber	PDMS
Thermoplastic elastomers[b]		TPE
Copolymers	Polyurethanes	PU
	Block co-polyesters	COPE
	Block co-polyamides	COPA
	Other block copolymers	
Alloys and blends	Elastomer–thermoplastic blend	
	Thermoplastic vulcanizate	TPV
Ionomers		

[a] Some polymers can be either thermosets or thermoplastics.
[b] Most thermoplastic elastomers are copolymers.

elastomers. A thermoplastic melts and flows prior to reaching its degradation temperature. Thermosets actually degrade, since the presence of cross-links do not allow bulk material flow (i.e., melting) prior to degradation. Cross-linked elastomers are obviously a subset of thermosets, but are considered separately due to their high extensibility. In addition, there are thermoplastic elastomers, which, with some exceptions, are thermoplastics that function as elastomers below the melt temperature or high strains. The four types may also be generally grouped with respect to their deformation or elongation properties with the nonelastomeric thermosets having the lowest elongation, the thermoplastics having medium elongation, and the elastomers, of course, having the greatest elongation properties.

1.6 IDENTIFICATION OF PLASTICS

Identification of plastics can be as simple as finding the resin identification code (typically on the bottom of the component) or as complex as separating the base resin from the filler, reinforcements, and colorants to analyze via infrared (IR) spectroscopy. The resin identification codes developed by the Society of Plastics Industry in 1998 are shown in Table 1.5. Despite the desire of many individuals to recycle, many communities recycle often only recycle plastics polyester (polyethylene terephthalate (PET)) and high-density polyethylene (HDPE). In fact, local recycling companies in the United States will often bury plastics in a landfill, either with the rest of the waste or in separate locations, ostensibly to mine at a later date. Collected recyclables are also shipped from the United States as baled

TABLE 1.5
Society of Plastics Industry Resin Identification Codes

Resin	Symbol	Alternative Symbol	Recycled Uses[2]
PET–Polyester	1	1 PET	Beverage bottles, shreded beverage bottles becomes carpet and mats
HDPE	2	2 HDPE	Fencing, bottles
PVC	3	3 V	Pipe, decking, fencing, electrical insulation
LDPE	4	4 LDPE	Bins, plastic lumber
PP	5	5 PP	Battery cases, bins, and pallets
PS	6	6 PS	Insulation, expanded packaging material
Other-resin other than 1–6 or combination of resins	7	7 OTHER	Plastic lumber

plastics to Asia. Products made from virgin resin have the better appearance and properties (mechanical and other) than those made with recycled content. Recycled content can go directly back into products that have the same performance demands as the original product from which the material was recycled. There are even limited cases of upcycling, where the new product is considered to have more value than the original. However, while this can occur, it does not necessarily imply that the new product has enhanced mechanical, electrical, or environmental performance. Thus, it is typical that most products containing recycled plastic are intended for applications that are considered to be downcycled, such that the new product has less value and usually engineering performance than the original. However, examples such as plastic lumber and fencing show that it is economically feasible and environmentally advantageous to create these products with recycled content. Due to aesthetic as well as performance considerations, recycled content is often <25%.

If you live in an urban area, visit a local plastics supplier and search through the scrap or off-cut bin. A great deal of insight into the difficulty of identifying a random plastic comes from asking one of the employees, what material you are holding in your hand. The usual response is "I don't know for certain, but I think it is …." If you are attempting to do perform a reverse engineering analysis of a commercial product, analyze a failed component, or ensure that the off cut is actually a polymer that will perform in your design, you usually want a rapid guess as to what the material actually is. The first test that one performs is to determine whether or not the material is a thermoplastic or thermoset. This can be performed with a heated rod (or soldering iron) at 260°C (500°F) or a hot plate. The second test is to determine density. One quick test of density is whether or not the polymer floats. If it floats it is either polypropylene (PP) or polyethylene (PE). The presence of chlorine can be detected by melting some of the polymer onto a heated copper rod. The rod is then placed into a flame, if the flame turns green, this indicates the presence of copper chloride. This test is a subset of a range of tests called the burn tests, where the polymer is placed into a flame and the response is noted. Some materials will self-extinguish when removed from the flame and others will continue to burn. The observations of these burn tests allow us to go through a flow chart of observations (smoke color and odor) to potentially identify the polymer. One of the first codifications of this methodology was documented in Lavioe.[3] While the original of this codification in chart form is almost impossible to obtain, it exists in many locations (virtually all inadequately referenced) on the Internet. Another version of this flowchart is found in Shah.[4] There are also texts on the identification of plastics, such as Braun.[5] Perhaps one of the best recommendations for identifying an unknown polymer is to have a series of known polymers for comparison as these tests are performed. Care must be taken since many of the materials release toxic gases when heated or burned (e.g., when burned, polyvinyl chloride (PVC) yields HCl gas).

The most reliable (time-consuming and costly) identification method is to use infrared spectroscopy measurements to determine the material. The Rapra Collection of Infrared Spectra of Rubbers, Plastics, and Thermoplastic Elastomers[6] can be used to compare the spectrum of a test material to reference spectra. The transmission spectra in this reference are obtained either from cast or molded thin film or in the case of cross-linked materials by pyrolysis of the material in a Pyrex tube.

1.7 RAW MATERIALS AND PRODUCTION OF POLYMERS

As shown in Figure 1.3, the raw materials or primary resources from which polymers are produced are crude oil, natural gas, chlorine, and nitrogen. These are refined into basic petrochemicals such as ethylene, propylene, styrene, and butadiene. These then are further processed into various polymer materials. During processing other non-petrochemical raw materials such as fillers or reinforcing fibers may be added for color, strength, and other purposes.

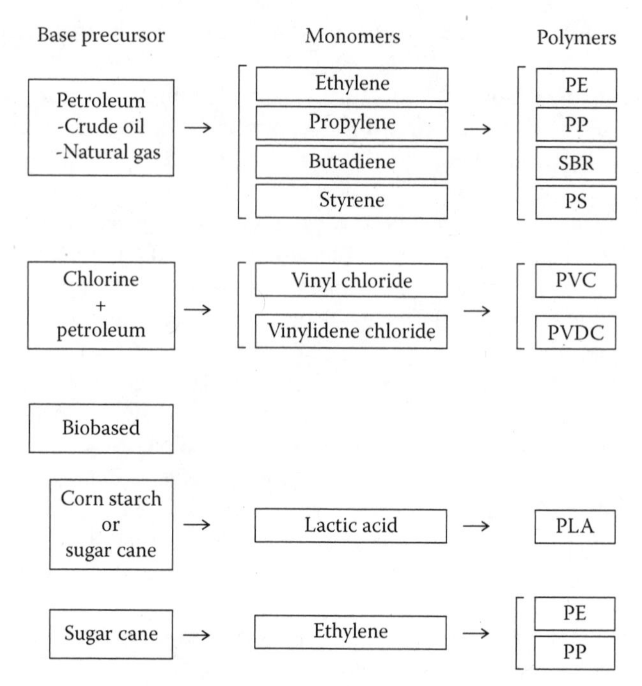

FIGURE 1.3 Production of polymer-based products.

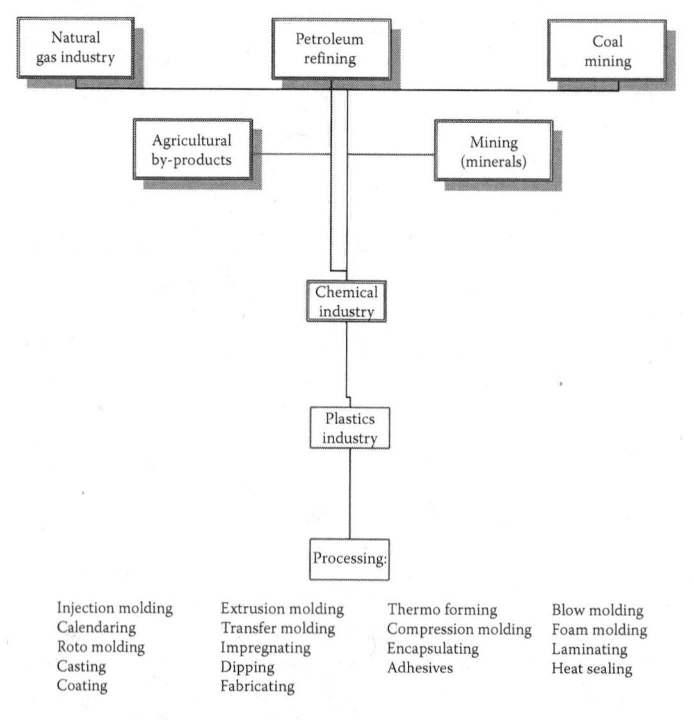

FIGURE 1.4 Organization of the plastics industry.

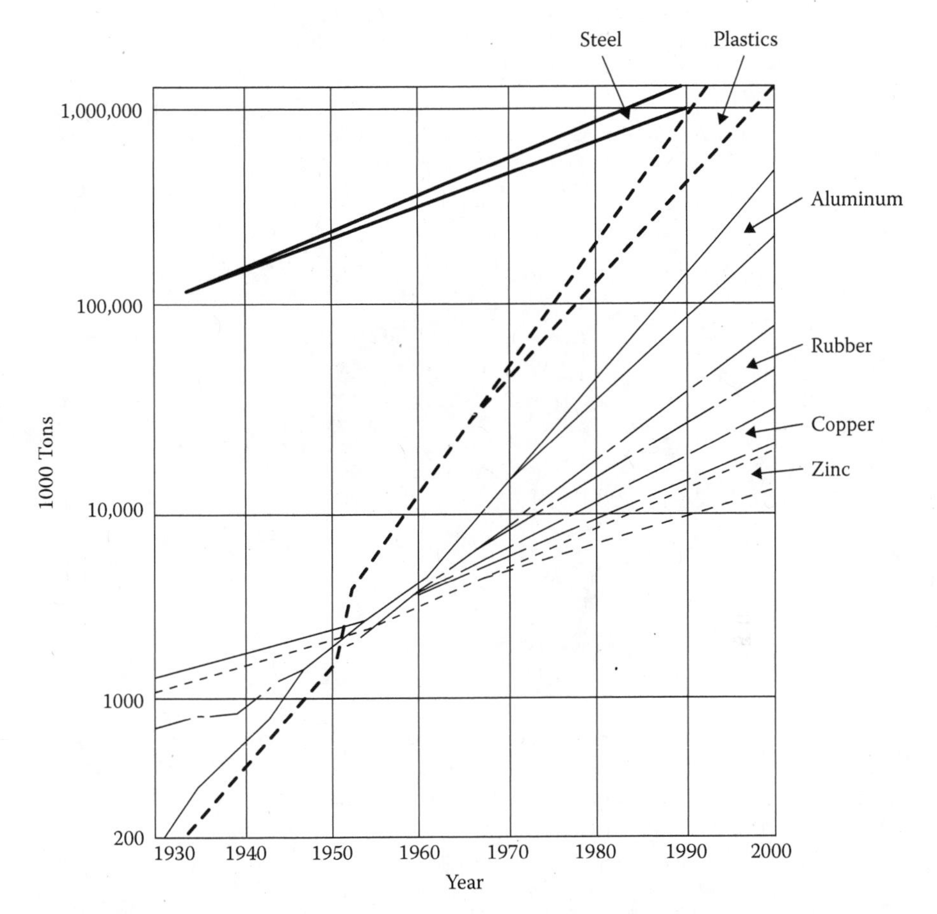

FIGURE 1.5 World consumption of plastics compared to iron and steel. (Adapted from Rosato, D. V., D. V. Rasato, and M. G. Rosato, *Plastics Design Handbook*. Norwell, MA: Kluwer, 2000.)

Figure 1.4 shows how the plastics industry is organized, starting with the mining of raw materials down to the final processing of the final polymer products. Figure 1.5 illustrates how the use of plastics has increased during the century and how in the middle 1980s that the use of plastics is about equal to that of iron and steel. One can recognize how important this is in the design, for example, of an automobile. In 2010, the average car had 1477 N (332 lb) of plastics and composites approximately 8.3% by weight.[8] One can recognize how an oil crisis could affect the availability of the raw materials and the future use of plastics versus metals.

The pie chart in Figure 1.6 shows the distribution of the use of thermoplastics in selected industries—the greatest uses being in building and construction, and packaging.[8] Figure 1.7 shows plastics sales in 2009 compared to 2008. As the chart in Figure 1.7 shows the production and sales were about 100 billion pounds in 2008 and 2009.[8]

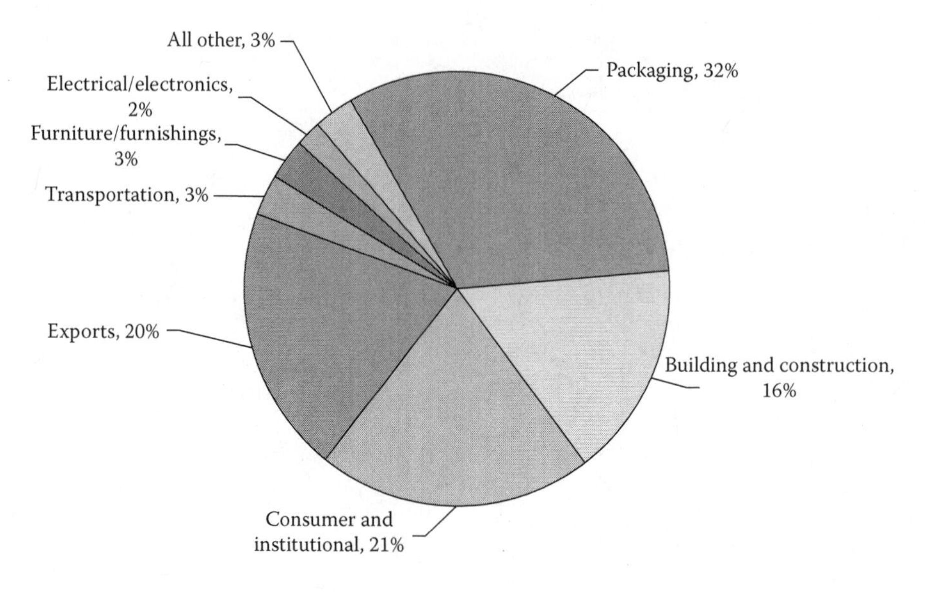

FIGURE 1.6 Use of plastics in different industries in 2009. (Data from http://www.americanchemistry.com/plastics.)

1.8 CHEMICAL STRUCTURES

Some polymers form a linear chain-type structure as shown under Section 1.1, for linear PE. There are other polymers that form linear chains, but where the repeating unit (the monomer) contains other atoms or hydrocarbon radicals. Examples of monomers for PVC, polystyrene (PS), and PP are shown below, where a hydrogen atom has been replaced by a chlorine (Cl) atom, a phenylene (C_6H_6) chemical radical, and a methyl (CH_3) chemical radical, respectively.

ACC resin statistics summary
2009 vs. 2008

U.S. production, sales and captive use
(millions of pounds, dry weight basis) (1)

Resin	Production			Total sales and captive use		
	2009	2008	% Chg 09/08	2009	2008	% Chg 09/08
Epoxy (2)	535	583	−8.2	497	603	−17.6
Other thermosets (5)	12,152	14,508	−16.2	12,108	14,443	−16.2
Total thermosets	12,687	15,091	−15.9	12,605	15,046	−16.2
LDPE (2)(3)	6667	7003	−4.8	6740	7143	−5.6
LLDPE (2)(3)	13,126	12,058	8.9	12,913	12,385	4.3
HDPE (2)(3)	16,956	16,247	4.4	17,010	16,823	1.1
PP (2)(4)	16,623	16,768	−0.9	16,754	17,235	−2.8
PS (2)(3)	4865	5220	−6.8	4969	5364	−7.4
Nylon (2)(4)	943	1148	−17.9	951	1164	−18.3
PVC (3)	12,754	12,789	−0.3	12,781	12,948	−1.3
Other Thermoplastics (6)	14,049	15,222	−7.7	15,282	16,054	−4.8
Total Thermoplastics	85,983	86,455	−0.5	87,400	89,116	−1.9
Grand total plastics	98,670	101,546	−2.8	100,005	104,162	−4.0

(1) Except phenolic resins, which are reported on a gross weight basis.
(2) Sales and captive use data include imports.
(3) Canadian production and sales data included.
(4) Canadian and Mexican production and sales data included.
(5) Includes: polyurethanes (TDI, MDI, and polyols), phenolic, urea, melamine, unsaturated polyester, and other thermosets.
(6) Includes: PET, ABS, engineering resins, SB latex, and other thermoplastics.

Sources: Plastics industry producers' statistics group (PIPS), as compiled by veris consulting, inc.; ACC

Year end 2009 statistics

FIGURE 1.7 Sales of plastic in 2009 and 2008. (Data from http://www.americanchemistry.com/plastics.)

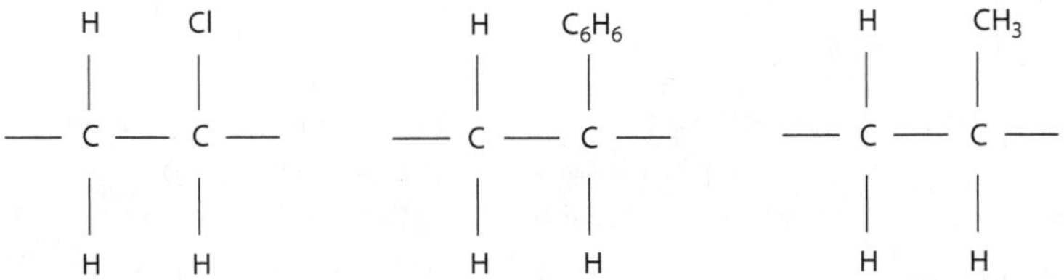

H	Cl		H	C_6H_6		H	CH_3
— C	— C —		— C	— C —		— C	— C —
H	H		H	H		H	H
PVC			PS			PP	

Atoms such as O, N, S, and Si can take the place of the carbon atom within the backbone of the chain. For example, the linear chain for the silicones is shown below where R represents a chemical radical or a pendent group. The R radical in the repeating unit may be a methyl or phenyl group.

```
            R
            |
  — C —  Si —
            |
            R
```

These linear chains in the polymers can develop branches. For example, a branch in the PE chain is represented as

```
    H    H     H     H
    |    |     |     |
  — C — C —  C —  C —
    |    |     |     |
    H    H     |     H
               |
          H — C — H
               |
          H — C — H
               |
               H
```

The branches can join and form a 3-dimensional network or a cross-linked polymer, and because the chains do not necessarily form straight lines, such a network can be illustrated schematically as

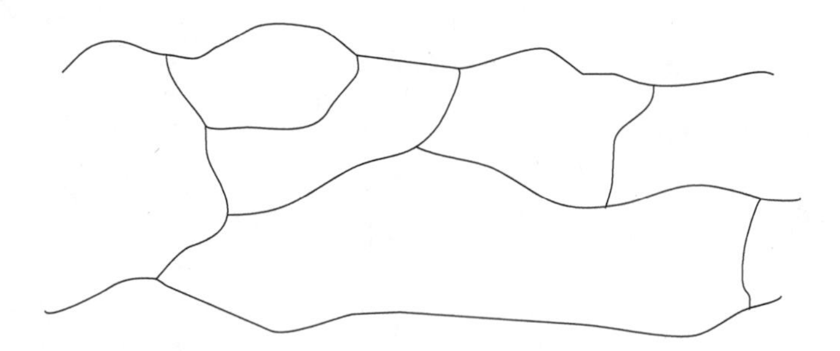

The chemical structures illustrated above will affect the mechanical properties of the polymers as will be described later.

As indicated in the network illustration, the linear chains do not necessarily form in straight lines. This is due to the attraction forces between atoms. For example, in methane (CH_4) the carbon atom likes to sit in the tetrahedral site to maintain its equilibrium, as shown:

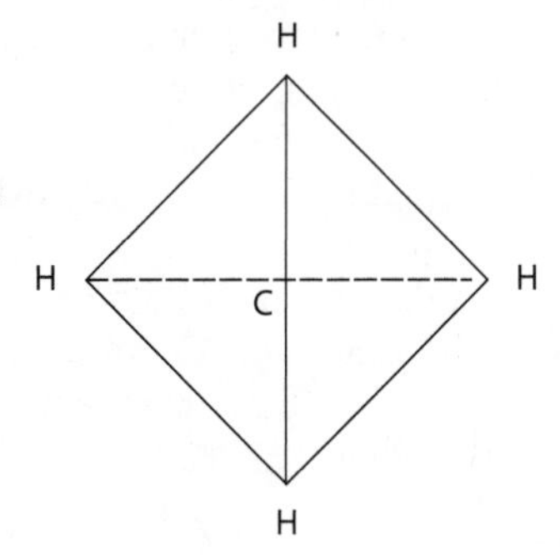

In ethane (CH_3-CH_3), there is a tendency toward this tetrahedral formation, but also a rotation about the C−C bond axis, as shown in Figure 1.8.

The tetrahedral geometry, and the rotation of the C−C axis, helps explain the zigzag pattern of a chain of polymethylene shown below where the hydrogen atoms (not shown) are located above and below the plane.

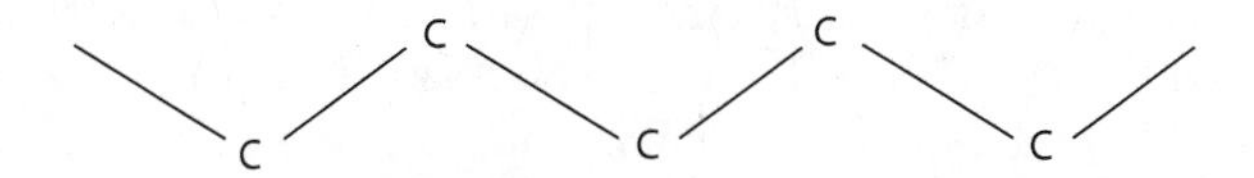

This possibility of different rotations of the C–C bond really causes the chain to twist into all kinds of three-dimensional configurations. In addition, the chain does not

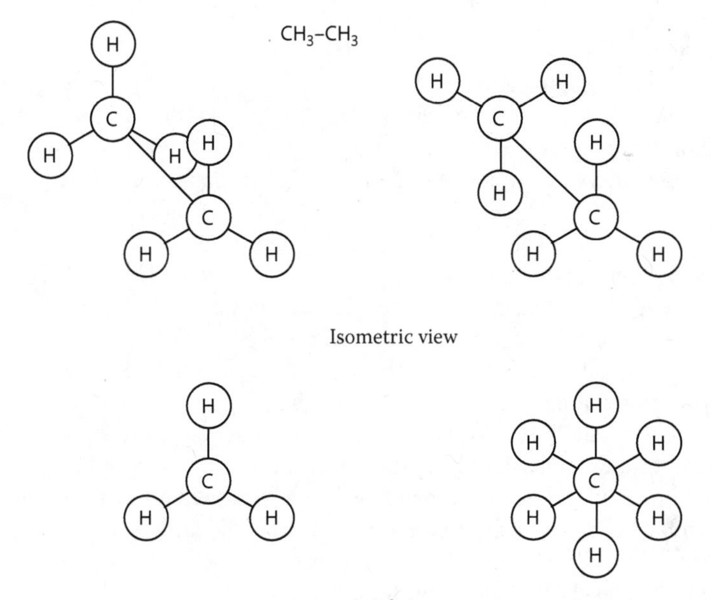

FIGURE 1.8 Rotation about the C–C bond in ethane.

remain stationary but is constantly wiggling around like a lively fish worm—more so with an increase in temperature. This motion can occur thousands of times in a second.

Zigzag chain formations can straighten out under stress, thus leading to larger elongations in elastomers. The temperature effects are also exhibited in the viscoelastic behavior of such polymers.

What are the sizes of these molecular chain structures? For example, linear PE can have from 1000 to 10,000 C atoms in a chain. A 1000 atom chain of PE has a length/thickness ratio equivalent to a 2 m length of string. The actual length of the 1000 atom chain is about 0.13 μm. The relative molecular mass of PE is

$$M = 14.02n + 2.02 \tag{1.1}$$

or when n is large,

$$M \approx 14.02n \tag{1.2}$$

where n = number of repeating CH_2 units in the chain.

The number of repeating monomer units in a chain is also referred to as the "degree of polymerization" or n. Then the molecular weight (M) of the entire polymer molecule is the n of the repeating unit multiplied by the molecular weight (m) of the repeating unit. In addition, repeating chains are often started and terminated by end groups that are not the repeat unit. Ignoring the starting and terminating groups, Equation 1.2 can be rewritten as

$$M \approx mn \tag{1.3}$$

where n = number of repeat units.

CH_2 linear polymer chains of shorter length with 30–40 C atoms constitute the structure of paraffin wax. Paraffin wax and PE look alike. Both are waxy, translucent, white solids that are easily melted. Both are electric insulators and both have the same density of $\rho = 900\ kg/m^3$. How can one tell the difference? They differ in their mechanical properties. The variation in density and melting temperature of PE (and paraffin wax), depending upon the chain length, is shown in Figure 1.9. The physical properties of PE as a function of molecular weight and crystallinity are shown in Figure 1.10.[5] Some polymers can also form a crystal structure where regular repeating units fit into a crystal lattice. For example, the zigzag shape of the PE chain can crystallize into the planar folded-chain structure shown in Figure 1.11. The PP polymer, on the other hand, crystallizes into a helical shape. Polymers whose chains are randomly made up of different units do not crystallize but remain amorphous.

Crystalline polymers do not completely crystallize, but there is a tendency to form grains similar to metals. This structure is known as the "fringed micelle" structure and is shown in Figure 1.12. Within the grains there is the regular crystalline phase. Between the grains there is the amorphous phase.

Instead of the planar structure, a spherical order of crystals may develop known as clusters of "spherulites." The fibrils of Nylon 6 can band together along radii of a sphere. PP can form into more "bushy" spherulites. Clusters of regular patterns of spherulites can also form in PS and PE.

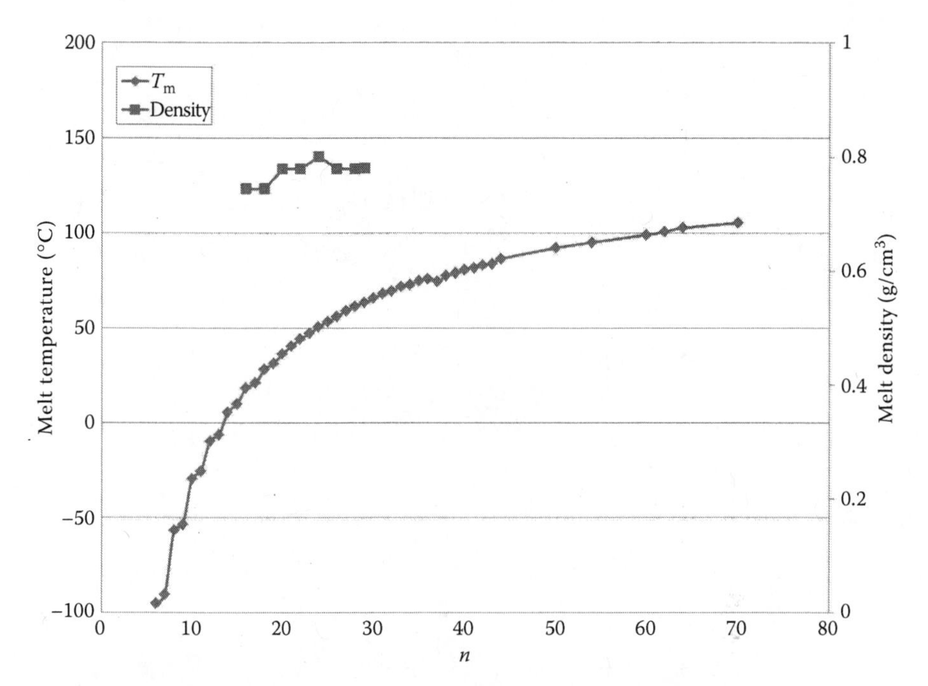

FIGURE 1.9 Liquid-state densities and melting temperatures of normal straight-chain hydrocarbons C_nH_{2n+2} (Densities at 20°C or closest available temperature). (Adapted from Seyer, W. F., R. F. Patterson, and J. L. Keays, *Journal of the American Chemical Society*, 66, 179–182, 1944.)

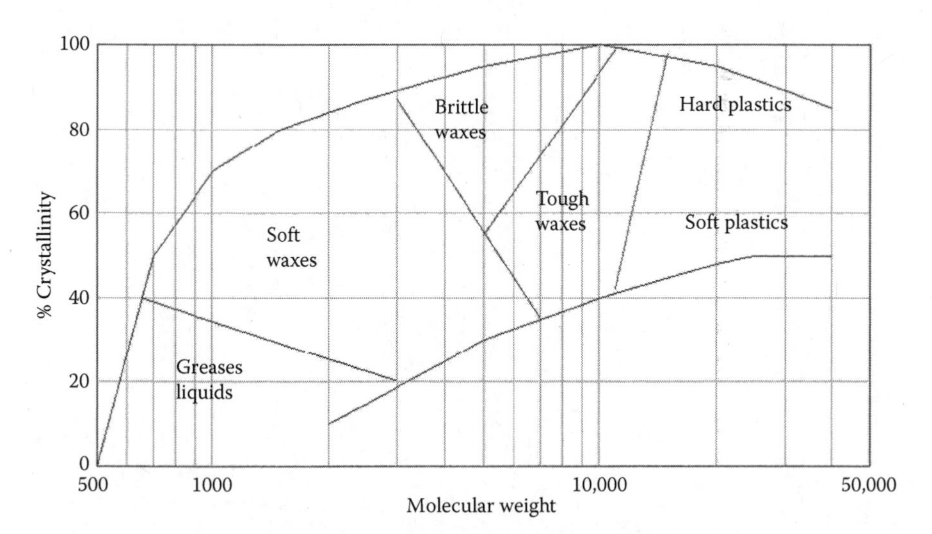

FIGURE 1.10 Physical properties of polyethylene as a function of molecular weight and crystallinity. (Adapted from Richards, R. B., *Journal of Applied Chemistry*, 1, 370, 1951.)

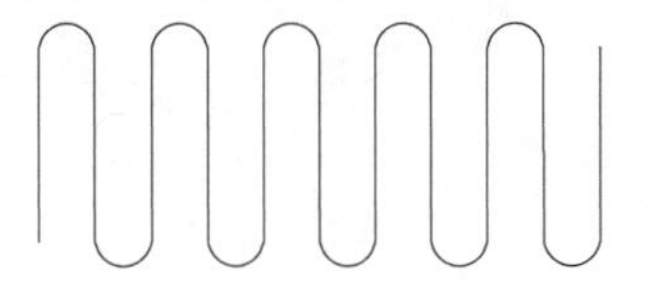

FIGURE 1.11 Folded-chain structure of PE crystals.

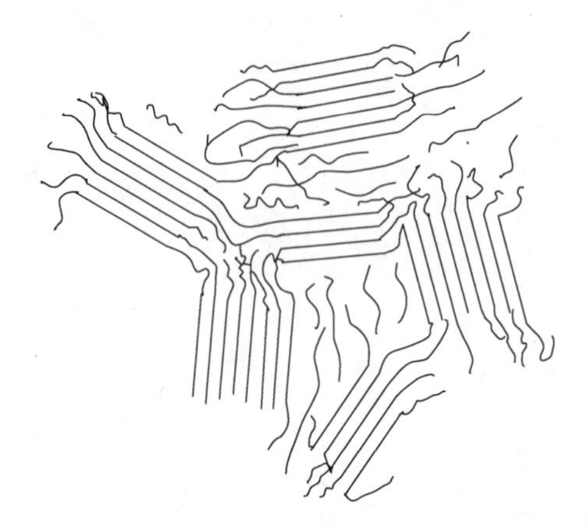

FIGURE 1.12 "Fringed-Micelle" structure of crystalline polymers.

TABLE 1.6
Crystalline and Amorphous Polymers

Crystalline	Borderline[11] and General	Amorphous
LDPE	NR[a]	Chlorinated PE
HDPE	PIB[a]	EPDM
PP	IR[a]	PS
PTFE	IIR[a]	ABS
PA	PVAL	PMMA
Cellulose	PCTFE	PF
PMP	PVC[b]	MF
PETP	MQ[a]	EP
PC	CR[a]	UP
PED		PUR
POM		PVC[b]
PPS		
PEEK		
PEK		
TPI		
PBT		

Note: The abbreviations of the polymers in this table are defined throughout the text and specifically in Chapter 6, Table 6.8.

[a] Some elastomers crystallize when extended to high strains.

[b] Some polymers have forms that are either crystalline or amorphous.

Table 1.6 lists some crystalline and amorphous polymers, and some borderline ones. The borderline cases, which are normally amorphous, may crystallize under high mechanical strain rates. Such is the case for NR, polyisobutylene (PIB), butyl (IIR), and polyvinyl alcohol (PVAL), silicone rubber (MQ), and chloroprene rubber (CR).

1.9 GLASS TRANSITION AND MELTING TEMPERATURES

Whether or not a solid polymer has a crystalline structure depends on whether it was supercooled or slowly cooled from its melting temperature T_m to below its glass transition temperature T_g, as shown in Figure 1.13. It also depends on whether or not T_g is above or below room temperature. Above T_m, a linear polymer is in a viscous liquid state (A). If it is rapidly cooled, it passes through a rubbery region (C). If it is rapidly cooler further to below a temperature T_g, it vitrifies into a glassy amorphous state (D). If, however, the same polymer is slowly cooled from T_m, crystallites form in a rubbery matrix (E), and if slowly cooled to below T_g, crystallites form in a glassy matrix—as shown previously in Figure 1.13.

The room-temperature properties of a polymer depend on its structure and its structure depends on whether the T_g is above or below room temperature. Table 1.7 gives glass transition temperatures for selected polymers. From the glass transition temperature, the reader can predict which of these polymers can be expected to be relatively hard and more rigid solids, or which can be expected to be relatively soft and more flexible solids at room temperature.

The mechanical properties of a polymer depend on the glass transition temperature. So logically we are led to ask how the glass transition temperature of a given polymer can be substituted to effect a change in its mechanical behavior. This can be accomplished through a change in chemical composition, mainly through the production of copolymers.

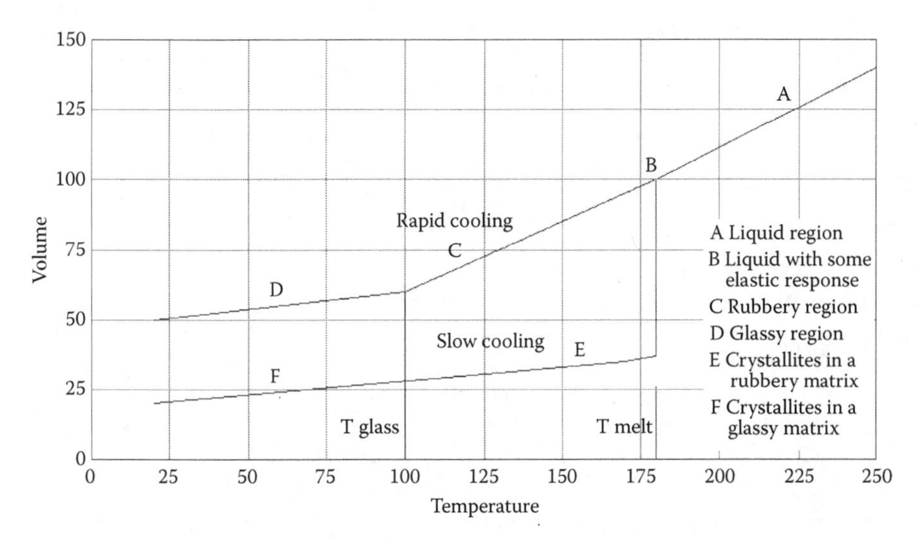

FIGURE 1.13 Volume–temperature curve for a crystalline polymer. (Adapted from Alfrey, T. and E. F. Gurnee, *Organic Polymers*. Englewood Cliffs, NJ: Prentice-Hall, Inc., 1967.)

TABLE 1.7
Glass Transition Temperatures and Melting Temperatures

Polymer	T_g (°C)	T_m (°C)
PE	−90 to −135	115–137
PP	−10	176
PS	95	240
PVC	85	212
PVF	−20–45	200
PVDC	−15	198
PA6	50	215
PA6/6	90	260
PMMA	105	175
PC	150	265
NR	−75	28
CR	−45	
NBR	−20	
ABS	100	230
PTFE	−65	327
BR	−90	154
PET	68–80	212–265

Note: The abbreviations of the polymers in this table are defined throughout the text and specifically in Chapter 6, Table 6.8.

Copolymers can vary in type depending on the structure developed. Therefore, it is convenient to characterize different classes of linear polymers as follows:

Class I: Chains with perfectly repeating units, ("the matched pearl necklace," e.g., PE).
Class II: Random copolymers.
Class III: Random D, L-"copolymers" (vinyl) and *cis–trans* copolymers (diene).
Class IV: Block copolymers.
Class V: Regular short-unit polymers (α-helix type).

Class-I-type structures can be made more flexible by lowering the glass transition temperature by substitution of other units sparsely but regularly along the chain. Typical "flexibilizing" units are oxygen and sulfur atoms, and ester or amide groups. The reverse effect can be achieved by substitution of a more rigid unit in the chain—for example, the more rigid benzene ring in PS.

Class-II random structures have an irregular substitution of the repeating units, and consequently do not fit into a crystal lattice as easily as the Class-I type and generally have a melting temperature lower than the monopolymers, as shown in Figure 1.14. The intermediate compositions, shown by the dashed line in Figure 1.15, correspond to a permanently amorphous structure with no crystallinity.

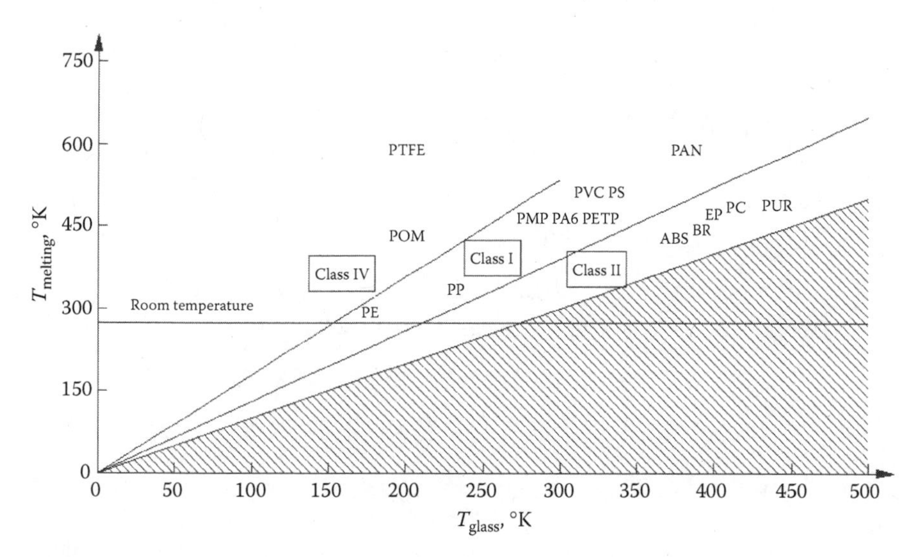

FIGURE 1.14 Schematic plot of melting temperature as a function of glass transition temperature.

Class-III structures are amorphous structure with low regularity, but its formation is different where substitution atoms as groups randomly sit above or below the plane of a zigzag chain referred to as a d- or 1-configuration. The configuration is generally not regular (e.g., dddddd), and is generally not perfectly alternating (e.g., dldldldl), but random (e.g., ddldllldlldd1d).

The dienes form a random structure because the more complicated pendent group itself (which forms the monomer or repeating unit) can have different structures, and

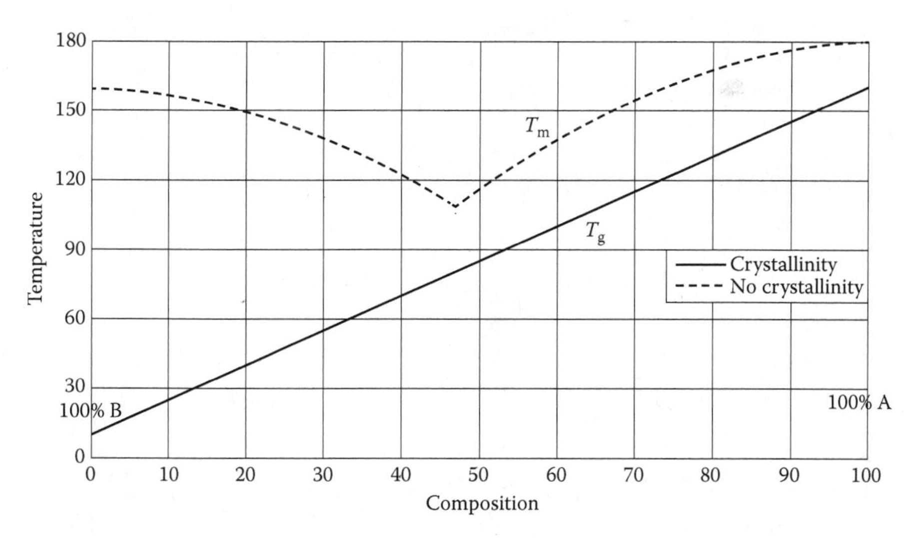

FIGURE 1.15 T_m and T_g as functions of composition for random copolymers composed of two Class-I polymers.

these different structures of the monomer itself occur randomly. For example, in butadiene three types of chain units can occur as shown:

$$-CH_2-\underset{\underset{HC=CH_2}{|}}{\overset{\overset{H}{|}}{C}}- \qquad -H_2C\diagdown_{CH=CH}\diagup^{CH_2}- \qquad -H_2C\diagdown_{CH=CH}\diagup^{CH_2}-$$

Class-IV structures consist of long sequences or blocks of one type (A) of polymer chain units alternating with long sequences of another type (B):

$$-AAAAAAAABBBBBBBBAAAAAAAA-$$

These long blocks of A units or B units, if large enough, can crystallize separately. With block copolymers, it is possible to achieve a wider range of mechanical properties than are possible with the Class I, II, or II types. For example, a poly A-block with a high melting temperature T_m and a low T_g as shown in Figure 1.14. Figure 1.14 also shows the Class I, and the Class II and IV Regions. This figure is very important in designing and selecting polymer materials with certain ranges of mechanical properties.

The (T_g, T_m) temperature point for the polymers must lie in the upper diagonal half of Figure 1.14 to satisfy $T_g < T_m$. The helical type, Class V, can form with some regularity if the adjacent helical chains are twisted the same relative amount to allow conformation. If the angular twist of each unit relative to its predecessor is a rational fraction of one revolution ($2\pi/n$), then the spatial orientation will exhibit a definite repeat distance. To learn more about the chemical structure and material behavior of polymers, the reader is referred to material science texts such as the one by Smith.[13]

Example Problem 1.1

Using the letters C and D to designate two types of repeating units, indicate representative segments of the following polymer chain architectures:

1. Linear chain polymer
2. Branched polymer
3. Block copolymer
4. Random copolymer

Answers:

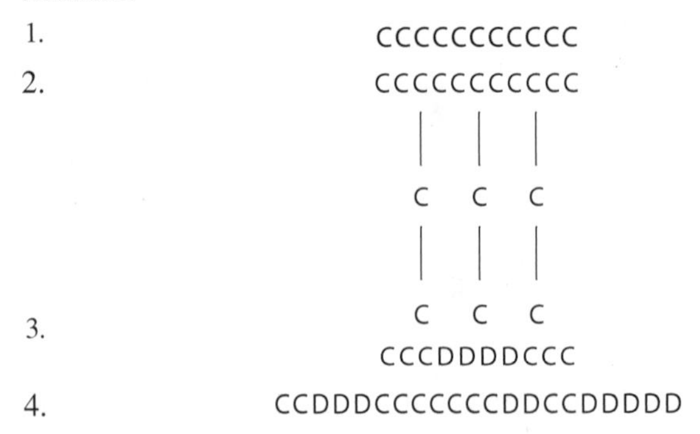

Example Problem 1.2

Which polymers would one expect to have a larger elastic modulus?

1. A linear chain polymer or a cross-linked polymer with the same length chain length as the uncrosslinked polymer (note that the molecular weight or length of the cross-linked chains would be different)?
2. A polymer with a high T_g or one with a low T_g?

Answers:

1. Cross-linked polymer
2. High T_g polymer

HOMEWORK PROBLEMS

Unlike most engineering homework, the following homework problems benefit greatly from an Internet search.

1.1. Open the plastic housing of a household appliance. Examples are vacuum cleaners, blenders, mixers, humidifiers, and so on. Describe the differences between modern-day appliance housing and one from the 1950s. What is the weight difference? What is the price difference in today's dollars based on current costs of materials? Discuss any differences in speed of manufacture.

1.2. Consider the evolution of the Chevrolet Corvette's body. What has been the evolution? Where there any other cars that either preceded or were concurrent with the Corvette body development?

1.3. What was the first composite pedestrian bridge designed for high volumes of foot traffic (i.e., in a city)? What were the design considerations?

1.4. What was the first composite automobile bridge? Compare the construction of a bridge for a rural two-lane highway with that of a 4+ lane urban highway. What were the design considerations?

1.5. Assume that you work in a chemical processing plant. How would polymers and composites be used in the plant that prior to the 1980s would have used steel and other materials?

1.6. Trace the development of downhill skis from 100 years ago to today.

1.7. Show that the "tetrahedral carbon angle" in methane is 109.47°. This is the angle every pair of C–H bonds form with each other.

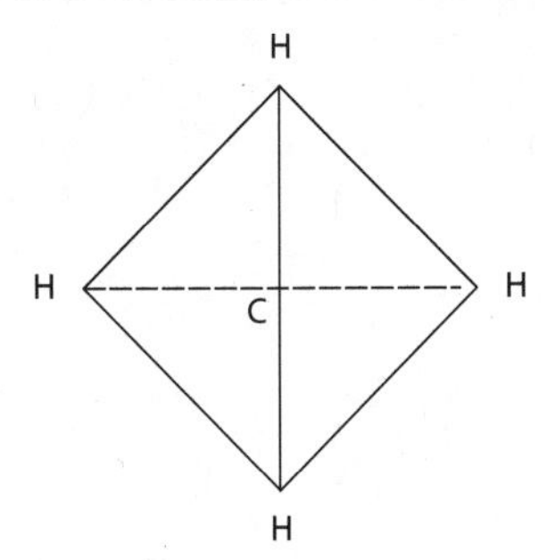

1.8. Calculate the relative molecular weight of linear polypropylene if the chain length is 1000 monomers. Is this polymer expected to be liquid and soft, or hard and solid?

1.9. What is the basic petrochemical from which the polymer PP is produced?

1.10. Identify the monomer in the following linear chain branched polymer.

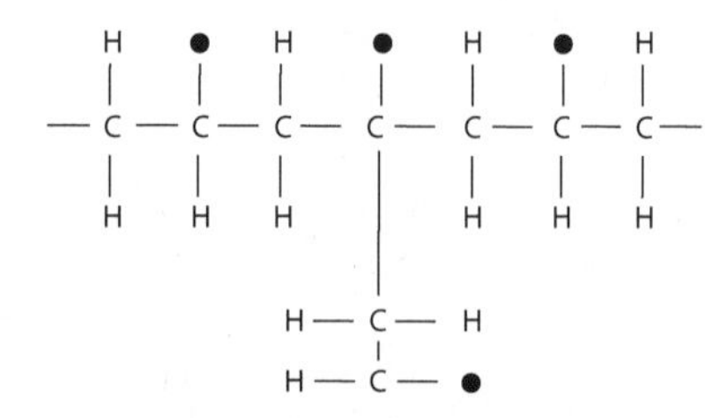

1.11. A Class-I polymer is injection molded from the hot melt into a cold metal mold. As it cools in the mold the material structure will vary across the mold. Sketch the structure that results at room temperature (T_{room}) if $T_{room} < T_g$ where T_g is the glass transition temperature.

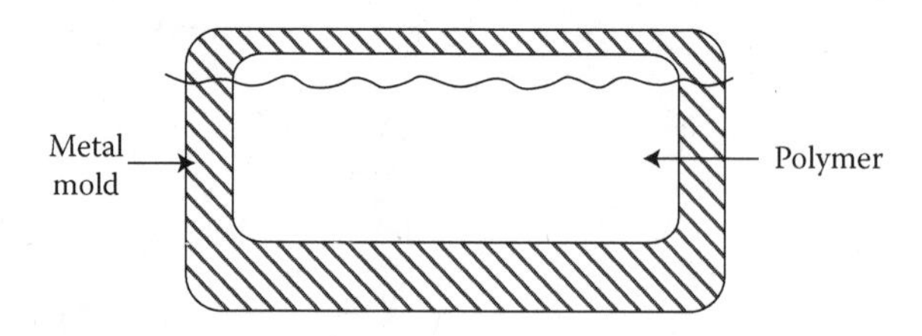

REFERENCES

1. http://www.nobelprize.org
2. Anonymous, *Plastic Packaging Resins*. American Chemistry Council.
3. Lavioe, F., ed., *Materials Selector Issue 2001, Materials Engineering*. Chicago: Penton/IPC, 2000.
4. Shah, V., *Handbook of Plastics Testing and Failure Analysis*. Hoboken, NJ:Wiley-Interscience, 2006.
5. Braun, D., *Simple Methods for Identification of Plastics*. Munich: Carl Hanser Verlag, 1986.
6. Sidwell, J. A., *Rapra Collection of Infrared Spectra of Rubbers, Plastics, and Thermoplastic Elastomers*. Shawbury: Rapra Technology Limited, 1997.
7. Rosato, D. V., D. V. Rasato, and M. G. Rosato, *Plastics Design Handbook*. Norwell, MA: Kluwer, 2000.
8. http://www.americanchemistry.com/plastics
9. Seyer, W. F., R. F. Patterson, and J. L. Keays, The density and transition points of the *n*-paraffin hydrocarbons, *Journal of the American Chemical Society*, 66, 179–182, 1944.
10. Richards, R. B., Polyethylene-structure, crystallinity and properties, *Journal of Applied Chemistry*, 1, 370, 1951.
11. Hall, C., *Polymer Materials*. New York, NY: John Wiley & Sons, 1981.
12. Alfrey, T. and E. F. Gurnee, *Organic Polymers*. Englewood Cliffs, NJ: Prentice-Hall, Inc., 1967.
13. Smith, W. F., *Foundations of Materials Science and Engineering*, 3rd ed., New York: McGraw-Hill, 2004.

2 Mechanical Properties of Polymers

2.1 INTRODUCTION

In the previous chapter, it was pointed out how the chemical structure of polymers influences their mechanical properties. The glass transition temperature and the rate of cooling from the melt determine whether the polymer will be a hard, stiff material, or a soft flexible material. These qualitative differences can be quantified by measuring standard mechanical properties. The student, generally, is introduced to the subject of mechanical behavior through the study of linear elastic metals that exhibit solid behavior at normal operating temperatures and conditions. However, this study of polymers reveals that polymers exhibit fluid as well as solid behavior and are viscoelastic and viscoplastic at room temperature. The mechanical properties of polymers are strain rate sensitive and highly temperature dependent.

2.2 TENSILE PROPERTIES

The standard tensile test,[1] is conducted on a uniaxial specimen with a reduced cross section. Standard tensile specimens have an overall length of 8 in. with a 2 in. gage length and cross section of 0.500 in. width and thickness t. Smaller specimens of one-half standard size are sometimes used for polymers. An extensometer is mounted on the central portion of the specimen to measure the elongation over the 2-in. gage length, and the conventional engineering strain is calculated from

$$\varepsilon = \frac{\delta L}{L_o} = \frac{(L - L_o)}{L_o}, \tag{2.1}$$

where L_o is the gage length and L is the length for any axial load P applied to the specimen. Equation 2.1 is satisfactory for small strains in relatively stiff materials. However, for large strains the true strain definition is recommended, where

$$\varepsilon_T = \int \frac{dL}{L} = \ln\left(1 + \frac{\delta L}{L_o}\right). \tag{2.2}$$

The conventional engineering tensile stress in the specimen is calculated from

$$\sigma = \frac{P}{A_o} \tag{2.3}$$

when strains are small. A_o is the original cross-sectional area. When strains are large, the true stress definition is recommended, where

$$\sigma_T = \frac{P}{A}, \tag{2.4}$$

where A is the current area. For volume constancy in the large strain range,

$$\sigma_T = \sigma(1+\varepsilon). \tag{2.5}$$

Tensile tests are conducted at a constant strain rate. Static data are determined at a slow strain rate of 10^{-4}((in./in.)/s). From the tensile data, a stress–strain curve is plotted as shown in Figure 2.1. From this curve, a number of important mechanical properties are determined.

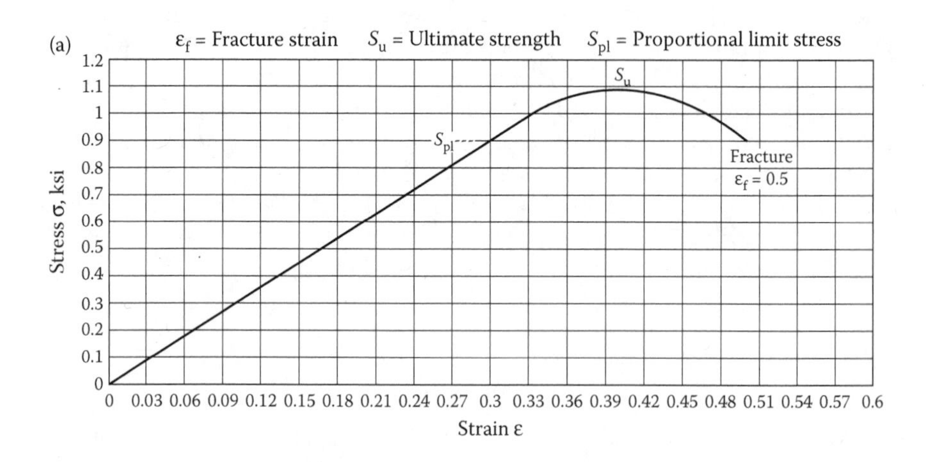

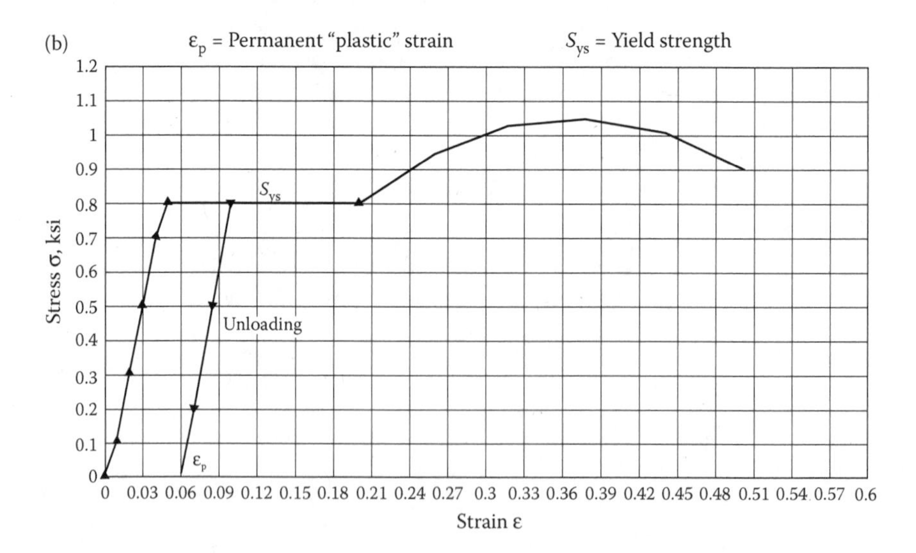

FIGURE 2.1 Stress (strain) curves and the effect of strain rate. (a) Engineering stress/strain curve, (b) ductile material with yield point, (c) stress/strain curves for different behavior, and (d) effect of strain rate.

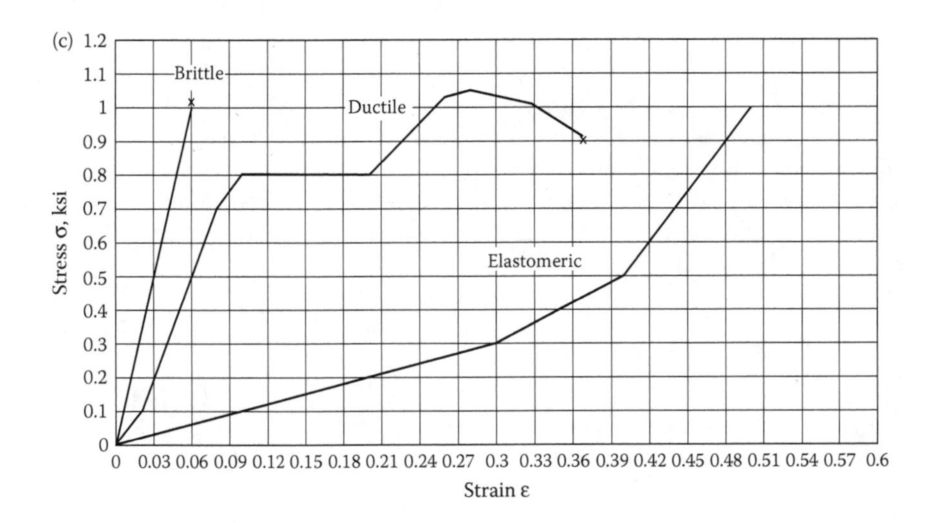

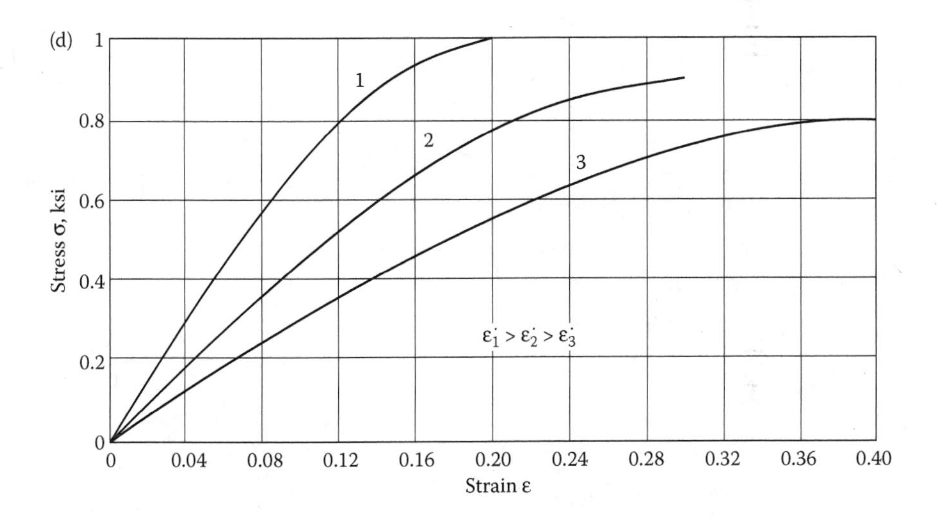

FIGURE 2.1 Continued.

2.2.1 Elongation

Elongation is defined as fracture in percent. It is a measure of ductility.

$\varepsilon_f < 10\%$ for a brittle material
10% $\varepsilon_f < 100\%$ for a ductile material
$100\% \leq \varepsilon_f$ for an elastomeric material (an elastomeric material is elastic up to 400–500% strain)

The percent strain limits are arbitrary relative divisions defining the behavior. The above definitions of a brittle and ductile material assume that the reference temperature is room temperature. It is best to refer to ductile behavior and brittle behavior, rather than define materials as ductile or brittle. A material that is brittle at room temperature will be ductile at some elevated temperature, and a material that is ductile at room temperature will exhibit brittle behavior at a low-enough temperature. Ductile and brittle are relative terms. The point is that mechanical properties, particularly for polymers, are highly dependent on temperature.

2.2.2 Elastic Modulus

The elastic modulus E is the slope of stress–strain curve which is a constant for a linear elastic material for stresses less than the proportional limit. If E is the slope of σ as a function of ε curve, then

$$E = \frac{d\sigma}{d\varepsilon} = \frac{\sigma}{\varepsilon}, \quad \text{for } \sigma < S_{pl},$$

where S_{pl} = proportional limit stress (based on original area A_o). For materials that have time-dependent creep strain, E is defined by

$$E = \left.\frac{d\sigma}{d\varepsilon}\right|_{t=0}.$$

2.2.3 Ultimate Tensile Strength

S_u is the ultimate tensile strength, the maximum stress the material can withstand before fracture. S_u is based on A_o.

2.2.4 Yield Strength

S_y is yield strength. For $\sigma > S_y$, permanent plastic strains ε_p occur after unloading. The notation of using S for strength and σ for stress is adopted here, because there is a difference. Strength is a material property. Stress is the result of a load applied to a material. Two materials may be subjected to the same stress, but have different strengths. In any case, the inequality holds, $\sigma \leq S$.

A ductile material with a pronounced yield point is shown in Figure 2.1b. For materials without a pronounced yield point, S_y is determined from a 0.2% offset line drawn parallel to the initial elastic line. Figure 2.1c shows the difference in appearance, in stress–strain curves for brittle, ductile, and elastomeric behavior. Figure 2.1d shows the effect of higher strain rates on the stress–strain(strain) curve.

There are some departures or exceptions to the above descriptions of stress (strain) behavior for viscoelastic polymers. For example, a stress–strain curve is shown in Figure 2.2a for an injection molded acetal copolymer. It shows strain softening with the yield and ultimate strengths approximately equal. It also shows an elastic unloading modulus much less than the initial elastic modulus when loaded. This results in a permanent strain that is much smaller than the ultimate strain. For example, an ultimate

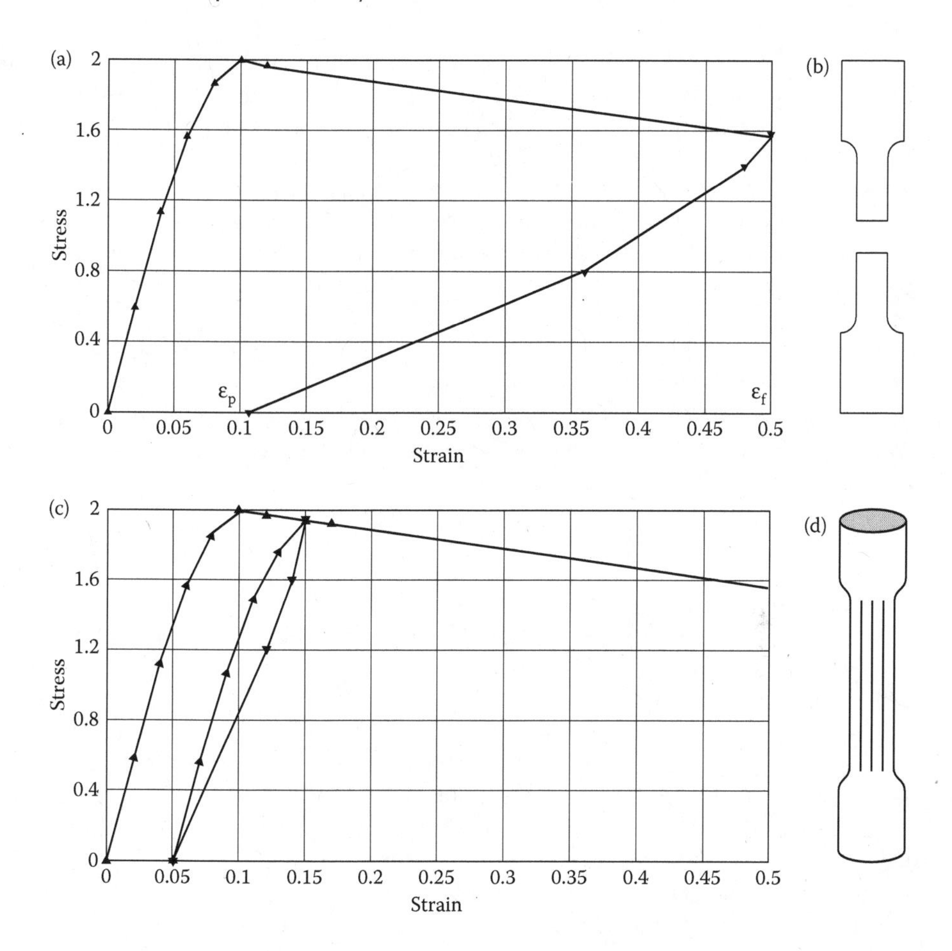

FIGURE 2.2 Different mechanical behavior of viscoelastic polymers under tension. (a) Stress–strain behavior of acetal copolymer, (b) normal fracture plane after 50% strain in acetal copolymer, (c) hysteresis in strain cycling acetal copolymer, and (d) longitudinal fracture of PE tube under tension to 400%.

strain at fracture of 50% has been found with a permanent strain after elastic springback of only 10.7%. Although this amount of ultimate strain indicates ductile behavior, the nature of the fracture surface is typical of brittle fracture being oriented normal to the maximum normal stress as shown in Figure 2.2b.

Figure 2.2c shows a hysteresis loop if the acetal copolymer is cycled to 5% strain, unloaded, and cycled to 10% strain. The strain rate was 0.2 min. in Figure 2.2b and c. There was a 0.1-min hold time at 5% and 10% strain, and it is evident that there was some small immediate relaxation of stress before the unloading at 0.2/min. occurred. (Modeling of this behavior is discussed in Chapter 3, where mechanical models and the dynamic modulus are described.)

Figure 2.2d shows 400% elongation under tension applied to a polyethylene tube. After removal from the tensile machine, the 400% was found to be permanent, but the tube fractured into many longitudinal slits between the linear chains that had been originally oriented longitudinally by extrusion of the tube during manufacture. Under tension the tube was contracting circumferentially (laterally) thus holding the bundle of chains together. As the tube extended, the kinks in the chains were straightening out and lateral bonds between the chains were evidently breaking. Poisson expansion after removal of the load revealed the broken lateral bonds. This example shows another case where the mechanical behavior of polymers is different from that of metals. Whereas a ductile metal has a 45° failure plane where the shear stress is maximum, the PE tube shows a failure orientation parallel to the direction of tension.

From the tensile test, another important property is measured and that is the Poisson ratio value μ, which is a measure of the contraction strain ε_t in the transverse direction when a strain ε is applied in the longitudinal direction. Stretching takes place in the load direction but shortening takes place in the other two directions for isotropic materials.

$$\mu = -\frac{\varepsilon_t}{\varepsilon} = \left|\frac{\varepsilon_t}{\varepsilon}\right| = \left|\frac{\varepsilon_W}{\varepsilon}\right| \tag{2.6}$$

If the material is not isotropic, the value μ measured in the thickness direction (t) will be different from the value determined in the width direction (w). Such materials are either orthotropic or anisotropic. These materials are considered in Chapter 8.

The mechanical properties under tension are not sufficient to determine total material behavior. The properties under compression, flexure, and shear can be different. For ductile materials, it is generally assumed that the behavior in compression is the same as that in tension, that is, that the compressive strength is the same and that the modulus E is the same as obtained in a simple tension test. Also, for ductile behavior, the maximum shear yield strength under tension is taken as a measure of the maximum shear strength, such that

$$S_{sy} = 0.577\,S_y \tag{2.7}$$

Furthermore, if the material is assumed isotropic, and if the stresses are within the proportional limit, the shear modulus G can be calculated from

$$G = \frac{E}{2(1+\mu)}, \tag{2.8}$$

where $G = \tau/\gamma$ in the linear elastic range, where γ is the engineering shear strain corresponding to shear stress τ. (The reader should be able to verify the validity of Equation 2.8.)

When direct measurement of the shear properties are desired, a round, cylindrical specimen with a circular cross section is used and a torque is applied to produce a pure shear stress state. The curve of τ as a function of γ yield the desired properties.

A brittle material is weak in tension and is difficult to even be tested in tension. Its compressive strength is much higher and a compression test is performed to determine its compressive strength. The tensile strength of a brittle material can be determined from a torsion test (if the material is isotropic) where failure will occur at 45° to the applied shear direction. The tensile strength of brittle materials and also of thinly laminated more ductile materials are also determined from flexural bending tests.[2] The compressive strain on one side of the bending specimen helps keep the material intact and a local failure begins on the tension side. Modulus of elasticity values, E_f, are also determined from flexure tests.

For polymers, the tensile strength S_u will generally be in the range of 1–10 ksi. The tensile strength can be increased by fiber reinforcement as discussed in Chapters 8 and 9 of this book. The modulus of elasticity values, E for polymers are in the range of $0.1–10 \times 10^6$ psi. Thus, the strength and modulus values are an order of magnitude less than those of metals. However, polymers are still being used from structural purposes. The main reason is their low density—also an order of magnitude less than metals. Based on their specific strength = strength/weight ratio, polymers are competitive. Other factors, however, are important such as relative cost and other mechanical properties yet to be considered.

2.3 STATIC FAILURE THEORIES

The traditional educational path to design and analysis with polymers and reinforced composites is typically first through exposure to metals. While metals do exhibit time dependency and thus viscoelasticity, this is usually ignored unless the temperature approaches the melt temperatures. For most design applications the creep of metals is rarely an issue at room temperature. Some exceptions are large civil engineering structures, and mechanically leaf springs on cars. At high temperatures, however, creep of metals is important. Examples include gas turbine blades and pressure vessels and piping in the utility industry. In these cases, high-temperature alloys must be used.

Let us summarize the majority of stress failures in metals at room temperature as either static (ductile or brittle) or fatigue. It has been common to treat most metals as isotropic. A general three-dimensional (3D) state of stress is expressed as the stress tensor

$$\sigma_{ij} = \begin{pmatrix} \sigma_{xx} & \tau_{xy} & \tau_{xz} \\ \tau_{yx} & \sigma_{yy} & \tau_{yz} \\ \tau_{zx} & \tau_{zy} & \sigma_{zz} \end{pmatrix} \tag{2.9}$$

where $\sigma_{ii} = \sigma_i$, the normal stress on the ith-face, $i = x$, y, and z, τ_{ij} = the shear stress on the ith face in the j direction, and $\tau_{ij} = \tau_{ji}$ for static equilibrium.

Note that the x, y, and z directions are arbitrary. However, it is beneficial and customary to choose them along the principal axes of the component to be analyzed. Failure theories are based on the principal stresses. Consider a general 3D state of stress on an infinitesimal cube, where all components of the stress tensor are nonzero. Principal stresses are the eigenvalues of the stress tensor. Mathematically, the eigenvalues are a stress tensor where the off-diagonal terms, that is, shear stresses are zero.

$$|\sigma_{ij} - \sigma_k I| = 0 \tag{2.10}$$

where σ_k = principal stresses k = 1, 2, and 3 with the convention that $\sigma_1 > \sigma_2 > \sigma_3$. Physically, this can be interpreted as rotating the infinitesimal cube around the three axes such that the shear stresses become zero. Traditionally, the principal stresses have been presented and calculated graphically as Mohr's circle. While this methodology is still used to present these concepts, there is actually no need to do so.

Mechanical design and analysis is based on either a deterministic or probabilistic approach. The overall probabilistic approach is beyond the scope of this text, but needless to say, it includes the distribution of both the loads and material strengths. The deterministic approach is usually referred to as the factor of safety approach. The first approach to a design is often to calculate the factor of safety η, which is in general defined as the ratio of allowable property over the applied loading. In mechanical terms, the allowable property is usually either an allowable load or strength over an applied load or stress.

$$\eta = \frac{P_{\text{critical}}}{P} \tag{2.11}$$

or

$$\eta = \frac{S}{\sigma}, \tag{2.12}$$

where P_{critical} = allowable load (e.g., critical buckling load), P = applied load, S = allowable strength, and σ = applied stress.

The static failure theory for brittle materials has historically been the Coulomb–Mohr failure theory. A simple definition of a brittle material is that the ultimate compressive strength is greater than the tensile strength, $S_{uc} > S_{ut}$, and the yield strengths are approximately equal to the ultimate strengths in tension and compression, $S_{yt} \approx S_{ut}$ and $S_{yc} \approx S_{uc}$. Cast iron, rock, and concrete are typical materials that are analyzed by the Coulomb–Mohr failure criterion. Actually, for metals such as cast iron, the modified Mohr failure criterion is used where there is a slight modification to the Coulomb–Mohr that results in a better fit to the experimental data. Both the Coulomb–Mohr and modified Mohr failure envelopes are shown in Figure 2.3.

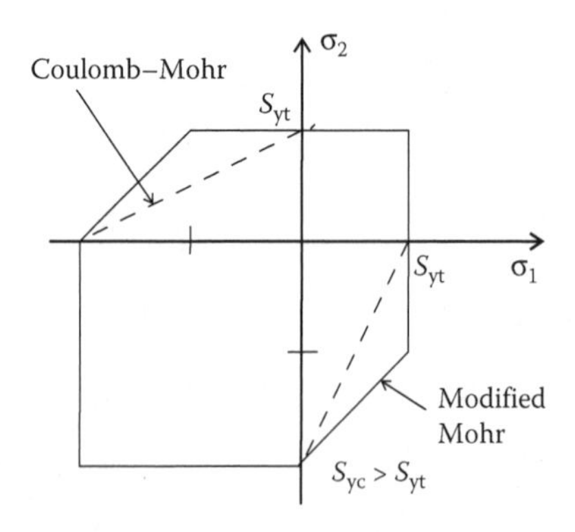

FIGURE 2.3 Coulomb–Mohr and modified Mohr static failure theories for brittle materials.

The static failure for ductile materials is the distortion energy, or von Mises criterion, which compares the yield strength to an equivalent tensile stress.

$$\sigma_e = \left(\frac{(\sigma_1 - \sigma_2)^2 + (\sigma_2 - \sigma_3)^2 + (\sigma_3 - \sigma_1)^2}{2} \right)^{1/2} \tag{2.13}$$

Finite-element analysis will often exhibit a 3D state of stress for a two-dimensional (2D) simple loading; however, the analysis will also directly calculate the von Mises stress from the results. Many, if not most, loading situations are macroscopically 2D with $\sigma_3 = 0$ and thus the von Mises equivalent stress is

$$\sigma_e = (\sigma_1^2 - \sigma_1\sigma_2 + \sigma_2^2)^{1/2} \tag{2.14}$$

In terms of x and y components of stress

$$\sigma_e = (\sigma_x^2 - \sigma_x\sigma_y + \sigma_y^2 + 3\tau_{xy}^2)^{1/2} \tag{2.15}$$

Furthermore, the majority of stress states for simple loadings additionally only have $\sigma_x \neq 0$ and $\tau_{xy} \neq 0$, and thus Equation 2.15 can be further reduced to

$$\sigma_e = (\sigma_x^2 + 3\tau_{xy}^2)^{1/2} \tag{2.16}$$

The von Mises criterion defines an ellipse in the 2D principal stress plane as shown in Figure 2.4. The maximum shear stress theory is also shown in Figure 2.4. However, for metals, while the maximum shear stress criterion is conservative, not only can the von Mises criterion be derived, but it also fits the experimental data better than the maximum shear criterion, and thus is the best estimation of the failure envelope.

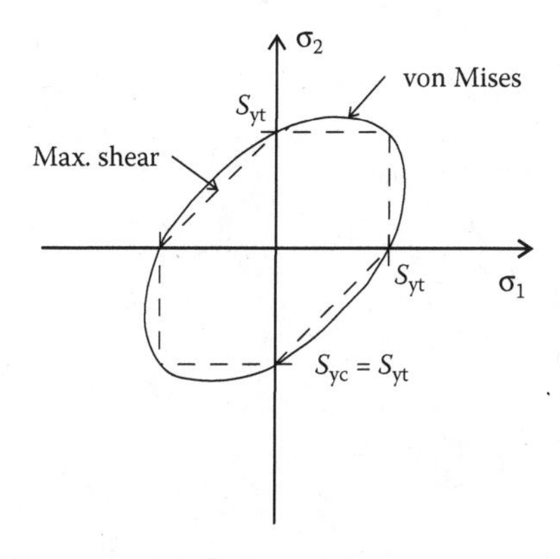

FIGURE 2.4 Static failure theories for ductile materials.

Example 2.1

Let us consider a simple example of a cantilever beam with an applied transverse load, F, and a torsion T as shown in Figure 2.5. This example will set the groundwork for future discussion on the factor of safety. Assume that the beam is circular.

At the wall, ignoring primary shear, the stresses are

$$\sigma_x = \frac{Mc}{I} = \frac{Flc}{I} = \frac{32Fl}{\pi d^3} \tag{2.17}$$

and

$$\tau_{xy} = \frac{Tc}{J} = \frac{16T}{\pi d^3}, \tag{2.18}$$

where F = transverse load, $c = d/2$ distance from neutral axis to outermost fibers, $I = \pi d^4/64$ = area moment of inertia, $J = \pi d^4/32$ = polar moment of inertia, d = diameter of beam, and l = length of beam.

Substituting Equations 2.17 and 2.18 into the equivalent, von Mises stress Equation 2.16 results in

$$\sigma_e = \left(\left(\frac{32Fl}{\pi d^3}\right)^2 + 3\left(\frac{16T}{\pi d^3}\right)^2\right)^{1/2} \tag{2.19}$$

The factor of safety depends on how the beam is loaded from the nominal design point. If both loads occur proportionally, then the factor of safety, η is applied to both loads and the resulting expression for factor of safety is

$$\eta = \frac{S_y}{\sigma_e} = \frac{S_y}{\left((32Fl/\pi d^3)^2 + 3(16T/\pi d^3)^2\right)^{1/2}} \tag{2.20}$$

Note that at this stage, Equation 2.20 can also be used to design, by selecting a factor of safety and calculating the diameter

$$d = \sqrt[3]{\frac{\eta\left((32Fl)^2 + 3(16T)^2\right)^{1/2}}{\pi S_y}}. \tag{2.21}$$

However, if only one of the loads will experience overloading, then the factor of safety is applied to that load only. Consider the case where only the

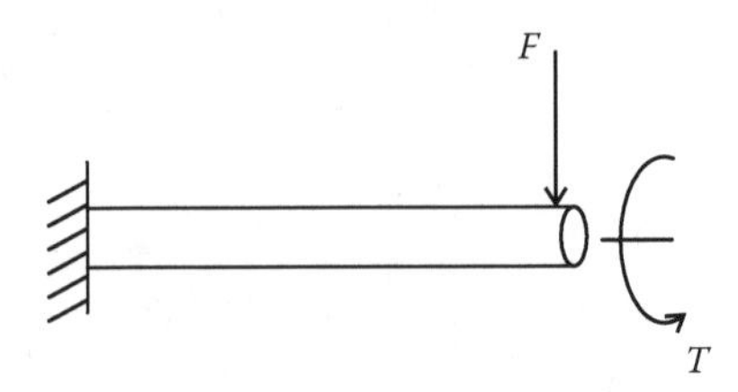

FIGURE 2.5 Cantilever beam with transverse load and torsion.

transverse load will increase and the torque remains constant. The factor of safety is applied directly to the transverse load and not to the torque.

$$1 = \frac{S_y}{\left((32\eta Fl/\pi d^3)^2 + 3(16T/\pi d^3)^2\right)^{1/2}} \tag{2.22}$$

The factor of safety for this case when the diameter is known is

$$\eta = \left(\frac{\pi d^3}{32Fl}\right)\sqrt{S_y^2 - 3\left(\frac{16T}{\pi d^3}\right)^2} \tag{2.23}$$

In the late 1960s and early 1970s, it became apparent that the yield failure of polymers (for both brittle and ductile polymers) had a hydrostatic pressure dependence that was neither present in metals, nor accounted for by the existing failure criterion. For polymers, where the compressive and tensile strengths are often different, Raghava et al.[3] found that inclusion of both the hydrostatic and differing compressive and tensile strengths could be accommodated in what can be considered a more general form of the distortion energy or von Mises criteria

$$(\sigma_1 - \sigma_2)^2 + (\sigma_2 - \sigma_3)^2 + (\sigma_3 - \sigma_1)^2 + 2(C - T)(\sigma_1 + \sigma_2 + \sigma_3) = 2CT \tag{2.24}$$

where C = compressive yield strength (always positive) and T = tensile yield strength (always positive).

Note that Equation 2.24 reduces to the von Mises equation when $C = T$. They found that this worked well at C/T ratios below 2. At a $C/T = 2.0$, the experimental data generated an elliptical failure envelope, the only discrepancy was that the curve fit to the data which determined C and T while conservative (i.e., the failure ellipse was smaller than the experimental ellipse) did not fit the data well. The authors attribute this to the sensitivity of the curve fit to the minimal data available in the quadrant where both σ_1 and σ_2 are negative. This failure criterion was applied to experimental data by the authors and others[4–6] on PVC, polycarbonate (PC), PS, and PMMA, which is a wide range of the commodity polymers. Figure 2.6 shows how the ellipse both shifts and grows when the C/T ratio goes from 1.0 to 1.5 with T = 70 MPa.

Sauer et al.[7] observed that both the compressive and tensile yield strength and modulus changed as a function of hydrostatic pressure. When the compressive and tensile strengths are determined at a given hydrostatic pressure, the result is incorporated into the experimentally determined failure envelope. However, when the strengths are determined at atmospheric or hydrostatic pressures different from the use of pressure, they must be correct for the difference in pressure. The relationships for tensile, σ_{yt}, and compressive, σ_{yc}, yield stress are a simple linear function of hydrostatic pressure.

$$S_{yt} = S_t^0 + kp \tag{2.25}$$

$$S_{yc} = S_c^0 + k'p \tag{2.26}$$

where S_t^0 = tensile yield strength at atmospheric pressure, S_c^0 = compressive yield strength at atmospheric pressure, p = pressure, k = experimentally determined constant

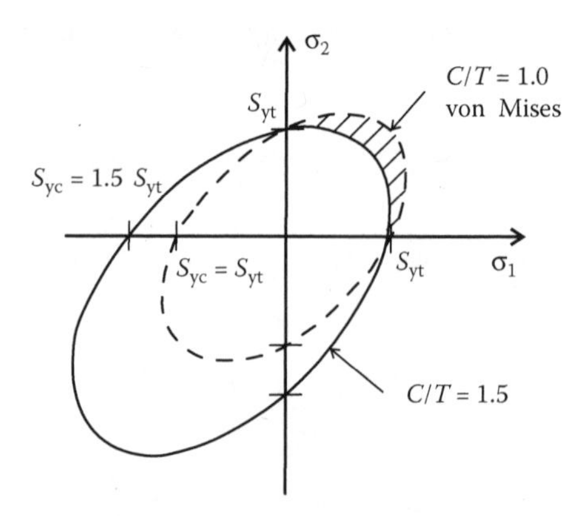

FIGURE 2.6 Comparison of von Mises to shear distortion with hydrostatic pressure effect 2D failure envelope (C = compression yield strength, T = 70 MPa tensile yield strength).

from tensile yield stress as a function of pressure, and k' = experimentally determined constant from compressive yield stress as a function of pressure.

Similarly, they found that modulus was also a function of pressure. However, the slope of change was related to the atmospheric Poisson's ratio

$$E = E^0 + mp \tag{2.27}$$

where E^0 = atmospheric modulus, $m = 2(5 - 4\upsilon_0)(1 - \upsilon_0)$, and v_0 = atmospheric Poisson's ratio.

The previous static failure theory discussions were intended to transition from the knowledge that a design engineer may bring from a background of metals or other geotechnical materials to that of polymers. Specific time-dependent failures such as creep rupture and fatigue are discussed in more detail in the applicable, subsequent chapters.

2.4 CREEP PROPERTIES

In contrast to the static elongation, the creep strain is time dependent. The standard creep test is to apply a stress constant with time and measure the resulting creep strain as a function of time as shown in Figure 2.7. If no plastic strains occur (the stress applied σ is less than a yield value which may be less than S_{sy}), the material is said to be viscoelastic and the total strain can be represented as

$$\varepsilon = \varepsilon_e + \varepsilon_c, \tag{2.28}$$

where

$$\varepsilon_e = \sigma / E = \text{initial elastic response} \tag{2.29}$$

and

$$\varepsilon_c = \varepsilon_c(t) = \text{creep response}. \tag{2.30}$$

The creep test is a standard test to determine whether a polymer is a solid. If the polymer exhibits solid behavior, the creep strain reaches an asymptote. If the polymer

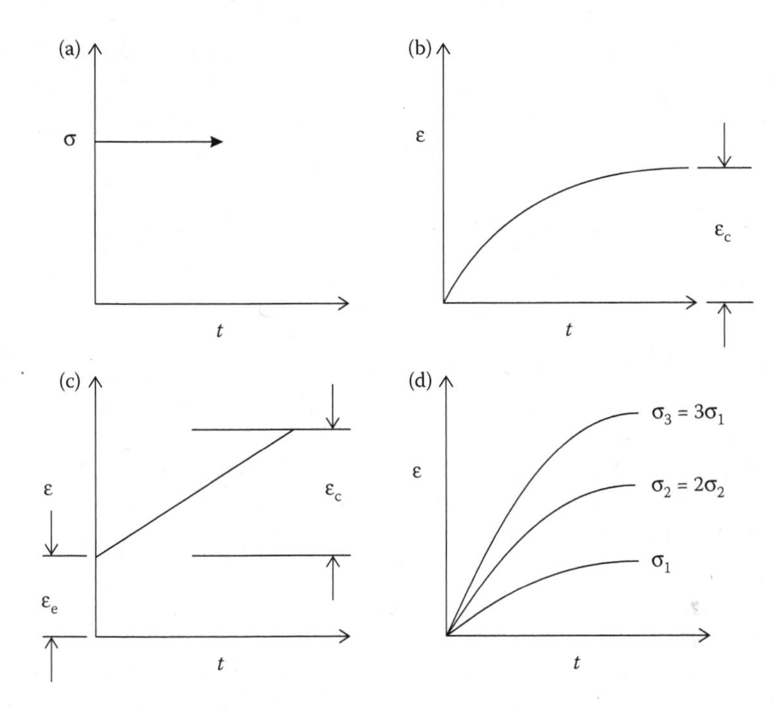

FIGURE 2.7 Standard creep test. (a) Stress, (b) creep, solid behavior, (c) creep, fluid behavior, and (d) creep as a function of stress.

exhibits fluid behavior, the creep strain is unbounded: $\varepsilon \to \infty$ as $t \to \infty$. The creep strains increase with stress as shown in Figure 2.7d. If the creep strain, as well as the elastic strain, is proportional to the applied stress, the polymer is a linear viscoelastic material, and it has a mechanical property called its creep compliance such that

$$\varepsilon(t) = \sigma J(t) \tag{2.31}$$

where σ = constant and $J(t)$ = creep compliance. The creep behavior of linear viscoelastic materials can be represented by one master curve as shown in Figure 2.8.

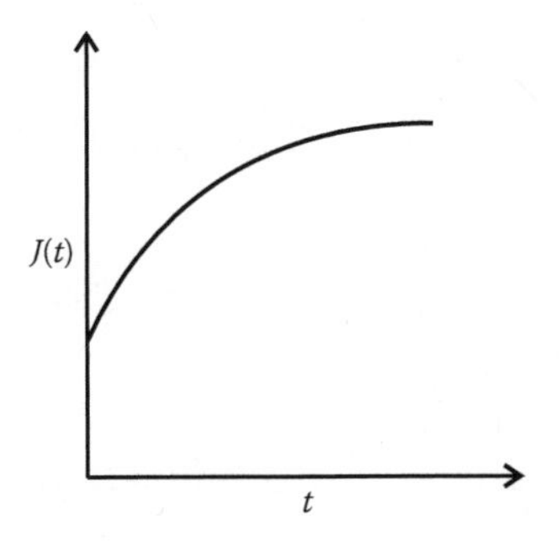

FIGURE 2.8 Master curve of creep compliance $J(t)$ for a linear viscoelastic polymer.

The configuration of the creep test specimen is usually made the same as the standard tensile test specimen for static testing. Creep data are generally determined in tension, but creep strains can be measured in compression, shear, or bending also.

2.5 RELAXATION PROPERTIES

The relaxation of polymers is another time-dependent property. The standard relaxation test is opposite that of the creep test; the strain is kept constant, and the relaxation in stress is measured as shown in Figure 2.5. The relaxation in stress is measured as shown in Figure 2.9. The relaxation test is a standard test to determine if a polymer can be classified as a fluid. Fluid behavior is characterized by the stress relaxing to zero, $\sigma \to 0$ as $t \to 0$.

The initial elastic response in this case is $\sigma = E\varepsilon$. Similar to the creep compliance, a linear viscoelastic polymer has a relaxation modulus $Y(t)$, which is a characteristic material property.

$$\sigma(t) = \varepsilon Y(t) \tag{2.32}$$

where ε = constant. Like $J(t)$, the relaxation modulus can be plotted on one master curve. In a subsequent chapter, analytical expressions will be derived for $J(t)$ and $Y(t)$, by using an analogy with mechanical models. First, dynamic properties need to be considered.

2.6 DYNAMIC PROPERTIES

It has been demonstrated that polymers can exhibit fluid as well as solid behavior. The viscous fluid behavior of polymers becomes evident under dynamic loading where the viscous damping property reduces the amplitude of free vibrations. It is also observed that the strain will lag the stress and that there can be an energy loss through heat dissipation during periodic loading.

2.6.1 DYNAMIC TESTS

Two common types of dynamic tests[3] are the torsional pendulum test and the vibrating reed test shown in Figure 2.10a and b. The first measures a shear response (a dynamic

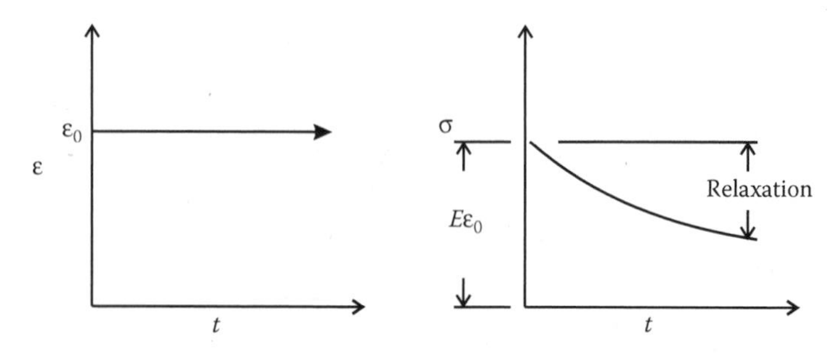

FIGURE 2.9 Standard relaxation test.

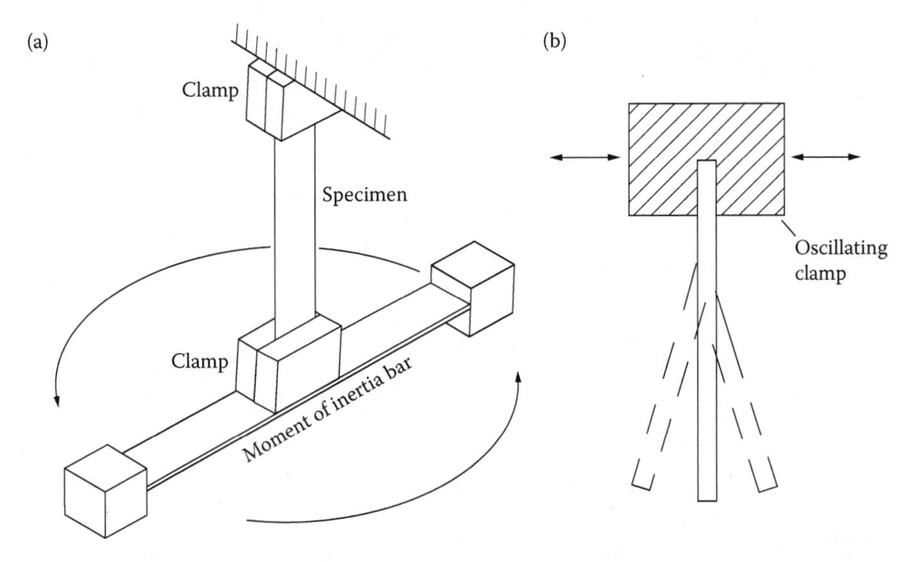

FIGURE 2.10 Two common dynamic tests. (a) Torsional pendulum test and (b) vibrating reed test. (From Nielsen, L. E., *Mechanical Properties of Polymers and Composites*, Vol. 1, pp. 11–17, Marcel-Dekker, Inc., New York, NY, 1974.)

shear modulus, G), and the second measures a longitudinal bending response (a dynamic tensile modulus, E).

Another test[10,11] is called the dynamic mechanical analysis (DMA) test, and is shown in Figure 2.11. A beam specimen is clamped between two parallel arms, and an electromagnetic motor drives one arm. A linear variable differential transformer senses the motion, and a feedback loop to the motor is used to control the motion.

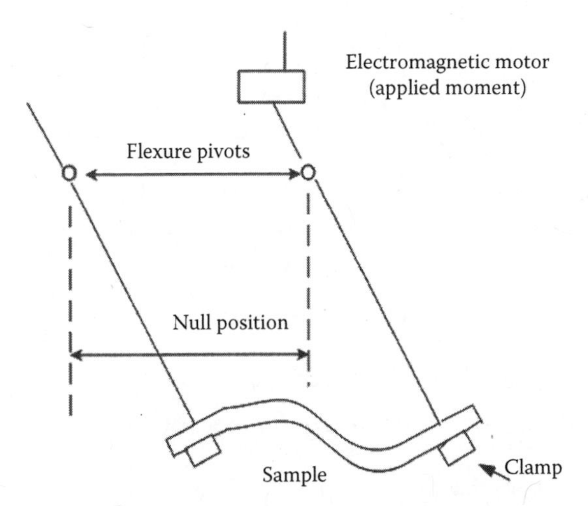

FIGURE 2.11 Electromagnetic dynamic mechanical analyzer. (From Sepe, M. P., *Advanced Materials and Processes Magazine*, 4, 32–41, 1992.)

Frequency and amplitude or force can be controlled. The sample can also be positioned in a temperature-controlled chamber to study the effects of temperature on mechanical properties. (A study of the parameters governing the DMA test is left to a homework problem.)

When experiments are necessary to determine mechanical properties, the tests should be designed and conducted according to ASTM standards. ASTM is an abbreviation for the American Society for the Testing of Materials. For example, the DMA test above is compared with the ASTM D.790 flexural modulus test in Sepe.[10]

2.6.2 Dynamic Modulus and Damping

From a dynamic test, for example, the vibrating reed test, or the DMA test, a complex modulus E^* is defined as

$$E^* = E' + iE'' = E_0 e^{i\delta t}, \tag{2.33}$$

where E' is the real part of the complex modulus ($E' = \mathrm{Re}(E^*)$), and E'' is the imaginary part ($E'' = \mathrm{Im}(E^*)$), and $i = \sqrt{-1}$. E^* is not to be confused with the mathematical use of the superscript * which indicates the complex conjugate. E' is the same as E measured in the static test if the strain rate and damping are small. E'' is the loss modulus representing the damping and energy dissipation. For a sinusoidal variation in stress

$$\sigma = \sigma_0 \sin(\omega t), \tag{2.34}$$

strain can be shown from $\varepsilon = \sigma / E^*$ to lag the stress by a phase angle δ where

$$\varepsilon = \varepsilon_0 \sin(\omega t - \delta) \tag{2.35}$$

and

$$\tan \delta = \frac{E''}{E'} \tag{2.36}$$

Thus, E'' is a measure of the phase angle lag in the strain cycle. If the energy loss is calculated for one cycle of stress and strain (i.e., for one hysteresis loop) for t varying from 0 to 2π, and for the stress and strain given by Equations 2.34 and 2.35, respectively, it can be shown that the energy loss per unit volume dissipated per cycle in the form of heat energy is

$$\delta H = \int \sigma \, d\varepsilon = E\varepsilon_0^2 \pi \sin \delta = E'' \varepsilon_0^2 \pi \tag{2.37}$$

Similarly, a complex creep compliance can be found, namely that

$$J^* = J' - iJ'' \tag{2.38}$$

It can also be shown that the following relationships exist:

$$E' = \frac{J'}{J^2}, \quad J' = \frac{E'}{E^2}, \tag{2.39}$$

And furthermore,

$$\frac{E''}{E'} = \frac{J''}{J'}, \tag{2.40}$$

where

$$E^2 = E'^2 + E''^2, \quad E^2 = E^{*2} \tag{2.41}$$

and

$$J^2 = J'^2 + J''^2. \tag{2.42}$$

A damping measure that is commonly used is the logarithmic decrement Δ. It is defined as

$$\Delta \equiv \ln\left(\frac{A_1}{A_2}\right) = \ln\left(\frac{A_2}{A_3}\right) = \ln\left(\frac{A_i}{A_{i+1}}\right), \tag{2.43}$$

where A_i are successive amplitudes shown in Figure 2.12. If the damping is low (see Section 2.8), then

$$\Delta \approx \pi\frac{E''}{E'}. \tag{2.44}$$

When measuring amplitudes of vibration to obtain a damping factor as shown in Figure 2.12, we have to be aware of possible errors. For example, if an accelerometer is mounted on the end of a vibrating cantilever beam as shown in Figure 2.10b, a decay in amplitude will be measured not only due to the internal damping in the material, but also due to friction in the clamp as well. One method to reduce this friction is to design a contoured clamp with a curvature varying according to the theoretically calculated curvature of a bent beam. Tests have been performed on beams suspended

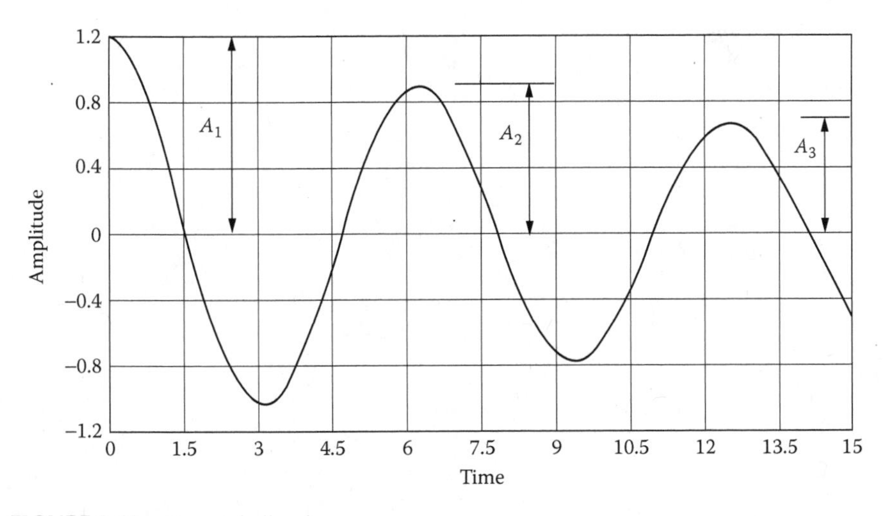

FIGURE 2.12 Damped vibration curve.

by fine wires or strings under three and four point bending where the points of suspension are considered as free boundary conditions. The measurement error can be minimized by using $\Delta = \ln(A_1/A_N)/(N-1)$, where $N > 2$.

2.6.3 Dynamic Property Data

As mentioned in Section 2.6.1, the DMA test can be used to determine dynamic properties as a function of temperature. The thermomechanical spectrum is a plot of the storage modulus, loss modulus, and tan δ as functions of frequency as will be shown later in both this chapter and the text. Let us examine the thermomechanical spectrum in more detail by considering the storage modulus temperature dependence as shown in Figure 2.13. Figure 2.13a shows the four regions of the spectrum; glassy, leathery, rubbery, and terminal (either melting or degradation). In addition, there are a few other points of importance. Note that the temperature difference between the glass transition temperature and the temperature associated with the terminal region is typically between 100°C and 150°C. In addition, the difference in modulus between the glassy and rubbery regions is usually of the order of 2–6 decades. Figure 2.13b

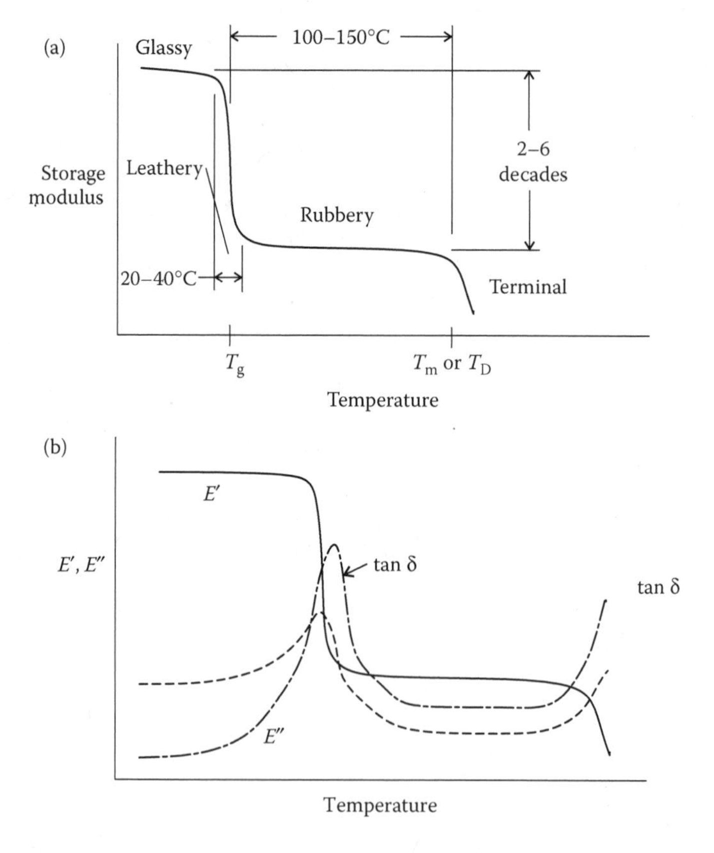

FIGURE 2.13 Thermomechanical spectra fundamentals. (a) Regions of thermomechanical spectrum, (b) viscoelastic parameters of the thermomechanical spectrum.

shows an idealized representation of the three viscoelastic properties, E', E'', and $\tan\delta = E''/E'$. It is useful to note that the glass transition temperature is estimated based on, in order of decreasing commonality, the peak in $\tan\delta$, the peak in E'', and last due to degree of difficulty in estimation, the inflection point of E'. Figure 2.14a shows three hypothetical polymers, A, B, and C. Note that polymer B is only shifted in temperature relative to polymer A. Polymer C is shifted with respect to storage modulus relative to A. Figure 2.14b shows that if the storage modulus is plotted relative to the glass transition temperature, the curves overlap with respect to temperature. The moduli could also be normalized and then for the idealized case, all three of the curves would coincide. A variety of polymer properties that have viscoelastic dependence can be normalized with respect to the glass transition temperature as shown in this hypothetical example. As an actual example, Figure 2.15 shows flexure modulus E' for epoxy, polyether sulfone (PES), and PET as a function of temperature.

Also indicated in Figure 2.15 are the values of deflection temperature under load (DTUL). This is the temperature at which excessive deflections start to occur and should correlate with a decrease in modulus E'. The DTUL as determined under static loads correlates with E' determined under dynamic loading for PES and PET,

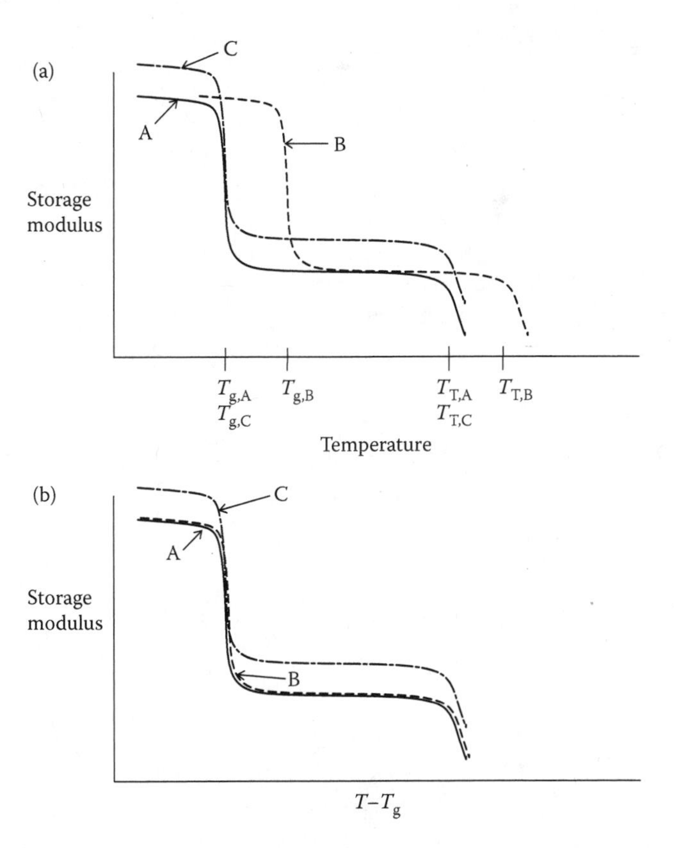

FIGURE 2.14 Thermomechanical spectra comparisons. (a) Comparison of three polymers, (b) thermomechanical spectra of polymers normalized by glass transition temperature.

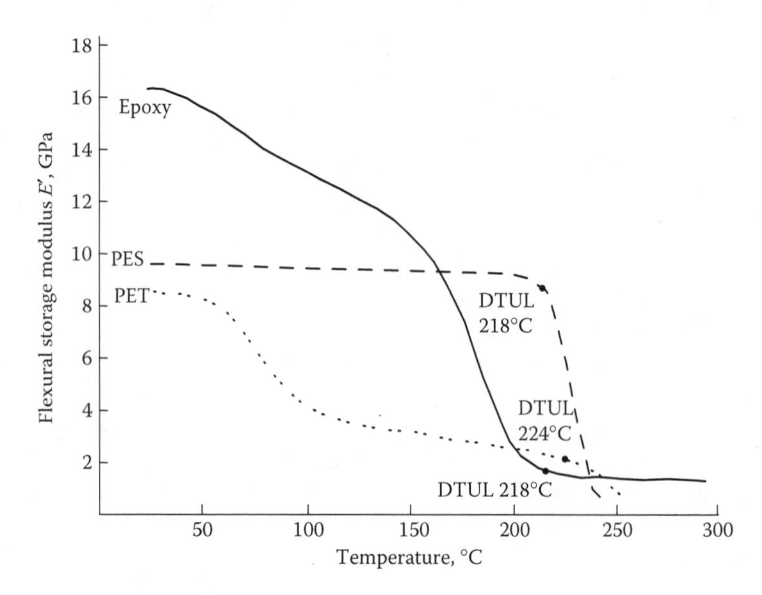

FIGURE 2.15 Flexure modulus E' for epoxy, PES, and PET as a function of temperature determined by the DMA test. (PES, polyether sulfone; PET, polyethylene terephthalate.) (From Sepe, M. P., *Advanced Materials and Processes Magazine*, 4, 32–41, 1992. Reprinted with permission of ASM International.)

but not for epoxy. Damping in the form of tan δ for polyertherether ketone (PEEK) is shown in Figure 2.16. For this semicrystalline thermoplastic, postbaking at 285°C for 1 h after molding increases the modulus E' but decreases the damping. The glass transition is defined in various ways from mechanical tests, one way is from the inflection point in the leathery region of E', the peak of E'' or the peak of tan δ.

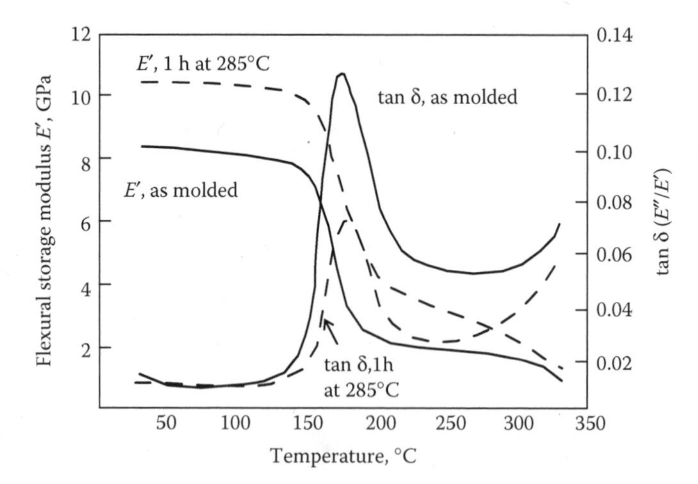

FIGURE 2.16 Modulus and damping properties of PEEK as changed by postmold baking. (PEEK, polyertherether ketone.) (From Sepe, M. P., *Advanced Materials and Processes Magazine*, 4, 32–41, 1992. Reprinted with permission of ASM International.)

More such dynamic property data can be found in Nielsen,[8] Sepe,[10] and Baer.[11] In later chapters in this text, Chapters 4 and 5, further attention will be given to some other dynamic properties, namely the impact and fatigue properties, respectively. In Chapter 3, mechanical modeling of polymers is considered with the purpose of deriving differential equations to represent the fundamentals of viscoelastic behavior. Before doing so, some additional special topics are considered here.

2.7 LARGE STRAIN DEFINITIONS

In addition to the true strain definition in Equation 2.2, there are other large strain definitions that are used in nonlinear continuum mechanics. These are the Lagrangian and Eulerian definitions depending on whether the original coordinates X_i or the deformed coordinates x_i are used as the reference coordinates.[9] The Lagrangian strain is defined by

$$E_{ij}^* = \frac{1}{2}\left[\frac{\partial u_i}{\partial X_j} + \frac{\partial u_j}{\partial X_i} + \left(\frac{\partial u_1}{\partial X_j}\right)\left(\frac{\partial u_1}{\partial X_i}\right) + \left(\frac{\partial u_2}{\partial X_j}\right)\left(\frac{\partial u_2}{\partial X_i}\right) + \left(\frac{\partial u_3}{\partial X_j}\right)\left(\frac{\partial u_3}{\partial X_i}\right)\right], \quad (2.45)$$

where u_i are the displacements in the X_i directions, respectively, with $i = 1, 2, 3$. It can be shown that for uniaxial strain that the Lagrangian strain is

$$E_{11} = \frac{1}{2}\left[\left(\frac{L}{L_0}\right)^2 - 1\right]. \quad (2.46)$$

This is called the Kirchoff strain by Nielsen.[8] The Eulerian strain is defined by

$$E_{ij}^* = \frac{1}{2}\left[\frac{\partial u_i}{\partial x_j} + \frac{\partial u_j}{\partial x_i} - \left(\frac{\partial u_1}{\partial x_j}\right)\left(\frac{\partial u_1}{\partial x_i}\right) - \left(\frac{\partial u_2}{\partial x_j}\right)\left(\frac{\partial u_2}{\partial x_i}\right) + \left(\frac{\partial u_3}{\partial x_j}\right)\left(\frac{\partial u_3}{\partial x_i}\right)\right]. \quad (2.47)$$

It can be shown for uniaxial strain that the Eulerian strain is

$$E_{11}^* = \frac{1}{2}\left[1 - \left(\frac{L_0}{L}\right)^2\right]. \quad (2.48)$$

This is called the Murnaghan strain.[8]

2.8 ANALYSIS OF DAMPING

An analysis is made with a mechanical spring/dash pot model as shown in Figure 2.17, where a mass M is restrained by a spring of stiffness k and a dashpot with viscosity η. The solution as illustrated in Figure 2.17 is an exponentially decaying cosine wave, that is, the displacement of the mass is given by

$$X = X_0 e^{-\alpha\omega t}\cos(\omega t). \quad (2.49)$$

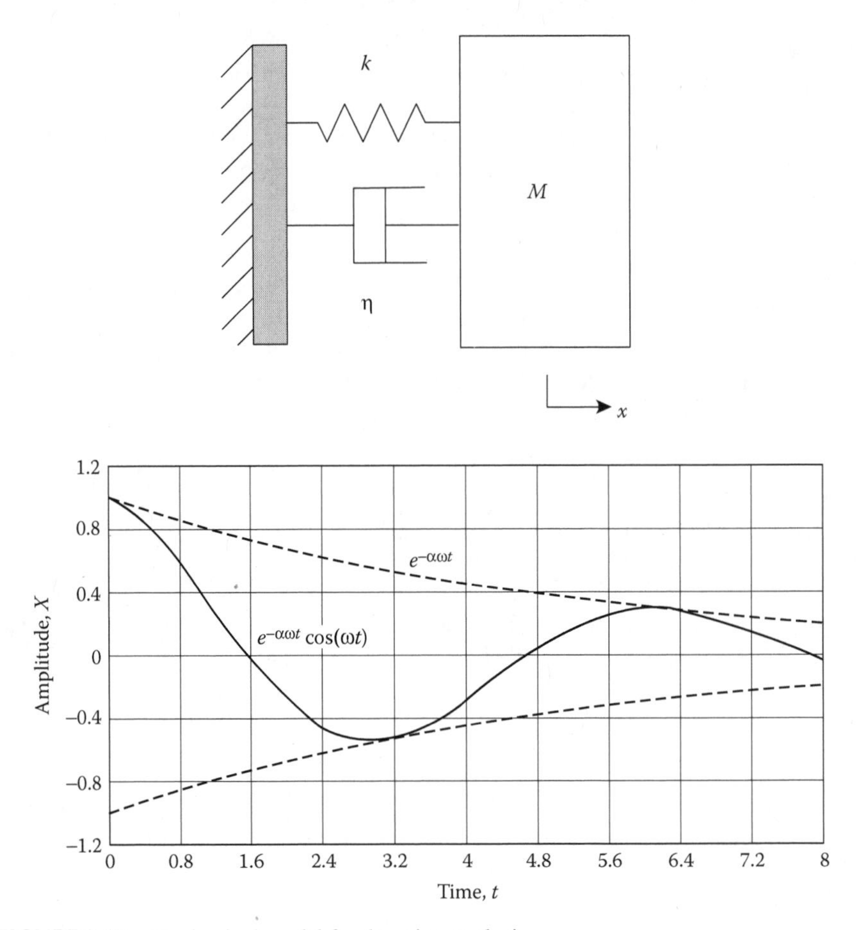

FIGURE 2.17 Mechanical model for damping analysis.

Analogous to the displacement X in the mechanical model, let the internal strain in a polymer material be represented by

$$\varepsilon = \varepsilon_0 e^{-\alpha\omega t}\cos(\omega t). \tag{2.50}$$

To determine α, use Figure 2.17 as reference and note that at $t = 0$,

$$\varepsilon = \varepsilon_0 = A_1 \tag{2.51}$$

and at $t = 2\pi/\omega$,

$$\varepsilon = \varepsilon_0 e^{-2\alpha\pi} = A_2. \tag{2.52}$$

Thus,

$$\frac{A_1}{A_2} = e^{2\alpha\pi} \tag{2.53}$$

and

$$\Delta = \ln\left(\frac{A_1}{A_2}\right) = 2\alpha\pi, \tag{2.54}$$

which shows how the log decrement is related to the exponential decay factor α.

Next a relation to actual material properties will be found. The equation of motion for the mechanical model in Figure 2.17 is

$$M\ddot{x} + \eta\dot{x} + kx = 0 \tag{2.55}$$

whose solution is

$$x = ce^{\beta t}, \tag{2.56}$$

where

$$\beta = \frac{(\eta/M) \pm \sqrt{(\eta/M)^2 - (4k/M)}}{2}. \tag{2.57}$$

For small damping, assume

$$\left(\frac{\eta}{M}\right) < 4\frac{k}{M}. \tag{2.58}$$

Thus,

$$\beta = \frac{(\eta/M) \pm i\sqrt{(4k/M) - (\eta/M)^2}}{2}. \tag{2.59}$$

This expression has the form

$$\beta = \alpha\omega \pm i\omega, \tag{2.60}$$

where

$$\omega = \sqrt{\frac{k}{M} - \left(\frac{\eta}{2M}\right)^2}. \tag{2.61}$$

For $(\eta/2M) \ll 1$,

$$\omega = \sqrt{\frac{k}{m}}. \tag{2.62}$$

From Equations 2.59 and 2.60,

$$\alpha\omega = \frac{\eta}{2M}. \tag{2.63}$$

Next, we make a relation to stress and strain. Because stress is derived from force and strain from displacement, the following expression can be written using Newton's second law:

$$\sigma = -M\ddot{\varepsilon}, \tag{2.64}$$

where a negative sign is taken since the stress represents an internal restraining force. The mass term M may need a factor to account for units, but this factor will cancel out as the analysis proceeds.

Let the strain be given by

$$\varepsilon = \varepsilon_0 e^{\beta t}. \tag{2.65}$$

Using the dynamic modulus from Equation 2.33 and using Equation 2.64 gives

$$\sigma = E_o e^{i\delta t}\,\varepsilon_o e^{\beta t} = -M\beta^2\varepsilon_0 e^{\beta t}. \tag{2.66}$$

Thus,

$$\begin{aligned} E' + iE'' &= -M\beta^2 \\ &= -M(\alpha^2\omega^2 - 2i\alpha\omega^2 - \omega^2). \end{aligned} \tag{2.67}$$

The real part yields

$$E' = -M\omega^2(\alpha^2 - 1) \approx M\omega^2 \tag{2.68}$$

for small damping, $\alpha^2 \ll 1$. The imaginary part yields

$$E'' = 2M\alpha\omega^2. \tag{2.69}$$

Eliminating $M\omega^2$ from Equations 2.68 and 2.69 gives

$$\alpha = \frac{E''}{2E'}. \tag{2.70}$$

Then from Equation 2.54

$$\Delta = \frac{\pi E''}{E'}. \tag{2.71}$$

Equation 2.71 is the same as Equation 2.44 given earlier that was the goal of this analysis.

2.9 TIME HARDENING CREEP

A time hardening creep law has been used is the Nutting equation.[8] This equation is represented here as

$$\varepsilon = K\sigma t^n, \tag{2.72}$$

where K and n are material constants. The time hardening exponent is n. For $n = 0$ the material is perfectly elastic. For $n = 1$, the material is a perfect fluid (Maxwell

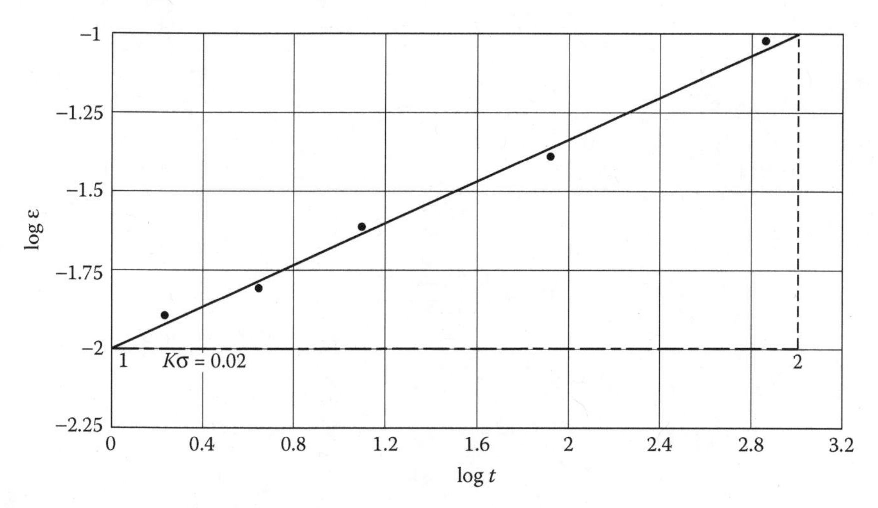

FIGURE 2.18 Creep data plotted on log–log scale.

fluid). The material constants can be found by plotting the data on a log–log scale as shown in Figure 2.18. The constant K is given by

$$K = \frac{\varepsilon}{\sigma} \quad \text{at } t = 0. \tag{2.73}$$

The constant n can be found from the slope on the log–log scale. Take the log of Equation 2.72 and obtain

$$\log\varepsilon = \log K + \log\sigma + n\log t. \tag{2.74}$$

When σ = constant, this is a creep test. A best straight fit line through the data is shown in Figure 2.19. Then from Equation 2.74, for the two points 1 and 2 (at an extreme distance apart for best accuracy), solve for n from

$$n = \frac{\log\varepsilon_2 - \log\varepsilon_1}{\log t_2 - \log t_1}. \tag{2.75}$$

2.10 ISOCHRONOUS CREEP CURVES

Literature data for creep are often replotted in a concise way in the form of isochronous stress as a function of strain curves as illustrated in Figure 2.19a and b. Creep data are obtained in the conventional way, as is Figure 2.7a–d, and are converted to constant time curves.

HOMEWORK PROBLEMS

2.1. Calculate and compare the values of engineering strain ε and the true strain ε_T as functions of $\delta L/L_0$ for $\delta L/L_0 = 0.0, 0.5, 1.0, 1.5$, and 2.0. Do

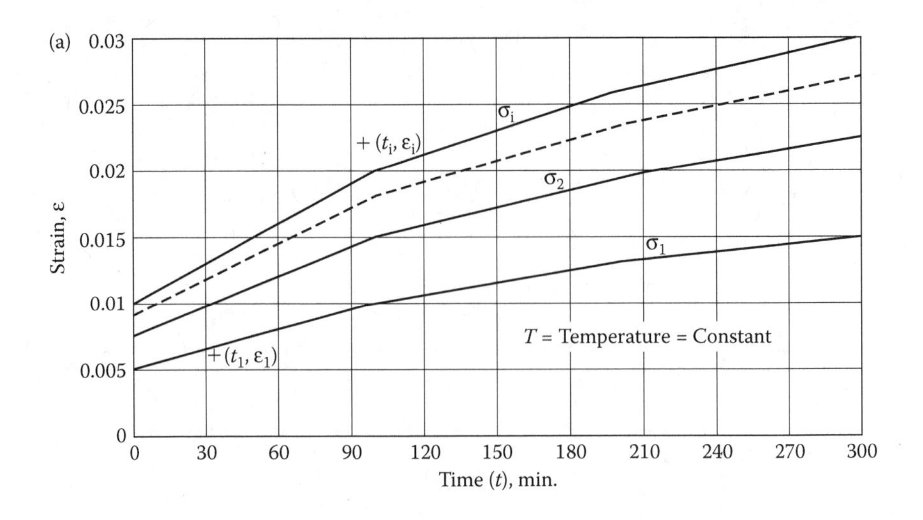

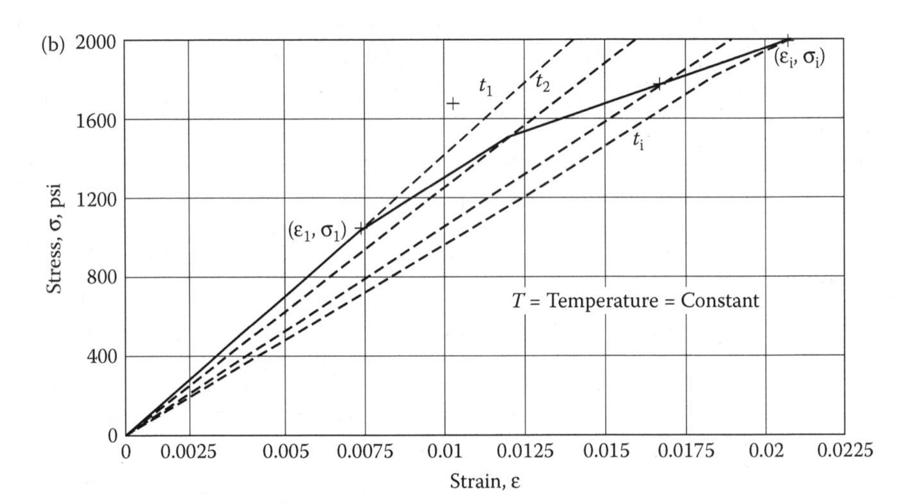

FIGURE 2.19 Isochronous (constant time) creep data. (a) Standard creep curves, (b) isochronous stress/strain curves.

the same for the engineering stress and the true stress. Assume $\sigma_T = K\,\varepsilon_T^N$ and $K = 2000$ and $N = 0.2$.

2.2. Derive Equation 2.8.

2.3. If a compression test is conducted on a brittle glassy amorphous polymer, at what angle would the fracture plane be oriented? (Review the "Mohr Coulomb" theory of failure.)

2.4. Verify Equations 2.19 through 2.42.

2.5. Show that the viscosity η of a linear viscous polymer element is equal to E''/ω, where $E^* = E' + iE''$ and ω is the frequency of vibration.

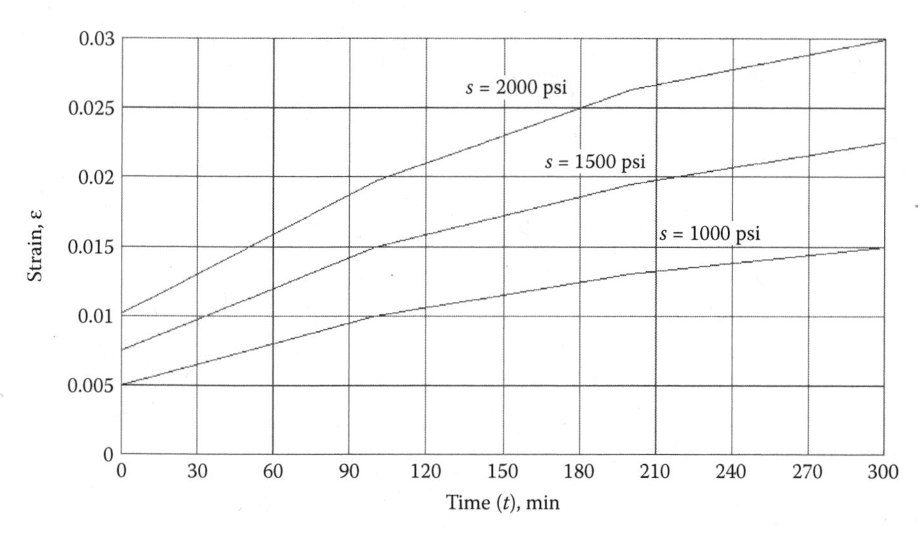

FIGURE 2.20 Creep data.

2.6. A polymer with a complex modulus E^* is tested in a vibrating beam test. Two different polymers A and B were tested. The periods of resonant oscillation were determined to be 0.05 and 0.06 s, respectively. Which polymer has the greatest modulus A or B? What part of E^* is determined by the period of oscillation E' or E''?

2.7. The creep data in Figure 2.20 were measured for a polymer at three different stress levels. Does the polymer have a creep compliance $J(t)$? If so, what is the $J(t)$ at time $t = 100$ min.? If the polymer has a $J(t)$, what kind of polymer is it?

2.8. Figure 2.21 gives some dimensions and forces for the DMA test, Figure 2.11.

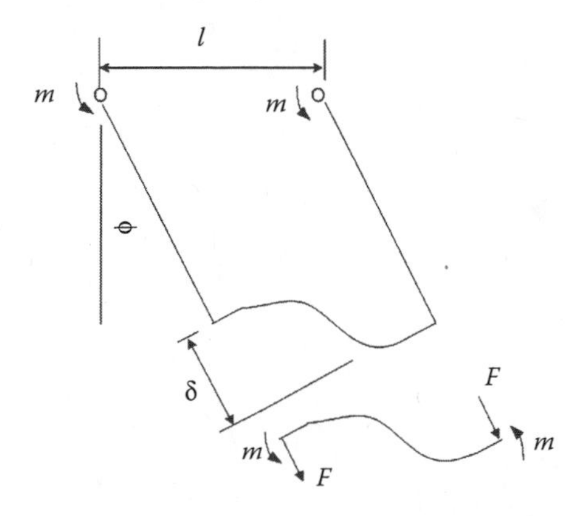

FIGURE 2.21 Free-body diagram for DMA homework Problem 2.8.

1. Show that the force F is approximately zero for small deflections.
2. Show that the deflection δ of the beam specimen is related to the moment by

$$\delta = \frac{ml^2}{6EI},$$

where

$$\delta = l \sin\theta.$$

3. Write the equation of motion of the system to control the motion. Use dynamics of rotation equations for rigid bars.

REFERENCES

1. ASTM (American Society for the Testing of Materials), Tensile Properties, Standard No. D638.
2. ASTM, Flexure Properties of Plastics, Standard No. D790.
3. Raghava, R., Caddell, R. M., and Yeh, G. S. Y. The macroscopic yield behaviour of polymers, *Journal of Materials Science*, 8, 225–232, 1973.
4. Whitney, W. and Andrews, R. D. Yielding of glassy polymers: Volume effects, *Journal of Polymer Science—Part C*, 16, 2981–2990, 1967.
5. Bauwens, J. C., Relation between the compression yield stress and the mechanical loss peak of bisphenol-A-polycarbonate in the β transition range, *Journal of Materials Science*, 7, 577–584, 1972.
6. Sternstein, S. and Ongchin, L. Yield criteria for plastic deformation of glassy high polymers in general stress fields, *ACS Polymer Preprint*, 10, 1117–1124, 1969.
7. Sauer, J. A., Mears, D. R., and Pae, K. D. Effects of hydrostatic pressure on the mechanical behaviour of polytetrafluoroethylene and polycarbonate, *European Polymer Journal*, 6, 1015–1022, 1970.
8. Nielsen, L. E., *Mechanical Properties of Polymers and Composites*, Vol. 1, pp. 11–17, Marcel-Dekker, Inc., New York, NY, 1974.
9. Malvern, L. E., *Introduction to the Mechanics of a Continuous Medium*, pp. 158–159, Englewood Cliffs, NJ: Prentice-Hall, 1969.
10. Sepe, M. P., Dynamic mechanical analysis pinpoints plastics temperature limits, *Advanced Materials and Processes Magazine*, 4, 32–41, 1992.
11. Baer, E., *Engineering Design for Plastics*, Polymer Science and Engineering Series, New York: Reinhold Book Corp., 1964.
12. TA Instruments, Inc., 109 Lukens Dr., New Castle, DE 19720.

3 Viscoelastic Behavior of Polymers

3.1 MECHANICAL MODELS

When a chemist or a chemical engineer looks at a polymer, he or she sees a molecular structure with atoms, C–C bonds, chains, branches, cross-links, and crystals. However, when a design engineer looks at a polymer, he or she sees a mechanism with springs, dashpots, and sliding blocks. For example, if we were to draw a box around an element of material, one can imagine that within that box there is a spring as in Figure 3.1a, or a dashpot (viscous damper) as in Figure 3.1b. These two models represent a linear elastic solid and a linear viscous fluid, respectively. For the first, the linear elastic solid, the $\sigma(\varepsilon)$ relationship is linearly dependent upon strain

$$\sigma = E\varepsilon. \tag{3.1}$$

For the linear elastic fluid, the $\sigma(\varepsilon)$ relationship is linearly dependent on strain rate

$$\sigma = \eta\dot{\varepsilon}, \tag{3.2}$$

where η is the viscosity and $\dot{\varepsilon} = (\mathrm{d}\varepsilon/\mathrm{d}t)$. The linear solid behavior may be expected for small strains and for temperature $T < T_g$ (the glass transition temperature). The linear fluid behavior may be expected at $T = T_m$ (the melting temperature).

We can easily imagine building more complicated mechanical models, using springs and dashpots in series or in parallel. One spring and one dashpot in series are called a Maxwell fluid, Figure 3.2a. If the spring and dashpot are in parallel, a Kelvin (or Voigt) solid is represented, Figure 3.2b. Three parameter models are shown in Figure 3.3.

By now, the student should be wondering why one model is called a "solid" and why the other is called a "fluid." Two physical rules will be given here and mathematically proven later. If there is a free spring in the chain of elements, then the behavior is characterized as a solid. This represents the behavior of a network polymer. If there is a free dashpot in the chain of elements, then the behavior is fluid behavior as in a pure linear chain polymer with irrecoverable flow.

Obviously, larger N-parameter models (Figure 3.4) can be constructed with $N > 3$, but the two rules still hold regarding solid or fluid behavior. More parameters may be desired to fit experimental data over a wider range of conditions.

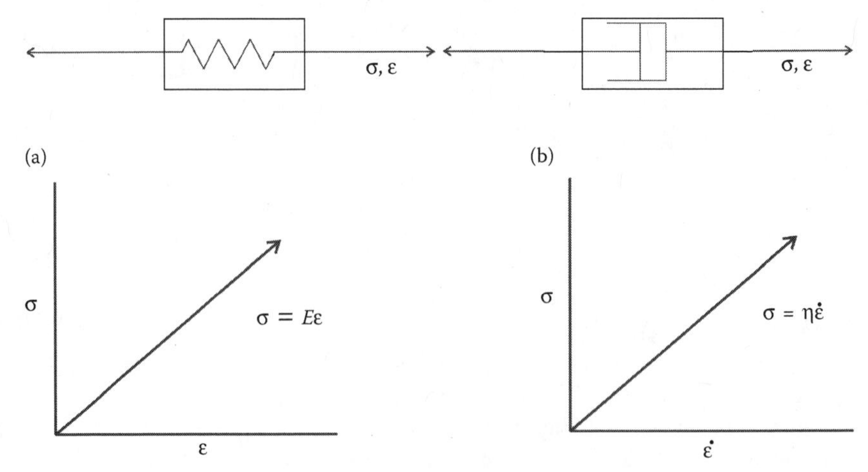

FIGURE 3.1 Mechanical models of the linear solid and the linear fluid. (a) Linear elastic solid ($\varepsilon < 1, T < T_g$) and (b) linear elastic fluid ($\varepsilon > 1, T > T_g$).

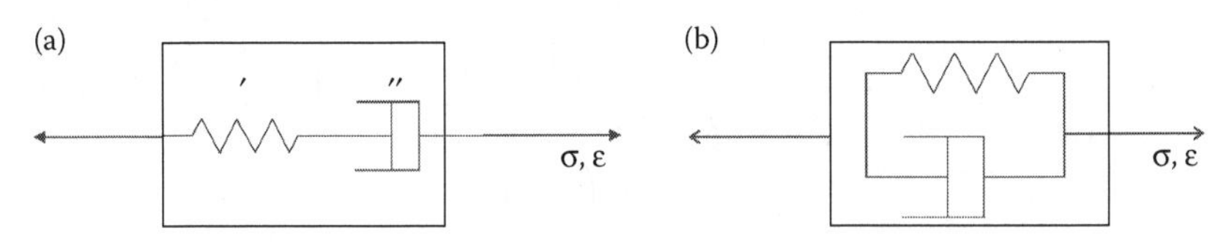

FIGURE 3.2 Two-parameter models. (a) Maxwell fluid and (b) Kelvin solid.

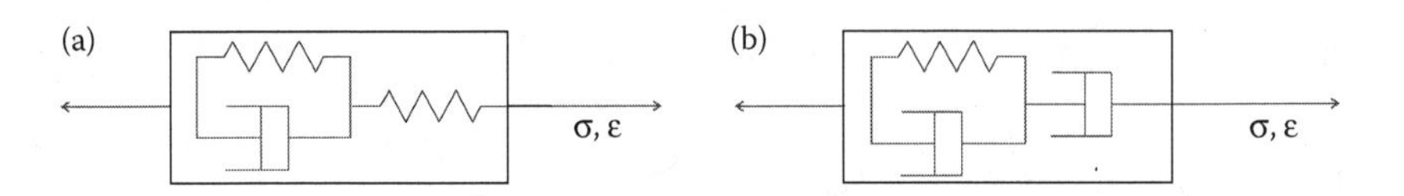

FIGURE 3.3 Three-parameter models. (a) 3-Parameter solid and (b) 3-parameter fluid.

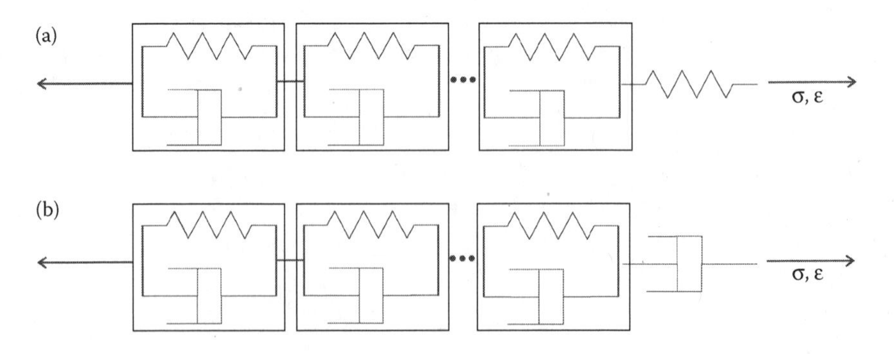

FIGURE 3.4 General N-parameter models. (a) N-parameter solid and (b) N-parameter fluid.

3.2 MATHEMATICAL MODELS

All of the above models represent linear viscoelastic behavior in that they are all assumed to be made up of combinations of linear springs and linear dashpots. In general, all such linear stress (strain) behavior can be described by a linear differential equation[1] of the type

$$p_0\sigma + p_1\dot{\sigma} + p_2\ddot{\sigma} + \cdots = q_0\varepsilon + q_1\dot{\varepsilon} + q_2\ddot{\varepsilon} + \cdots \tag{3.3a}$$

or

$$\Sigma p_n \frac{d^n\sigma}{dt} = \Sigma q_m \frac{d^m\varepsilon}{dt}. \tag{3.3b}$$

For the linear elastic solid (Figure 3.1a),

$$p_0 = 1; \quad p_n = 0, \quad n > 0, \tag{3.4}$$

$$q_0 = E; \quad q_m = 0, \quad m > 0.$$

For the linear viscous fluid (Figure 3.1b),

$$p_0 = 1; \quad p_n = 0, \quad n > 0, \tag{3.5}$$

$$q_0 = 0; \quad q_1 = \eta, \quad q_m = 0, \quad m > 2.$$

How do we derive the differential equations for the higher order polymers, that is, how do we find the values for p and q? The principles of statics and kinematics are applied to the composite elements, and a creep test and a relaxation test are applied.

3.3 THE MAXWELL FLUID

In the following analysis, the spring in Figure 3.2a is identified by a prime (′) and dashpot by a double prime (″) as shown. Continuity of stress (statics or Newton's law of action and reaction) requires that

$$\sigma = \sigma' = \sigma''. \tag{3.6}$$

Kinematics, the analysis of the deformation of the element, require that

$$\varepsilon = \varepsilon' + \varepsilon''. \tag{3.7}$$

Differentiation of Equation 3.7, because Equation 3.2 has $\dot{\varepsilon}$, gives

$$\dot{\varepsilon} = \dot{\varepsilon}' + \dot{\varepsilon}''. \tag{3.8}$$

Substitution of Equations 3.1 and 3.2 into Equation 3.8 gives

$$\dot{\varepsilon} = \frac{\dot{\sigma}}{E} + \frac{\sigma}{\eta}.$$

This equation is rewritten in standard form as

$$\sigma + \frac{\eta}{E}\dot{\sigma} = \eta\dot{\varepsilon}. \tag{3.9}$$

Comparison with Equation 3.3 shows that for the Maxwell fluid,

$$p_0 = 1; \quad p_1 = \frac{\eta}{E}, \quad q_0 = 0, \ q_1 = \eta$$

for the standard form:

$$\sigma + p_1\dot{\sigma} = q_1\dot{\varepsilon} \tag{3.10}$$

Consider a creep test, Figure 2.3a, applied to the Maxwell fluid. For $t < 0$, $\sigma = 0$, and for $t > 0$, $\sigma = \sigma_0$ = constant. Thus for $t > 0$, Equation 3.9 gives

$$\sigma_0 = \eta\dot{\varepsilon}, \quad \varepsilon = \frac{\sigma_0}{\eta}t + C_1 \tag{3.10a}$$

To find the constant of integration C_1, note that as $t \to 0^+$, that the strain will be the initial elastic response since there has been no time elapsed for creep to occur. Thus,

$$\varepsilon = \varepsilon_0 = \sigma_0/E = C_1 \quad \text{for } t \to 0^+, \tag{3.10b}$$

and the creep response for the Maxwell fluid becomes

$$\varepsilon = \frac{\sigma_0}{\eta}\left(t + \frac{\eta}{E}\right). \tag{3.11}$$

(The above is an engineer's solution to a differential equation. The more mathematically inclined, who are worried about the singularity in σ at $t = 0$ (Figure 2.3), can consult Flugge,[1] where the same solution is obtained by integrating the step function in σ.)

Solution 3.12 is plotted in Figure 3.5a. The Maxwell fluid has an initial elastic modulus

$$\varepsilon = \frac{\sigma_0}{E}, \tag{3.12}$$

as noted above, but as $t \to \infty$, $\varepsilon \to \infty$ which is the characteristic of a fluid.

Next, consider a relaxation test, Figure 2.4a, applied to the Maxwell fluid. It is more realistic to consider relaxation after some time t_1, after some creep has occurred as shown in Figure 3.5b. Thus, for $t > t_1$, $\dot{\varepsilon} = 0$, $\varepsilon = \varepsilon_1$ = constant, and from Equation 3.9,

$$\frac{\eta}{E}\dot{\sigma} + \sigma = 0, \tag{3.13}$$

$$\sigma = C_2\, e^{\left(\frac{-Et}{\eta}\right)}. \tag{3.14}$$

At $t = t_1$, $\sigma = \sigma_0$. Therefore,

$$\sigma = C_2\, e^{\left(\frac{-E(t-t_1)}{\eta}\right)}. \tag{3.15}$$

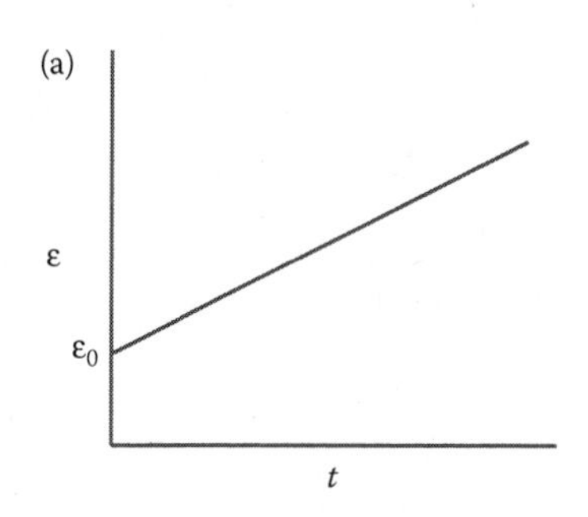

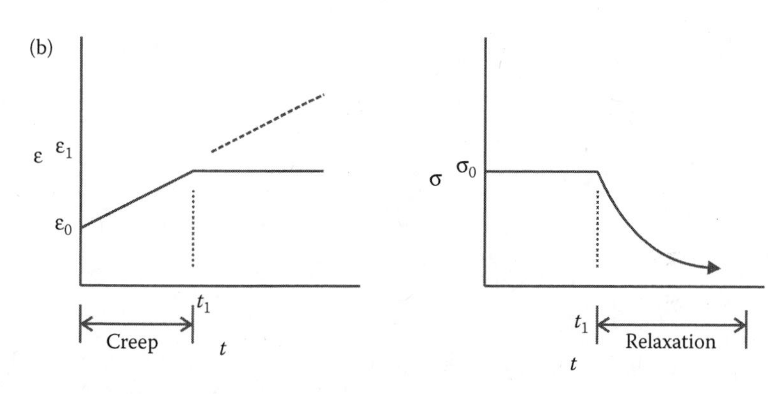

FIGURE 3.5 Creep and relaxation of the Maxwell fluid. (a) Creep curve for a Maxwell fluid and (b) relaxation of a Maxwell fluid.

This equation represents the relaxation of a Maxwell fluid. As $t \to \infty$, $\sigma \to 0$. The stress completely relaxes out, which is also characteristic of a fluid.

In viscoelastic behavior of polymers there is another mechanical property which is characteristic. It is the "relaxation time" denoted as τ. It is the time (arbitrarily defined for convenience) when the stress has relaxed to (1/e) of its original value. For the Maxwell fluid, the relaxation time is

$$\tau = \frac{\eta}{E}. \tag{3.16}$$

Example 3.1

Problem: Given $E = 10{,}000$ psi and $\eta = 5000$ psi-min.

1. Find the relaxation time.
2. Find the creep rate if $\sigma_0 = 500$ psi.

SOLUTION

1. $\tau = \frac{5000}{10{,}000} = 0.5\,\text{min}$
2. $\dot{\varepsilon} = \frac{\sigma_0}{\eta} = \frac{500}{5000} = 0.1\frac{1}{\text{min}}$

3.4 THE KELVIN SOLID

In order to derive the differential equation describing the stress (strain) behavior of the Kelvin solid, the same principles of mechanics are applied, except that now account must be made of the parallel configuration of the spring and dashpot element shown in Figure 3.2b. From a force balance,

$$\sigma = \sigma' + \sigma''. \tag{3.17}$$

From kinematics,

$$\varepsilon = \varepsilon' = \varepsilon''. \tag{3.18}$$

Substitution of Equations 3.1 and 3.2 into Equations 3.17 and 3.16, gives

$$\sigma = E\varepsilon + \eta\dot{\varepsilon}. \tag{3.19}$$

The standard form is

$$p_0\sigma = q_0\varepsilon + q_1\dot{\varepsilon} \tag{3.20}$$

where now $p_0 = 1$, $q_0 = E$, and $q_1 = \eta$. When a creep test, $\sigma = \sigma_0 =$ constant, is applied for $t > 0$, the solution of Equation 3.19 is

$$\varepsilon = \frac{\sigma_0}{E} + C_1 e^{(-Et/\eta)}. \tag{3.21}$$

To evaluate the constant of integration C_1, again engineering reasoning is used. At time $t = 0$, the dashpot locks up the Kelvin element, because it takes time for the dashpot to move. Thus the initial condition is $\varepsilon = 0$ at $t = 0$. This relation yields

$$C_1 = -\frac{\sigma_0}{E} \tag{3.22}$$

and the creep solution for the Kelvin solid becomes

$$\varepsilon = \frac{\sigma_0}{E}\left(1 - e^{\left(\frac{-Et}{\eta}\right)}\right). \tag{3.23}$$

This equation is plotted in Figure 3.6a. The Kelvin solid has no initial elastic response ($\varepsilon_0 = 0$), but has delayed elasticity with an asymptotic modulus, namely as

$$t \to \infty, \quad \varepsilon \to \frac{\sigma_0}{E} \tag{3.24}$$

or the asymptotic modulus can be determined from

$$\frac{1}{E_\infty} = \frac{\varepsilon_\infty}{\sigma_0}. \tag{3.25}$$

From the creep solution Equation 3.21, the time factor η/E in the exponential is called the "retardation" time τ, such that

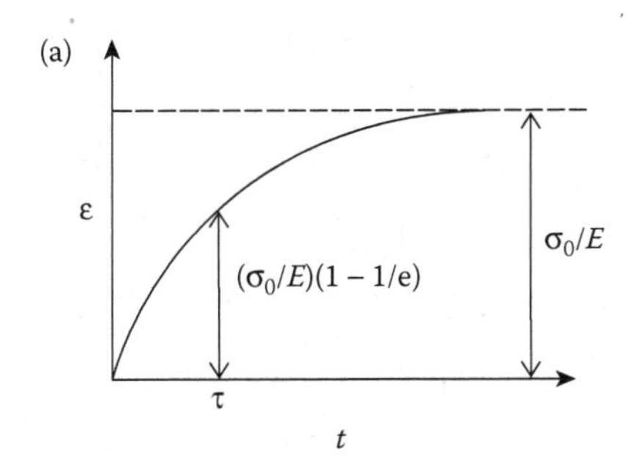

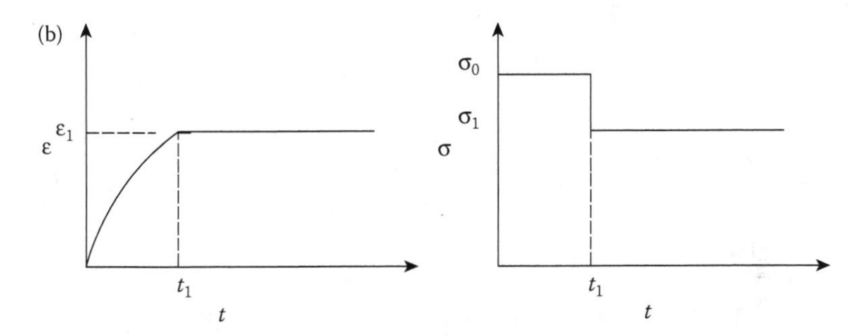

FIGURE 3.6 Creep and relaxation of the Kelvin solid. (a) Creep curve for the Kelvin solid and (b) relaxation of the Kelvin solid.

$$\tau = \frac{\eta}{E}. \tag{3.26}$$

It is the time when 63% of the creep has occurred, or $t = \tau$, when

$$\varepsilon = \varepsilon_\infty \left(1 - \frac{1}{E}\right) \tag{3.27}$$

as shown in Figure 3.6a. Knowledge of τ is important when conducting experiments. Let t_e = time of the experiment. If $t_e << \tau$, the experiment will measure only small strains and determine primarily a creep response. If $t_e >> \tau$ and the time steps are too large, the experiment will determine only the elastic response σ_0/E. The experiment must be designed with measurements in both ranges $t_e < \tau$ and $t_e > \tau$. At $t_e \approx \tau$, both fluid and solid responses are present.

Example 3.2

Problem: Determine material specifications for a design of a shock absorber for a design that requires 63% of the elastic strain to be delayed (retarded) 2.0 s. Assume a Kelvin solid-type material behavior and assume the maximum elastic strain is limited to 0.01 when a constant stress of 1000 psi is applied.

SOLUTION

From Equation 3.26, $t = \tau = 2.0\ \text{s} = \eta/E$.

The modulus E required is $E = \sigma_0/\varepsilon = 1000/0.01 = 10^5$ psi.

The viscosity η required is $\eta = E\tau = 10^5(2.0) = 2(10^5)$ psi s.

The characteristic time, the retardation time is a common measure in the original literature on this subject. However, for engineering purposes, a measure of the time for creep to reach an asymptote is more useful. Two additional times are introduced here. The first comes from considering an analogy with shell theory and corresponds to the decay length in the edge zone. Bending stresses decay by 95% when the exponential is three times the characteristic length.[2]

Let

$$\tau_1 = \tau = \eta/E \tag{3.28}$$

Thus, define

$$\tau_3 = 3\tau \tag{3.29}$$

For the subject at hand, at this time 95% of the creep will have occurred in a Kelvin solid-type material.

The second time is taken from analogy with electronics theory where the steady state time is taken as five times the characteristic time. Thus, define

$$\tau_5 = 5\tau \tag{3.30}$$

At this time 99% of the creep will have occurred in a Kelvin solid. The engineer can make a choice as to how close to estimate the asymptotic creep time.

Next, a relaxation test is applied after creep to a time t_1 as shown in Figure 3.6b. For $t > t_1$, $\dot{\varepsilon} = 0$, so that Equation 3.17 has the solution

$$\sigma = \sigma_1 = E\varepsilon_1 \tag{3.31}$$

or from Equation 3.21

$$\sigma_1 = E\varepsilon = \sigma_0(1 - e^{(-Et_1/\eta)}) \tag{3.32}$$

Thus, the Kelvin solid relaxes immediately to a constant stress, $\sigma_1 < \sigma_0$. This behavior is characteristic of a solid—that as $t \to \infty$, $\sigma > 0$ (i.e., the stress does not fully relax).

Example 3.3

Given $E = 70$ kPa, $\eta = 35$ kPa min, and $\sigma_0 = 3.5$ kPa for a Kelvin solid material.

1. Find the time for 99% of the creep to occur.
2. Find the initial creep rate at time $t = 0$.
3. What is the relaxation stress if the creep strain is held constant after 99% creep has occurred?

SOLUTION

1. $\tau = \frac{35}{70} = 0.5$ min. $\tau_5 = 5\tau = 5(0.5) = 2.5$ min for 99% of the creep to occur.
2. From differentiating Equation 3.23 when $t = 0$, $\dot{\varepsilon} = \frac{\sigma_0}{\eta} = \frac{3.5}{35} = 0.1\frac{1}{\text{min}}$.
3. From Equation 3.32

$$\sigma_1 = 3.5(1 - e^{(-70(2.5)/35)}) \cong 3.5(0.99) = 3.465\,\text{kPa},$$

that is, there is little or no relaxation. (This answer shows that after reaching asymptotic creep, a stress proportional to the elastic response is required to hold the strain constant.)

3.5 THE FOUR-PARAMETER MODEL

On the basis of the knowledge of the behavior of the one- and two-parameter models, it is now possible to approximate any *N*-parameter linear chain model by a four-parameter model as shown in Figure 3.7. The differential equation of the four-parameter model is

$$\sigma + p_1\dot{\sigma} + p_2\ddot{\sigma} = q_1\dot{\varepsilon} + q_2\ddot{\varepsilon}. \tag{3.33}$$

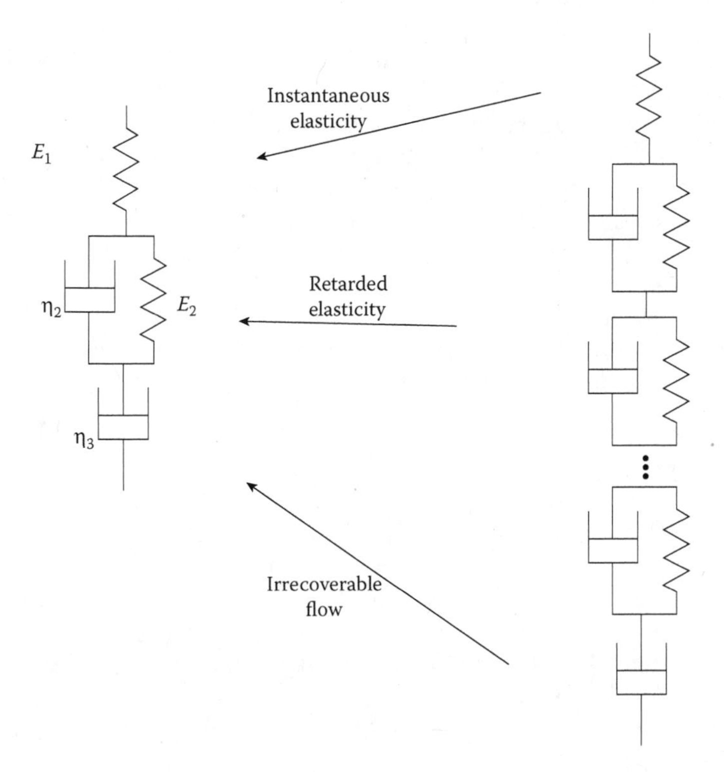

FIGURE 3.7 The four-parameter model.

Instead of solving this differential equation, the creep solution is more easily constructed from physical considerations by combining the solutions already available for the linear elastic solid, the Kelvin solid, and the linear viscous fluid. Thus,

$$\varepsilon(t) = \varepsilon_1(t) + \varepsilon_2(t) + \varepsilon_3(t), \tag{3.34}$$

where

$\varepsilon_1(t) = \dfrac{\sigma_0}{E_1}$ = instantaneous elastic response,

$\varepsilon_2(t) = \dfrac{\sigma_0}{E_2}(1 - e^{(-E_2 t/\eta_2)})$ = retarded elastic response, and

$\varepsilon_3(t) = \dfrac{\sigma_0}{\eta_3} t$ = irrecoverable flow.[1,2]

The four unknown parameters to be determined from experiments are E_1, E_2, η_2, and η_3. If η_3 is large, $\eta_3 \to \infty$ the model reduces to a three-parameter solid.

3.6 THE BOLTZMANN SUPERPOSITION PRINCIPLE

When a mechanical part is made from a polymer, and when it is to be used as a load-carrying component, obviously it is not necessarily always going to be subject to a constant stress as in the creep test. It generally has to be designed to withstand some history of stress variation. How will the polymer respond to the stress history? Can its response be predicted? Fortunately, for linear viscoelastic behavior, predicting the response is possible, because of the principle of superposition of solutions to linear differential equations. The student, of course, remembers that if $y_1(x)$ and $y_2(x)$ are both solutions of an ordinary differential equation for $y(x)$, then the sum $y_1(x) + y_2(x)$ is also a solution. This is the basis of the Boltzmann Superposition Principle for linear polymer behavior.

Consider the creep test in Figure 3.8 where the stress is changed by a step increase σ_1 at time t_1. As a result, the creep strain also changes. From the superposition principle, the total creep strain can be written as

$$\varepsilon = \sigma_0 J(t), \quad \text{for } t \le t_1 \tag{3.35}$$

and

$$\varepsilon = \sigma_0 J(t) + \Delta\sigma_1 J(t - t_1), \quad \text{for } t > t_1. \tag{3.36}$$

This principle can be extended to arbitrary stress histories as shown in Figure 3.9. Let $d\sigma_i$ be a change in stress at any time t_i, and then sum up all the changes in an integral over time. Thus, the so-called hereditary integral for linear viscoelasticity is derived[1,2]:

$$d\sigma_i = \frac{d\sigma_i}{dt_i} dt_i,$$

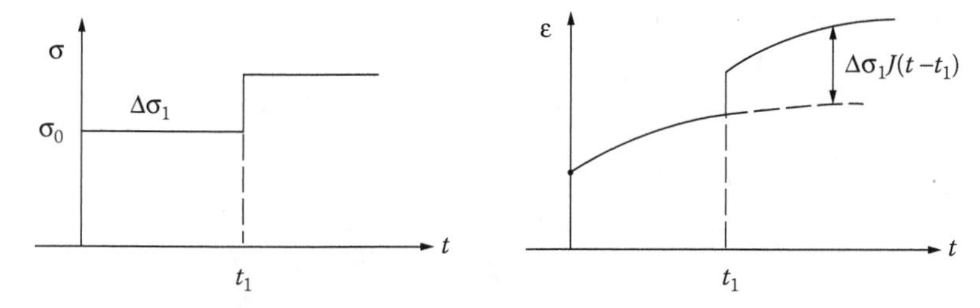

FIGURE 3.8 Creep test with step increase in stress.

and

$$\varepsilon(t) = \sigma_0 J(t) + \int_0^t J(t - t_i) \frac{d\sigma_i}{dt_i} dt_i \tag{3.37}$$

where $t_i < t$. This gives the total creep strain accumulated at any time t. Thus $J(t)$ from a standard creep test can be used for arbitrary stress histories for linear polymer behavior.

The superposition principle can easily be applied graphically as illustrated by the examples in Figure 3.10. In the first example when the stress is doubled after time t_1, the creep curve from $0 < t < t_1$, can be graphically superimposed on top of the original curve from $t_1 < t < 2t_1$. In the second example when the stress is taken off at time t_1, the creep curve from $0 < t < t_1$, can be graphically subtracted from the original curve from $t_1 < t < 2t_1$ (i.e., the creep caused by a negative stress is subtracted). The stress states in Figure 3.10d are equivalent to the stress state in Figure 3.10c.

Often a conservative approximation to the maximum deflection can be obtained by approximation of a general stress history by a series of step and linear functions. This estimation has the advantage that the displacement response to step and linear functions can be determined analytically.

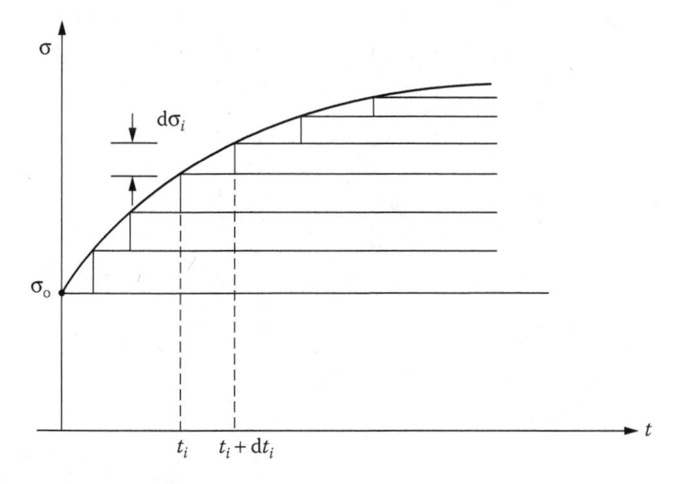

FIGURE 3.9 General stress history.

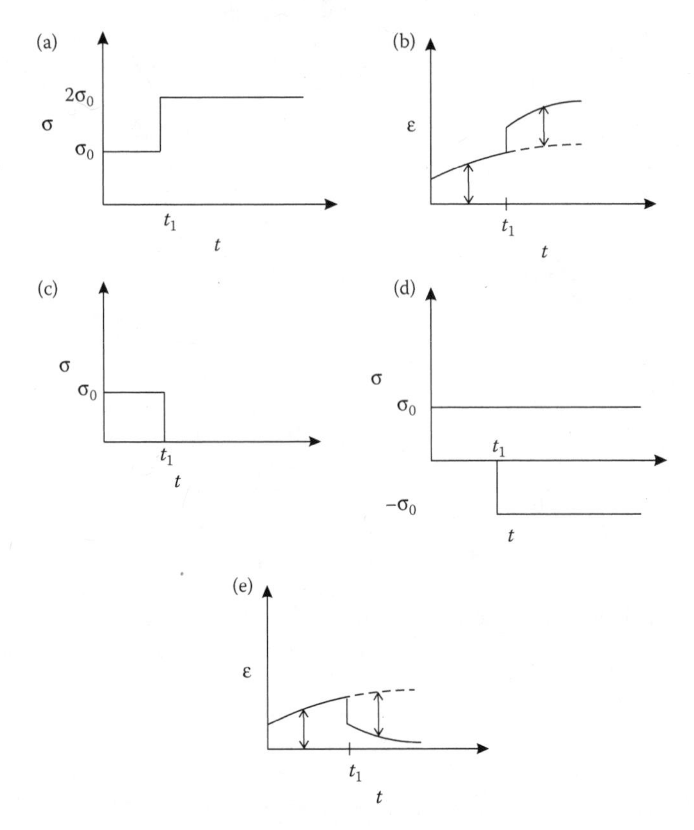

FIGURE 3.10 Example applications of the Boltzman superposition principle.

Example 3.4

Problem: Consider a stress history such as that in Figure 3.10a. Assume a Kelvin solid with properties and an initial stress level the same as Example 3.2. Assume the stress is doubled at time $t_1 = 2.0$ s. What is the total creep strain at $t_1 = 3$ s.

SOLUTION

A creep strain ε_1 from the original stress is calculated for $t = 3$ s:
From Equation 3.21

$$\varepsilon_1 = \frac{1000}{10^5}(1 - e^{(-3/2.0)}) = 7.77(10^{-3})$$

An additional creep strain is calculated for $t = 1$ s for the additional stress.

$$\varepsilon_2 = \frac{1000}{10^5}(1 - e^{(-1/20)}) = 3.93(10^{-3})$$

The total creep strain after 3 s is $\varepsilon = \varepsilon_1 + \varepsilon_2 = 7.77(10^{-3}) + 3.93(10^{-3}) = 11.7(10^{-3})$.

3.7 ADVANCED VISCOELASTIC MODELS

The Maxwell fluid and Kelvin solid model with one spring and one dashpot actually fits very little experimental data well. Experimental results of creep and stress relaxation often fits with a Prony series that is a series of Maxwell fluid elements in parallel or a series of Kelvin solids in series. It is typical to utilize one element per decade of time. While this is not always necessary, it has historically been standard convention.

The Prony series for the Maxwell fluid and Kelvin solid are

$$J(t) = \sum_{i=1}^{n} \left(\frac{1}{E_i} + \frac{t}{\eta_i} \right) \quad (3.38)$$

and

$$J(t) = \sum_{i=1}^{n} \frac{1 - e^{(-E_i t/\eta_i)}}{E_i}, \quad (3.39)$$

respectively.

For example, examination of the data in homework problem 3.1 shows that a single element does not result in an adequate fit. The underlying explanation for this is that the material responds with multiple timescales. The Prony series result in a pair of parameters E_i and η_i for each decade of time fit.

While there can be some rationalization of the Prony series to the effect that the response to mechanical stimuli has different time scales it is somewhat of a brute force methodology both mathematically and philosophically. There are more advanced viscoelastic predictions based on both more in depth mathematics and physics rational. If one takes experimental results and attempts to fit only a single Maxwell or Kelvin viscoelastic element it is seen that the fit is neither good nor adequate. They will often also occur with the Prony series. The first model to step beyond the Prony series is that which uses a stretching exponent for the creep or stress relaxation modulus.

One step in sophistication and certainly complexity is to utilize a fractional calculus-based model. In the fractional calculus model the dashpot is replaced by an element that has the properties of both a spring and dashpot. This element is called a spring-pot with the following constitutive relation

$$\sigma(t) = E^{1-\beta}\eta^{\beta}D^{\beta}\varepsilon(t), \quad (3.40)$$

where E and η are the modulus and viscosity of a spring and damper, β is a memory parameter. The range of β is $0 \le \beta \le 1$. When $\beta = 0$ the spring-pot constitutive equation is that of a spring and when $\beta = 1$ it is a dashpot. Note that when $\beta \neq 0$ and $\beta \neq 1$ or $0 < \beta < 1$, D^{β} becomes fractional, not integer, differentiation. There are two important aspects to this development. First, there is some indication from Bagley and Torvik[3] that rheological properties of viscoelastic materials have a response that can be described with fractional not integer calculus. Second, the response of this element is not over a single decade of time, but over multiple decades. This allows for the implementation of models with few viscoelastic elements to fit data over multiple decades of time.

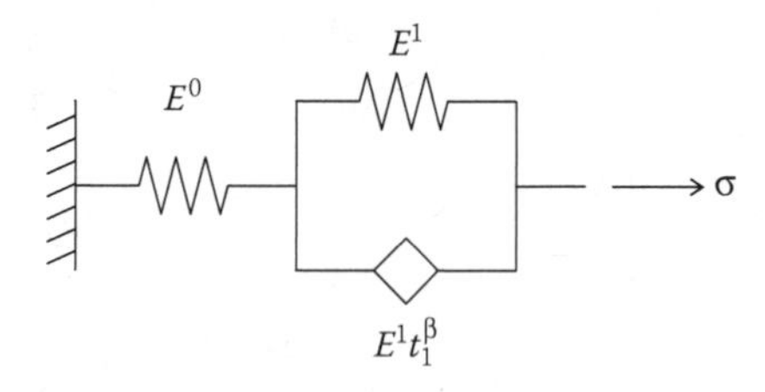

FIGURE 3.11 Four-parameter solid model with spring pot.

Welch et al.[4] used a four-parameter model as shown in Figure 3.11. The model results in the modulus

$$G(t) = G_0 + G_1 E_\beta \left[-\left(\frac{t}{t_1}\right)^{\beta} \right]. \tag{3.41}$$

$E_\beta[x]$ is the Mittag–Leffler function and

$$E_\beta[x] = \sum_{n=0}^{\infty} \frac{(x)^{\beta}}{\Gamma(1+\beta)^{n}} \tag{3.42}$$

where $\Gamma(1+\beta n)$ is the gamma function. Due to issues with convergence it is necessary to utilize the asymptotic approximation to evaluate the Mittag–Leffler function

$$E_\beta[x] = \sum_{n=1}^{\infty} \frac{(x)^{-n}}{\Gamma(1+\beta n)}. \tag{3.43}$$

Both the four-parameter fractional solid model from Figure 3.11 and an eight-decade Prony model were fit to the experimental data of Gottenberg and Christensen.[5] The result is shown in Figure 3.12. In addition, Table 3.1 shows the values for the

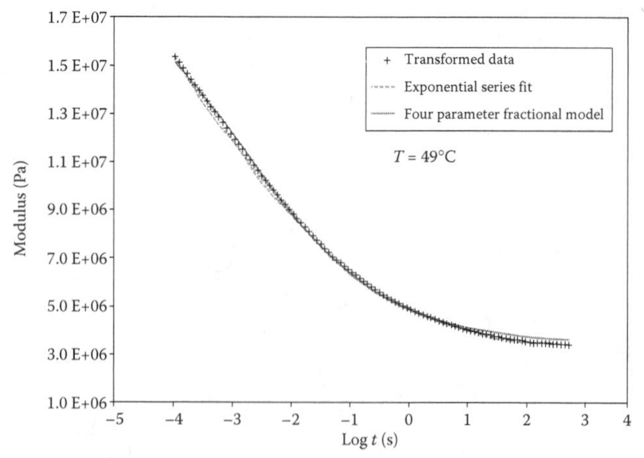

FIGURE 3.12 Four-parameter fractional solid and seven-decade series exponential model fit to polyurethane data. (Adapted from Welch, S. W. J., R. A. L. Rorrer, and R. G. Duren, Jr, *Mech Time-Dependent Matter*, 3, 279–303, 1999.)

TABLE 3.1
Parameters for Fractional Model Fit to Gottenberg and Christensen Polyurethane Data

Parameter	Value
G_0 (Pa)	3.4475(10^6)
G_1 (Pa)	1.6893(10^7)
β	0.336
t_1 (s)	1.92(10^{-3})

Sources: Adapted from Welch, S. W. J., R. A. L. Rorrer, and R. G. Duren, Jr, *Mech Time-Dependent Matter*, 3, 279–303, 1999; Gottenberg, W. G. and R. M. Christensen, *Int J Eng Sci*, 2, 45–57, 1964.

four-parameter fractional solid with G substituting for E in Equation 3.41. Christensen[6] required 17 parameters to fit the seven decades of time with a series exponential model.

3.8 THE VISCOELASTIC CORRESPONDENCE PRINCIPLE

The superposition principle presented above enables the prediction of the time-dependent creep elongation of a uniaxial member under tension. In an analogous way, the shear deformation can be predicted for a time-dependent shear stress. In both of these cases, a homogeneous stress state (independent of the coordinates x, y, t) has been assumed. How can this analysis be extended to nonhomogeneous stress states such as bending of beams, and torsion of shafts? For small deformations where the geometry change does not significantly affect the equilibrium equations for forces and moments, the creep solution for linear viscoelastic structural members can be found by analogy with the elastic solution by use of the *viscoelastic correspondence* principle, where the elastic modulus E is replaced by $1/J(t)$ where $J(t)$ is the viscoelastic creep compliance.

As an example, consider here the deflection of a beam under arbitrary loading. For small deformations, the deflection δ at some point is a product of three functions:

$$\delta = f_1 \, f_2 \, f_3, \tag{3.44}$$

where f_1 is a function of the loads on the beam, f_2 is a function of the material properties of the beam, and f_3 a function of the geometry of the beam (assumed not to change and not to be coupled with f_1 and f_2).

If f_1 and f_2 change with time, these changes can be handled using the Boltzmann superposition principle. The load history is analogous to the stress history considered in Figure 3.9. The material properties can be incorporated into a time-dependent creep compliance $J(t)$ or in a time-dependent relaxation modulus $Y(t)$.

Example 3.5

Problem: A cantilever beam with transverse load P on the end is shown in Figure 3.13. The elastic solution for the deflection $y(x)$ (from *Mechanics of Materials* book) is

$$y(x) = -P\frac{f(x)}{6EI}, \tag{3.45}$$

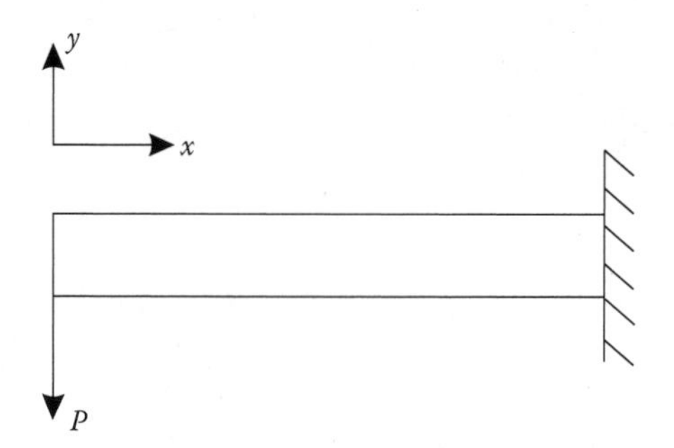

FIGURE 3.13 Cantilever beam.

where

$$f(x) = (x^3 - 3x^2 l). \tag{3.46}$$

By use of the viscoelastic correspondence principle, the linear viscoelastic solution for a polymer can be immediately written as

$$y(x,t) = -P \frac{f(x)}{6I} J(t), \tag{3.47}$$

where $J(t)$ replace $(1/E)$. Thus, for a Kelvin solid,

$$y(x,t) = -P\, f(x) \left(\frac{1 - e^{(-t/\tau)}}{6EI} \right) \tag{3.48}$$

where τ is the retardation time for beam deflection.

The student should realize the importance of viscoelastic correspondence principle. All of the mechanics of materials solutions for linear elastic materials have a corresponding solution for linear viscoelastic materials! New solutions do not need to be derived. Only the transformation $(1/E) \rightarrow J(t)$ needs to be made.

What if the load on the cantilever beam changes with time? Let $P = P(t)$ and use the hereditary integral formulation. Thus,

$$y(x,t) = -\left(P_0\, J(t) + \int_0^t J(t - t_i) \left(\frac{dP}{dt_i} \right) dt_i \right) \frac{f(x)}{6I}, \tag{3.49}$$

where $P_0 = P(t)$ when $t = 0$.

For the Kelvin solid, for example, and for $P(t_i) = Kt_i$ for $0 \le t_i \le t$

$$y(x,t) = -f(x) \int_0^t (1 - e^{(-t/\tau)}) \frac{K}{6EI} dt$$

or

$$y(x,t) = -f(x)(t - \tau(1 - e^{(-t/\tau)})) \frac{K}{6EI} \tag{3.50}$$

Example 3.6

An aluminum cantilever beam is replaced by a polymer beam of the same dimensions. The maximum defection of the aluminum beam was 0.01 in. What will be deflection of the polymer beam after 2 s if $J(t) = 0.5(10^{-5})$ psi^{-1} when $t = 2$ s?

SOLUTION

Assume for aluminum that $E = 10(10^6)$ psi.

$$\frac{y_{\text{polymer}}}{y_{\text{aluminum}}} = \frac{J(t)}{(1/E)} = 0.5(10^{-5}) \times 10(10^6) = 50$$

$$y_{\text{polymer}} = 50(0.01) = 0.50 \text{ in.}$$

3.9 THE TIME–TEMPERATURE EQUIVALENCE PRINCIPLE

The viscoelastic properties of polymers depend strongly on temperature. In rubber, the elastic modulus is nearly directly proportional to the absolute temperature T. Likewise, the initial elastic modulus (of the spring element) in network polymers can be taken as proportional to T. On the other hand, the viscous properties of amorphous network polymers are changed by an order of magnitude by moderate changes in temperature. However, it is found that this change can be accounted for by a "shift factor" when plotted on a log timescale.[7] For example, if storage compliance plotted as a function of log ωa_T, as in Figure 3.14, similar-shaped curves are found. These curves can be represented by one master curve if a shift factor a_T is defined. Thus, if the same function $J'_{pp} = J' \frac{T\rho}{T_0\rho_0}$ is plotted versus ($\log t - \log a_T$), a master curve as shown in Figure 3.15 results. The empirically determined shift factors are shown in Figure 3.16.

The shift factor is philosophically based on the concept that the viscoelastic response is a result of the ability of the polymer chains to respond to stress or deformation. Largely this response is temperature or conversely rate dependent. However, one of the factors often overlooked is that the ability of a polymer to respond is also a function of the volume available for the polymer chain to deform, which is ultimately related to density. This results in an additional vertical shift of the raw experimental data. This shift is often overlooked for two reasons. The first reason is that many practitioners are unaware that it exists and the more valid reason is that the time–temperature transformation (TTT) of the raw data is at best often only an order of magnitude predictor of the response over long times.

For Kelvin solid-type materials, the viscosities can be represented by the Arrhenius equation, and a_T can be calculated from an equation of the form

$$a_T = e^{\left(b\left(\frac{1}{T} - \frac{1}{T_0}\right)\right)}, \tag{3.51}$$

where b is a constant and T_0 is a reference temperature.

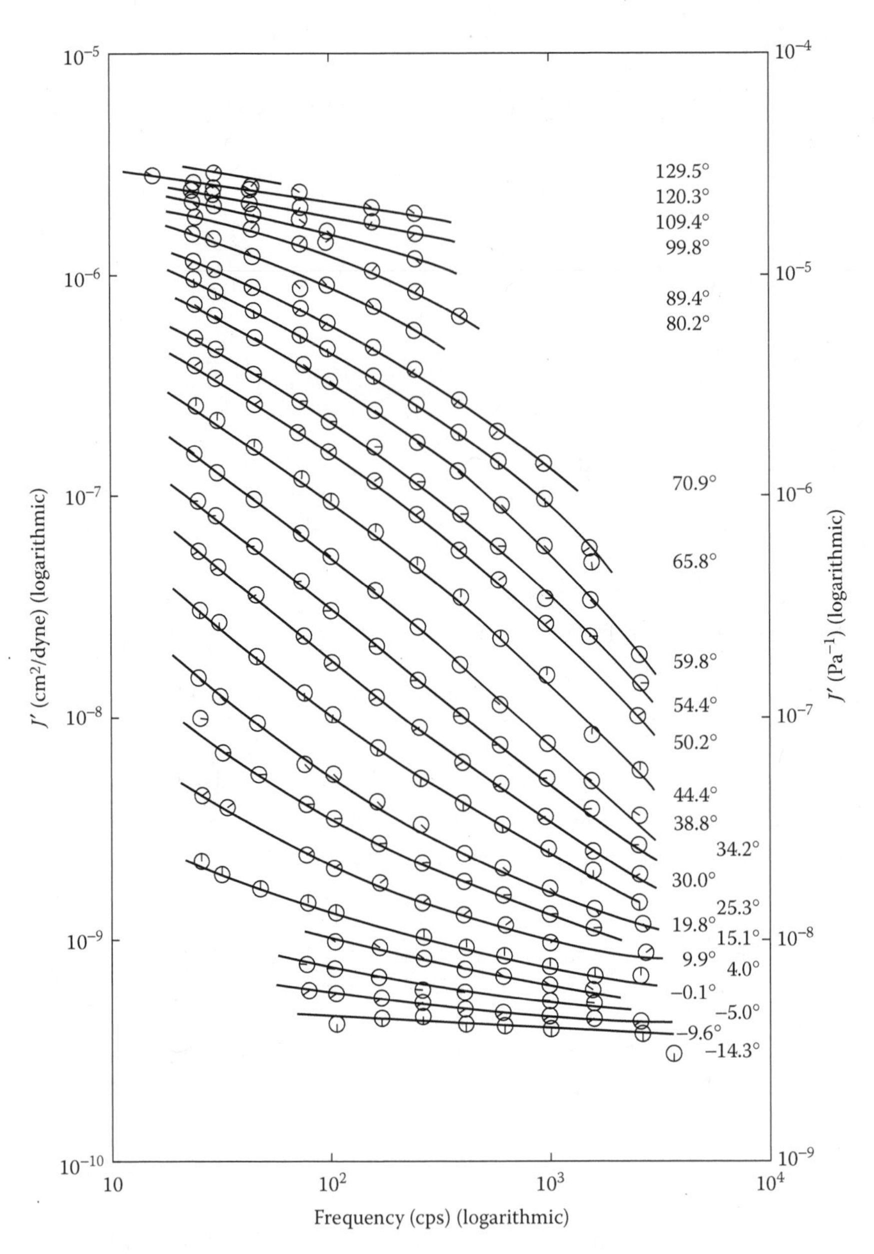

FIGURE 3.14 Storage compliance of poly(*n*-octyl methacrylate) determined under isothermal conditions. (Adapted from Ferry, J. D., *Viscoelastic Properties of Polymers*, New York, NY: John Wiley & Sons, 1980.)

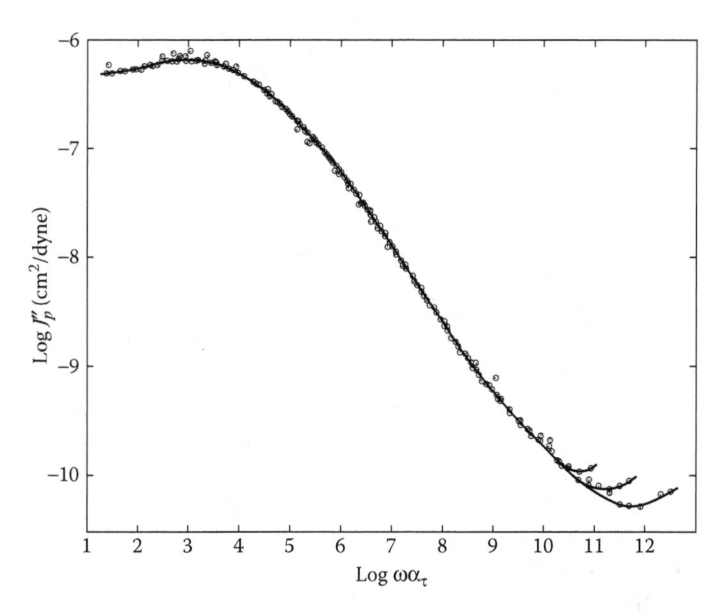

FIGURE 3.15 Master curve of data shown in Figure 3.14 (reference temperature $T_0 = 100°C$). (Adapted from Ferry, J. D., *Viscoelastic Properties of Polymers*, New York, NY: John Wiley & Sons, 1980.)

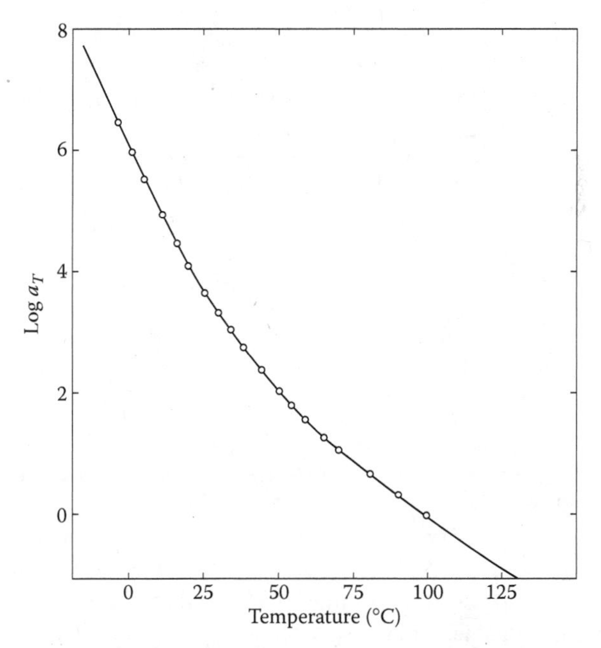

FIGURE 3.16 Shift factors for Figure 3.14. (Adapted from Ferry, J. D., *Viscoelastic Properties of Polymers*, New York, NY: John Wiley & Sons, 1980.)

Another shift factor that has been used is based on the work of Williams, Landel, and Ferry (WLF), and referred to as the WLF shift factor

$$\log a_T = \frac{-C_1(T - T_0)}{(C_2 + T - T_0)} \tag{3.52}$$

where the constants C_1 and C_2 are fitted to experimental data. There are universal constants ($C_1 = 8.86$ and $C_2 = 101.6$) that cover a wide range of polymers and can be used to calculate initial estimates of the effect of temperature changes.

Different investigators have used different expressions for a_T to empirically fit experimental data over wider temperature ranges. This use of a shift factor a_T to develop a master curve to represent creep as a function of temperature T is called the time–temperature equivalence principle. It is generally used to interpolate creep data between two temperatures or to extrapolate creep data to an unknown temperature. The product $J(t)xT$ is plotted to factor out the proportionality of the initial elastic modulus in temperature. Originally this phenomenon was observed experimentally by Williams et al.[7] Since that time the behavior is predicted theoretically from a variety of starting assumptions. The time–temperature superposition principle is only valid in the range $T_g < T < T_g + 150°C$. However, it is often used outside of that range, especially for $T < T_g$. While not strictly valid in this region, it is useful for estimating behavior.

HOMEWORK PROBLEMS

3.1. A creep test is made on a polyethylene specimen that has a length of 4 in., a width of 0.50 in., and a thickness of 0.125 in. A load of 62.5 pounds is applied to the specimen, and its length as a function of time is given by

Time (min)	Length (in.)
0.1	4.033
1	4.049
10	4.076
100	4.110
1000	4.139
10,000	4.185

Plot the creep compliance as a function of time using a logarithmic time scale.

3.2. Would the curve of Problem 3.1 show the upward curvature on a linear time scale? (Hint: Let $t' = \log t$ and show that the slope on the t scale decreases by $1/t$ or by $1/10^{t'}$ times the slope on the log scale.)

3.3. Use a Kelvin model for creep and show that the inflection point on the log scale plot corresponds to the creep retardation time τ. Then the asymptotic time can be estimated at 3–5τ even though the experimental creep data is obtained to times just beyond τ. This can save time in conducting creep experiments.

3.4. Assuming that the Boltzmann superposition principle holds for the polymer in Problem 3.1, what would the creep elongation be from 100 to 10,000 min if the load were doubled after 100 min? (Show graphically on a plot.)

3.5. Assuming the Boltzmann superposition principle holds and that all of the creep is recoverable, what would the creep recovery curve be for the polymer in Problem 3.1 if the load were removed after 10,000 min? (Show graphically on a plot.)

3.6. A material has two relaxation times—10 s and 100 s. Plot its relaxation curve from 1 s to 1000 s.

3.7. A horizontal cantilever beam is made of an idealized material which has only two retardation times, 10 s and 1000 s. The beam is bent downward for 100 s. Then it is bent upward for 1 s and released without any vibrations taking place. Describe the motion of the beam for the next 10,000 s. (Show graphically on a plot.)

3.8. The maximum deflection of a simply supported beam made of steel with $E = 30(10^6)$ psi is found to be 0.010 in. What would the deflection of the beam be if the same beam was made from the polymer of Problem 3.1, and if the load were left on the beam for 10 min.

3.9. Given the digitized load history shown in Figure 3.17 and provided in the accompanying computer files, perform two estimates of the maximum displacement using numerical summation. The estimates should utilize the following: (a) one-step function only and (b) summation of step functions based on the digitized points. For both methods be conservative in your estimation. Plot the displacement that you can estimate from these data(time) for both methods on the same figure for

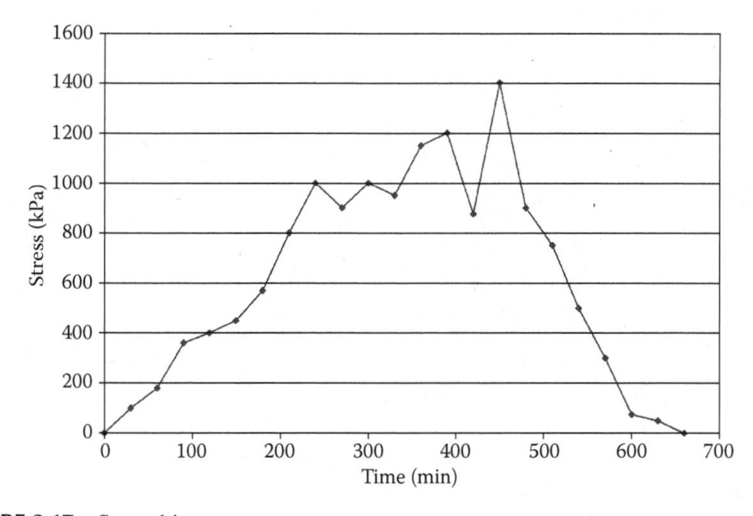

FIGURE 3.17 Stress history.

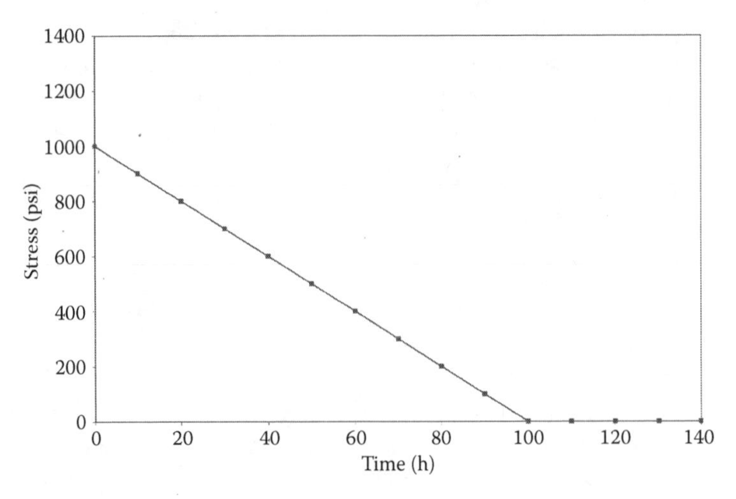

FIGURE 3.18 Decreasing ramp stress history.

comparison. Assume a vertical beam (25.4 mm in diameter, 0.25 m long) mounted in tension for a polymer selected by the instructor.

3.10. Estimate the result to Problem 3.9 analytically with a combination of ramp and step functions.

3.11. Shift the thermomechanical data for nylon obtained at 1 Hz provided on the Taylor & Francis website www.taylorandfrancis.com for 0.1, 10, 100, and 1000 Hz.

3.12. Solve for the strain for stress given in Figure 3.18 via superposition for $\varepsilon(t)$ for a generalized Kelvin solid and a Maxwell fluid.

3.13. Assume that an ideal solid or fluid model will fit the data shown in Figure 3.19. What type of polymer is this? What is the model? What

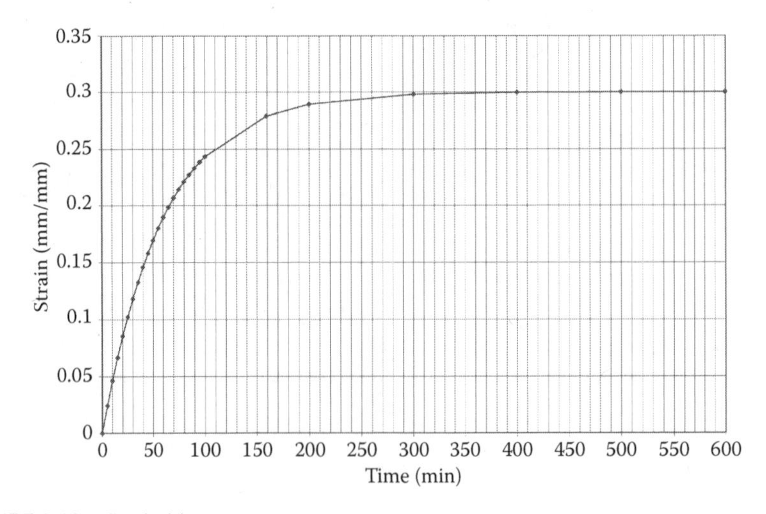

FIGURE 3.19 Strain history.

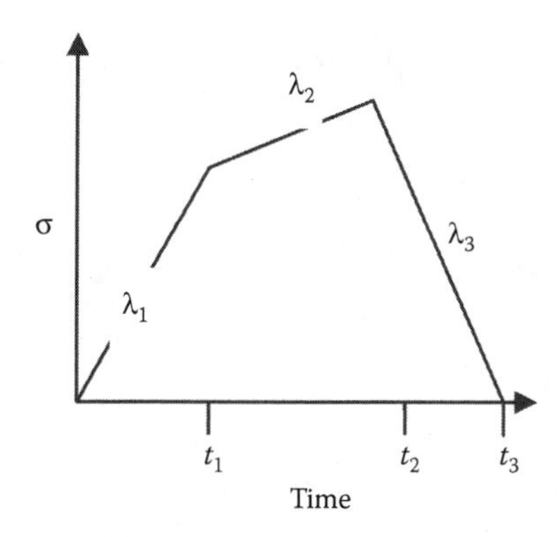

FIGURE 3.20 Triple ramp stress history.

are the values of E and η that you can estimate from these data? (σ = 1500 Pa).

3.14. How would you solve the creep problem in Figure 3.20? Set up the solution of displacement as a function of all time for a Maxwell fluid in equation form without solving.

3.15. How would you solve the creep problem in Figure 3.21? Set up the solution of displacement as a function of time for a Kelvin solid in equation form without solving.

3.16. For a beam loaded in compression under 27.5 lb (d = 1 in., L = 6 in.) and having the following Kelvin parameters (E = 2000, $\eta = 1(10^7)$) answer the following questions.
What is P_{cr} = at t = 0 s?
What is P_{cr} = at t = 10,000 min?
Will this column fail at t = 10,000 s? Prove by solving problem.

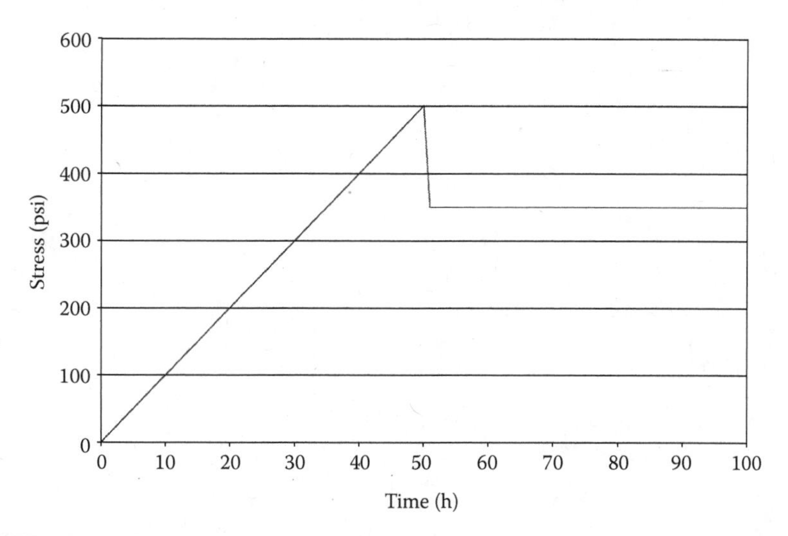

FIGURE 3.21 Ramp with step offset stress history.

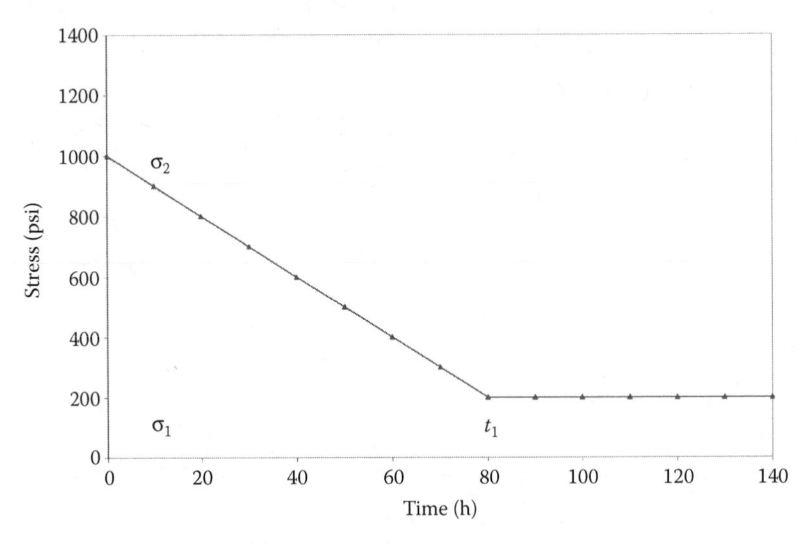

FIGURE 3.22 Decreasing ramp with step offset stress history.

3.17. For the stress history in Figure 3.22, set up the solution of $\varepsilon(t)$ for the Maxwell fluid with substitutions for J, and so on for $t > t_1$ ($t_1 = 80$ h). Do not completely solve this problem! What would change if this polymer was a Kelvin solid? Explicitly write down the change.

3.18. Demonstrate how the problem in Figure 3.22 would be solved via superposition. Without actually solving, estimate and sketch $\varepsilon(t)$ for a Kelvin solid (Figure 3.23).

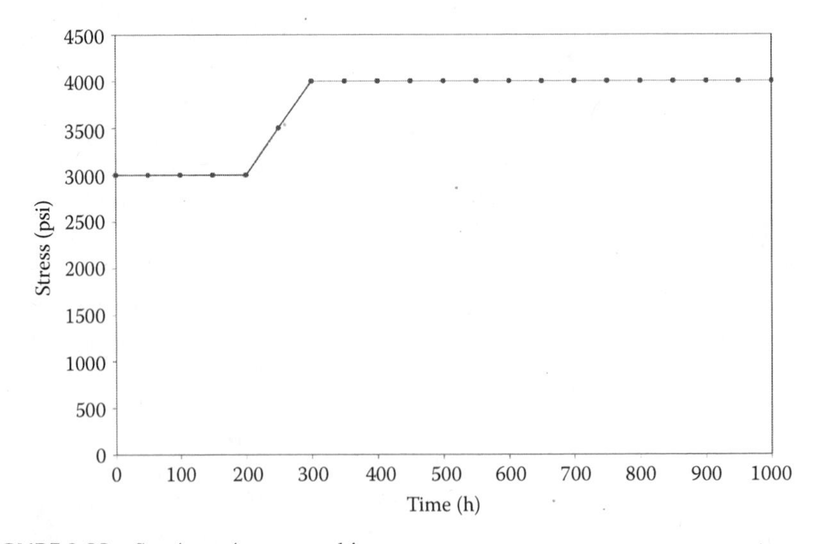

FIGURE 3.23 Step/ramp/step stress history.

REFERENCES

1. Flugge, W., *Viscoelasticity*, Waltham, MA: Blaisdell Publ. Co., 1967.
2. Alfrey, T. and E. F. Gurnee, *Organic Polymers*, New York, NY: Prentice-Hall, 1967.
3. Bagley, R. L. and P. J. Torvik, On the fractional calculus model of viscoelastic behavior, *J of Rheol*, 30(1), 133–155, 1986.
4. Welch, S. W. J., R. A. L. Rorrer, and R. G. Duren, Jr, Application of time-based fractional calculus methods to viscoelastic creep and stress relaxation of materials, *Mech Time-Dependent Matter*, 3, 279–303, 1999.
5. Gottenberg, W. G. and R. M. Christensen, An experiment for determination of the mechanical property in shear for a linear, isotropic viscoelastic solid, *Int J of Eng Sci*, 2, 45–57, 1964.
6. Christensen, R. M., *Theory of Viscoelasticity, An Introduction*, 2nd ed. New York, NY: Academic Press, 1982.
7. Williams, M. L., R. F. Landel, and J. D. Ferry, The temperature dependence of relaxation mechanisms in amorphous polymers and other glass-forming liquids, *Int Am Chem Soc*, 77, 3701–3707, 1955.
8. Ferry, J. D., *Viscoelastic Properties of Polymers*, New York, NY: John Wiley & Sons, 1980.

4 Creep and Fatigue Failure

4.1 CREEP FAILURE UNDER TENSION

As defined in Section 2.2, creep is the time-dependent deformation under a constant load. The constitutive relations that describe the stress–strain behavior have been derived form mechanical models in Section 3.1. Here attention is given to the creep strength. Unlike the static ultimate strength S_u, there is no unique creep rupture strength. The ultimate strength in creep, S_{uc} depends on time (t), temperature (T), and other ambient conditions, that is,

$$S_{uc} = S_{uc} \quad (t, T, \text{condition}) \tag{4.1}$$

This behavior is illustrated in Figure 4.1. Theoretically, a polymer will fail at any stress level if one waits long enough. Therefore, the designer needs to know the time under load for each application. The question that logically comes up next is how to estimate the creep strength when the load is not constant for all times. One way of estimating the creep strength under variable load is to use "Miner's Rule," which is well known in fatigue analysis.

Consider the load history shown in Figure 4.2. Estimate the creep damage for each time interval as a fraction of the strength used up for that interval and add these fractions together. That, in essence, is Miner's Rule. Thus, from Figure 4.2, if

$$\frac{t_1 - t_0}{t_{c_1}} + \frac{t_2 - t_1}{t_{c_2}} + \cdots = \sum_{N=1}^{N} \frac{t_n - t_{n-1}}{t_{c_n}} < 1, \tag{4.2}$$

where $(t_i - t_{i-1}/t_{ci})$, is an increment of creep damage, and if t_{c_1}, t_{c_2}, and t_{c_3} are critical times at the respective load levels; then the load history is estimated as safe—no failure is predicted; but if

$$\sum_{n=1}^{N} \frac{t_n - t_{n-1}}{t_{c_n}} \geq 1 \tag{4.3}$$

then creep rupture is predicted at some intermediate time t_n, when the sum is equal to one. Despite the fact that Miner's Rule was developed for metals, with experimental verification, it can also be applied to polymers and elastomers.

Figure 4.3 shows creep failure curves for PVC. The upper curve is for ductile failure due to necking, normally identified as the start of creep rupture. The dashed line indicates the initiation of crazing. In order to prevent the initiation of failure the strain must be kept below 0.010.

One must be aware of strength reducing ambient conditions similar to stress corrosion of metals, for example, steel in the presence of rust-producing salt. Placing a polymer in the presence of a solvent must be avoided. For example, a polymer-bearing

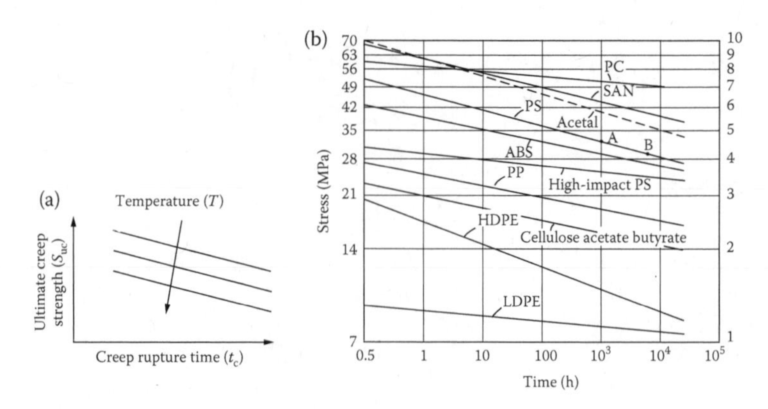

FIGURE 4.1 Creep rupture curves. (a) Effect of temperature and (b) creep strength data for some polymer, at 20°C. (From *Engineering Materials Handbook, Volume 2, Engineering Plastics*, ASM International, 1988, Figure 17, p. 666. Reprinted with permission of ASM International, all rights reserved, www.asminternational.org.)

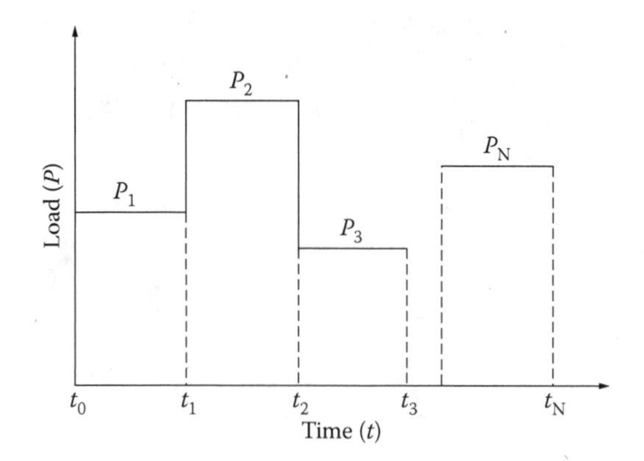

FIGURE 4.2 Variable load history.

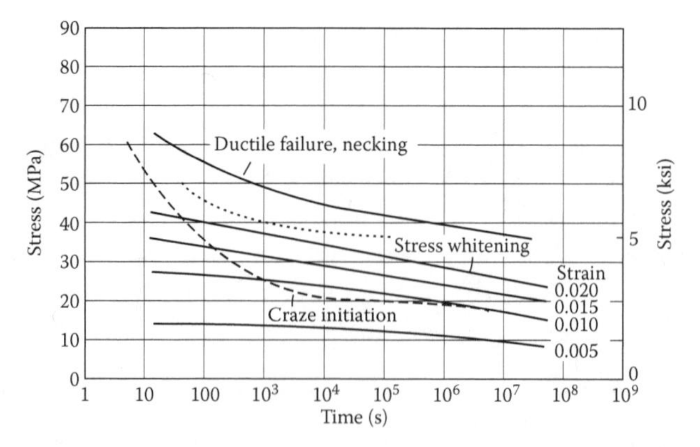

FIGURE 4.3 Creep rupture of PVC at 20°C, 50% relative humidity. (From *Engineering Materials Handbook, Volume 2, Engineering Plastics*, ASM International, 1988, Figure 25, p. 671. Reprinted with permission of ASM International.)

surface contacting a lubricant that contains a solvent will greatly reduce the service time. For the PVC in Figure 4.3, the creep rupture strength will be reduced by 50–60% in the presence of iso-octane or isopropanol. Also, some polymers such as PE are subject to deterioration from sunlight (ultraviolet rays). Much research is being done, however, on additives for polymers to counteract such deteriorating effects.

Example 4.1

At one time in recent history, an automobile manufacturer was using a plastic head gasket for the engine on some of its vehicles. The gasket was failing early and the vehicles were subject to a recall. Was the cause of this problem corrosion due to oil, gas or exhaust, or was it a creep problem at engine operating temperatures, or was it a combination of both causes?

4.2 CREEP FAILURE UNDER COMPRESSION

Compressive loading introduces the possibility of another mode of failure—buckling. It is not sufficient to say that an axial member under compression or a shell type configuration under external pressure can merely buckle, but one must say that it will buckle at some critical time, if the material is viscoelastic like a polymer. If the instantaneous Euler elastic buckling load for a column is

$$P_{cr}^{E} = \frac{n\pi^2 EI}{L^2} \tag{4.4}$$

(where n = constant depending on boundary conditions), then under creep buckling,

$$P = P_{cr} \quad \text{at } t = t_{cr},$$

where

$$P_{cr} < P_{cr}^{E} \tag{4.5}$$

as shown in Figure 4.4.

Let us estimate the creep buckling time for a column whose material is modeled by a Kelvin solid. The differential equation for the buckling deflection of an ideally elastic column (from any *Mechanics of Materials* book) is

$$\frac{d^2 y}{dx^2} + \frac{P}{EI} y = 0 \tag{4.6}$$

where y is the sideways deflection. For pinned ends, the solution

$$y = A \sin(mx), \quad P_{cr}^{E} = \frac{\pi^2 EI}{L^2}, \tag{4.7}$$

where $n = 1$, $mL = \pi$, and $m^2 = P_{cr}/(EI)$.

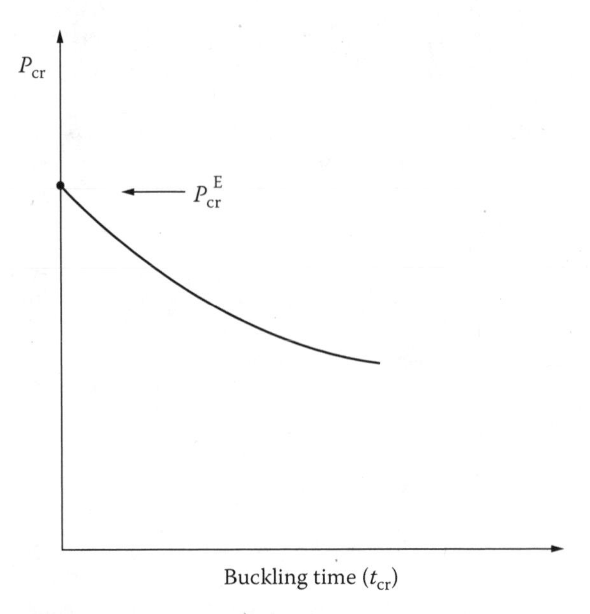

FIGURE 4.4 Creep buckling of columns.

Using the viscoelastic correspondence principle (Section 3.8), valid for small deflections on which Equation 4.6 is based, the buckling load for a viscoelastic column can be calculated by replacing $(1/E)$ by $J(t)$, thus

$$P_{cr} = \frac{\pi^2 I}{L^2 J(t)} \tag{4.8}$$

for a pinned–pinned viscoelastic column. For the Kelvin solid,

$$J(t) = \frac{1 - e^{-\frac{t}{\tau}}}{E} \tag{4.9}$$

Thus, the Kelvin solid approaches the Euler load as $t \to \infty$, which can be seen from Equation 4.9 and as shown in Figure 4.5. However, for a Maxwell fluid,

$$J(t) = \frac{1 + \frac{t}{\tau}}{E} \tag{4.10}$$

and the buckling load decreases from the Euler load at $t = 0$ as time increases as shown in Figure 4.5.

The fact that the buckling load decreases with time for viscoelastic materials is a very important design consideration.

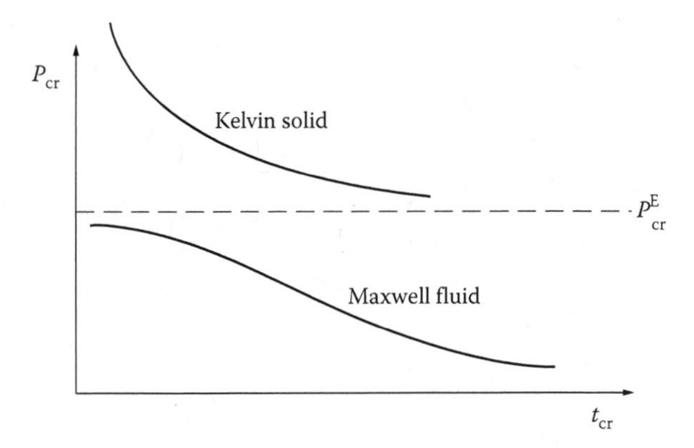

FIGURE 4.5 Buckling loads for Kelvin solid and Maxwell fluid.

4.3 FATIGUE OF POLYMERS

Polymers can fail in fatigue under repeated loading (shown in Figure 4.6a) in a similar manner as metals fail, but also in a different and unique manner. Generally, the causes of fatigue failures in polymers are (1) mechanical crack initiation and propagation (similar to metals) and (2) heating due to internal energy dissipation and consequent softening of the polymer (sometimes causing localized melting and thermal decomposition). Because of the heating effect, as the temperature goes up, the stiffness (the effective modulus E) of the polymer decreases, and consequently the strain of the polymer under constant stress cycling increases with time as shown in Figure 4.6b.

In the discussion on dynamic properties (Section 2.4), it was pointed out that there is an energy loss per cycle of stress and strain proportional to the loss creep compliance J''. This is represented by a hysteresis loop on the stress–strain diagram, Figure 4.7.

Let the stress cycle by represented by

$$\sigma = S\sin(\omega t) \tag{4.11}$$

and the strain cycle by

$$\varepsilon = \varepsilon_0 \sin(\omega t - \delta) \tag{4.12}$$

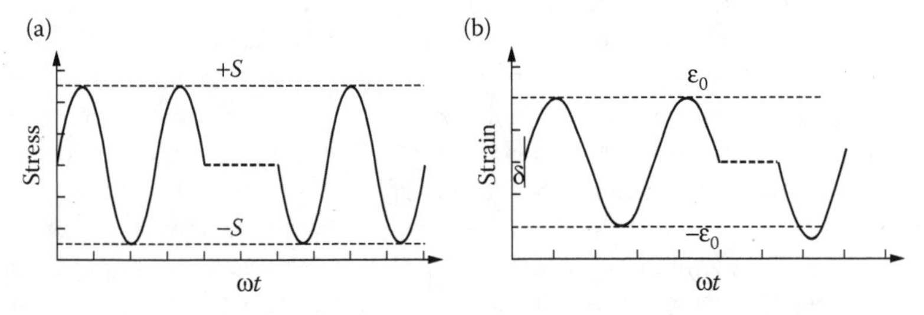

FIGURE 4.6 Nonisothermal fatigue behavior of polymers. (a) Stress input and (b) strain response.

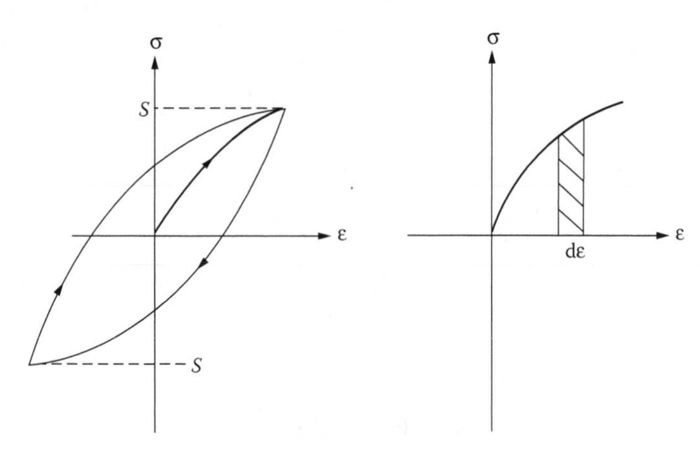

FIGURE 4.7 Hysteresis loop and energy loss.

where δ is a phase angle by which the strain lags the stress as shown in Figure 4.6. It can be shown that

$$\delta = \frac{1}{\omega\tau} \tag{4.13}$$

where τ is the creep retardation time. Let U_c be the strain energy loss per cycle where

$$U_c = \int \sigma \, d\varepsilon = S^2 \pi J^* \sin\delta = S^2 \pi J'' \tag{4.14}$$

Then the energy loss per unit time is

$$U = S^2 \pi J'' f \tag{4.15}$$

where f is the frequency of the test, cycles/time.

What Equation 4.15 shows is that the fatigue failure of polymers depends on the frequency of the test (f) $f$and the complex compliance of the material J''. Thus, a plot of cycles-to-failure (N) versus frequency f (f) shows a lower fatigue life for nonisothermal tests as shown in Figure 4.8. This is an important material characteristic that influences design. If nonisothermal fatigue test data are used for isothermal design applications, the design will end up conservative. If too conservative, the design engineer will waste the company's money by using too much material. For example, a laborer's plastic lunch box latch may see 4–6 cycles per day of opening and closing—an intermittent cycling that does not allow much heat buildup, that is, an isothermal application. Fatigue data at 1000 cpm, however, would predict a low fatigue life and the designer would lower the allowable stress and increase the size of the latch. However, in a dynamic application, that is, an engine operating at 2000 cpm, fatigue data from a 1000 cpm test are nonconservative.

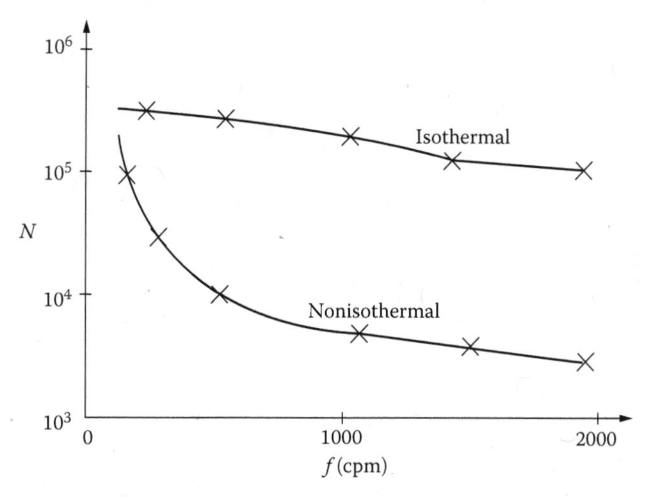

FIGURE 4.8 Cycles to failure versus test frequency. (Glass reinforced PP, initial stress = 8500 psi.) (Adatpted from Cessna, L. C. and J. A. Levens, *Proc. 4th Annual Plastics Conf.*, Eastern Michigan University, September, 1969, pp. 46–73.)

Example 4.2

One of the authors (Gerdeen) had a research project in the 1960s to analyze the reason for failure of plastic milk bottle handles. People were losing their milk occasionally when the handles broke carrying the glass bottles out of the store or lifting them out of the refrigerator. The glass bottles and handles were recyclable.

A fatigue study was made by designing an experiment with stainless-steel models of milk bottle necks mounted to an oscillating crank shaft. Several plastic handles could be tested in each run of the apparatus. The crank shaft and handles were all immersed in water which was circulated and kept at 40°F. The results of this test help locate the critical points of failure under certain loads and the handles were redesigned geometrically to avoid failure. Of course, today glass bottles are no longer used, and integral handles and bottles are blow molded together as a unit. (Blow molding is discussed in Chapter 10.)

It is desirable to expedite laboratory fatigue tests. No one can afford fatigue tests that take a year! Therefore, if one desires isothermal results at high frequency, a bath of cooling liquid can be provided, (as illustrated by Example 4.2). But beware! Because of solvent (or corrosive) effects, there can be a stress interaction. For example, vegetable oil causes a deleterious effect on PS by reducing the fatigue strength by 20–50%.[2]

Referring back to Equation 4.15, let us consider the effect of J''. The energy loss depends on J''. Because of this energy dissipation, heat will build up in the polymer with time. But J'' increases with temperature. These effects are shown in Figure 4.9 for PE, for example.

Many fatigue testing apparatuses are designed to run constant stress amplitudes. For metals, this does not matter since the stiffness of the metal does not change during the test. However, for polymers, the stiffness of the polymer decreases during the test because of internal heating and a constant strain cycling test gives quite different results than a constant stress cycling

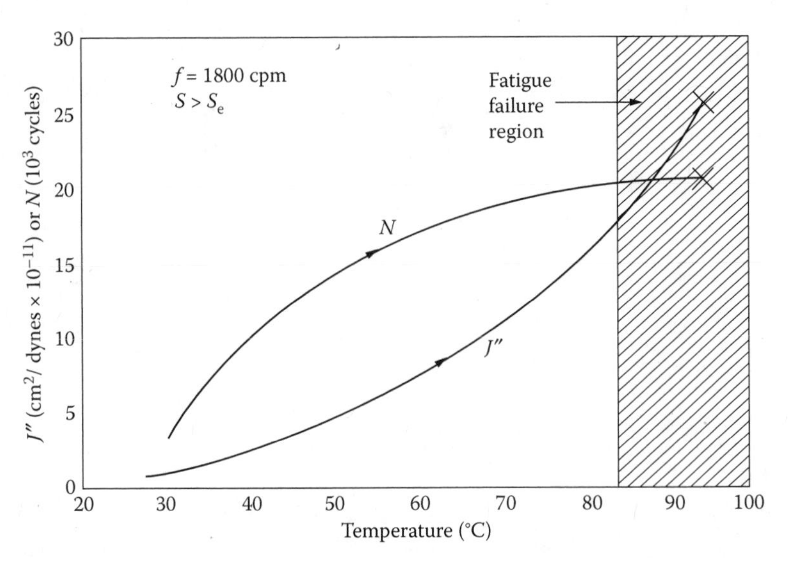

FIGURE 4.9 Temperature rise and increase in J'' in fatigue test of PE. (Adapted from Riddle, M. N., *Proc. 4th Annual Plastics Conf.*, Eastern Michigan University, September, 1969, pp. 1–18.)

test. For example, Figure 4.10 shows the results of a constant amplitude deflection test of cantilever specimens of a glass reinforced polypropylene. The "apparent stress" is plotted versus cycles to failure. The apparent stress is less than the initial stress because of the reduction in stiffness with time due to the temperature rise. A great reduction in stress occurs at about 2–3000 cycles. Whereas the initial stress was 5000 psi, the stress at failure was approximately 1000 psi.

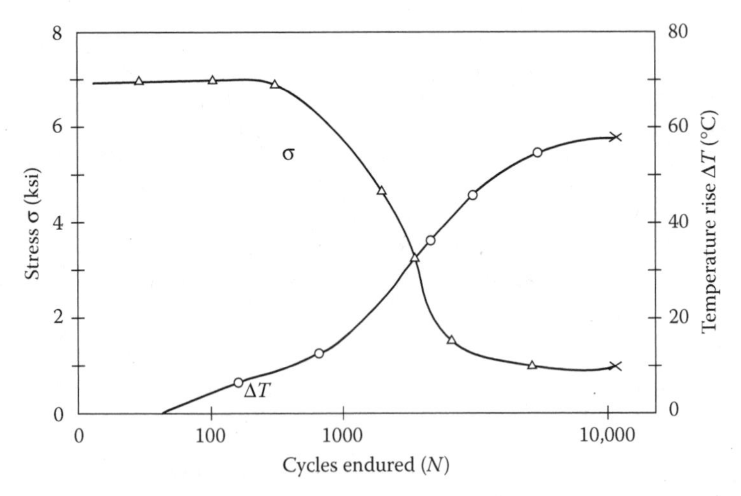

FIGURE 4.10 Apparent stress and temperature rise during constant deflection fatigue testing of a glass-reinforced PP at 1200 cpm. (From Cessna, L. C. and J. A. Levens, *Proc. 4th Annual Plastics Conf.*, Eastern Michigan University, September, 1969, pp. 46–73.)

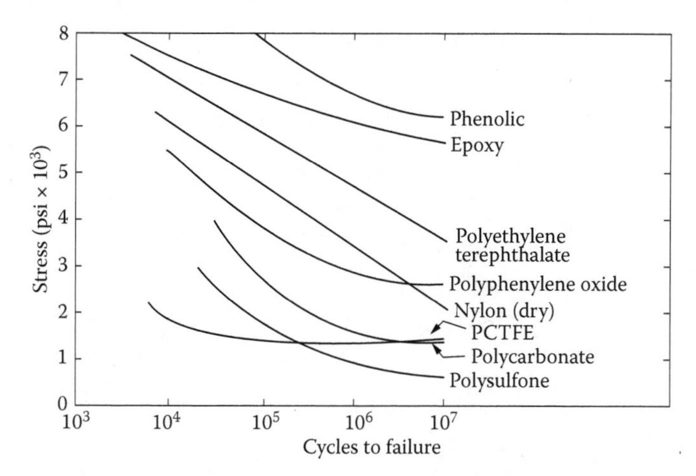

FIGURE 4.11 Fatigue life (S–N) curves for some polymers. (Test frequency: 1800 cpm.) (Adapted from Riddle, M. N., *Proc. 4th Annual Plastics Conf.*, Eastern Michigan University, September, 1969, pp. 1–18.)

How do the relative fatigue strengths of polymers compare? Figure 4.11 shows some representative $S(N)$ curves for a number of polymers from isothermal tests. Some polymers show an endurance limit (S approaching an asymptote S_e at 10^6–10^7 cycles), but many do not.

Another factor in design against fatigue failure is the probability of failure. Most fatigue data are given as average data corresponding to a 50% survival rate. If one wants a greater probability of survival, then one must lower the allowable design stress as shown in Figure 4.12 for glass-reinforced epoxy laminates.

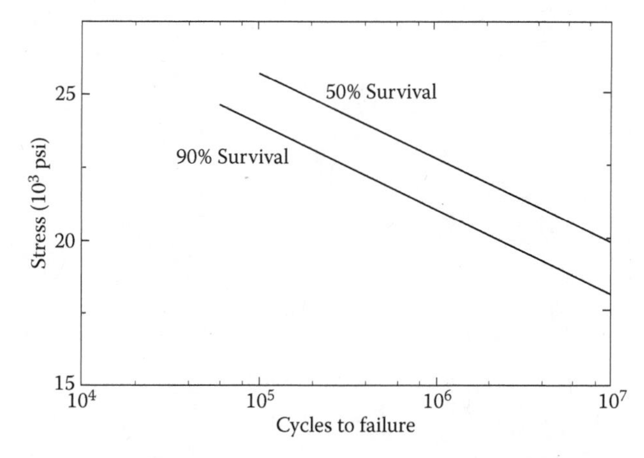

FIGURE 4.12 Flexural fatigue curves for 90% and 50% probability of survival of glass reinforced epoxy laminates (181 Style S-901, room temperature, saturated humidity, 30 cps). (Adapted from Lockwood, P. A., *Proc. 4th Annual Plastics Conf.*, Eastern Michigan University, September, 1969, pp. 74–99.)

4.4 NOTCH SENSITIVITY UNDER FATIGUE

Polymers can be sensitive to stress concentrations or notches under fatigue loading as metals are, but generally they are less sensitive. The notch sensitivity factor (q) is defined as follows:

$$q = \frac{K_f - 1}{K_t - 1}, \tag{4.16}$$

where

q = notch sensitivity factor, $0 < q < 1$, material dependent

K_t = theoretical or geometrical stress concentration factor under stress loading

K_f = stress concentration factor under dynamic loading (includes geometric effect and material dependence)

The stress concentration factor K_f is used in two essentially equivalent ways. The endurance limit of the material is reduced by dividing the nominal value by the stress concentration factor or the applied stress is magnified as shown

$$\sigma = K_f \sigma_{\text{nominal}} \tag{4.17}$$

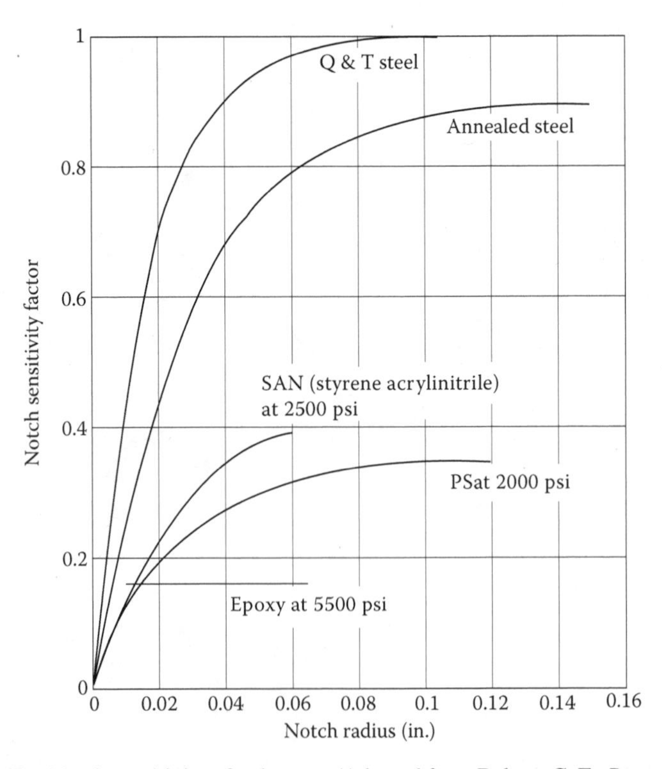

FIGURE 4.13 Notch sensitivity of polymers. (Adapted from Balazs, C. F., *Proc. 4th Annual Plastics Conf.*, Eastern Michigan University, September, 1969, pp. 20–45.)

Notch sensitivity factors have been determined for a number of polymers from rotating beam fatigue tests and they are plotted versus the notch radius in Figure 4.13. The notch sensitivity was determined at one stress level for each material. The frequency of the tests was 30 cps. The conditions were nonisothermal.

4.5 CREEP BUCKLING OF SHELLS

A review article[7] gives a thorough review of creep instability of steel and other metal shells and pressure vessels at elevated temperatures. Some of the same analyses can be adopted here for polymers at lower temperatures, for example, room temperature. For example, consider the axisymmetric buckling mode of a cylindrical shell under compression. (The beam column mode can be included under column buckling, Section 4.2.) For the axisymmetric mode, where axisymmetric buckling waves occur for $R/h < 33$, the critical buckling stress in compression is given by

$$\sigma_{cr} = \left(\frac{E_t E_s}{3(1-\mu^2)} \right)^{1/2} \left(\frac{h}{R} \right), \tag{4.18}$$

where E_t is the tangent modulus given by $E_t = \partial\sigma/\partial\varepsilon$, E_s is the secant modulus given by $E_s = \sigma/\varepsilon$, $\mu = 1/2 - (1-\nu)\,E_s/E$, ν = Poisson's ratio, E = the elastic modulus, h = thickness, and R = radius.

Example 4.3

Find the critical time and the critical strain for creep buckling for an axisymmetric mode of a cylindrical shell under compression for σ_{cr} = 1600 psi, where h = 0.1 in., R = 3.0 in., ν = 0.4. The solution involves an iterative method. For a stress of 1600 psi the moduli are found from isochronous curves in Figure 2.12b.

The elastic modulus is E = 700/0.005 = 140,000 psi.
The critical time is between t_2 and t_3.
The critical strain is $\varepsilon_{cr} = 0.0125 + 0.4(0.01500125) = 0.0135$.
The secant modulus is E_s = 1600/0.0135 = 118,500 psi.
The tangent modulus is E_t = (1800 – 1600)/(0.017 – 0.0135) = 57,140 psi.
The value $\mu = 1/2 - (1 - 0.4)118{,}500/140{,}000 = -0.0079$.

From Equation 4.17 the calculated buckling stress is

$$\sigma_{cr} = \left(\frac{57{,}140(118{,}500)}{3} \right)^{0.5} \left(\frac{0.1}{3.0} \right) = 1584\,\text{psi}$$

This result is close for $h/R = 0.1/3$. For other values of h/R iteration on estimated stress and modulus is required.

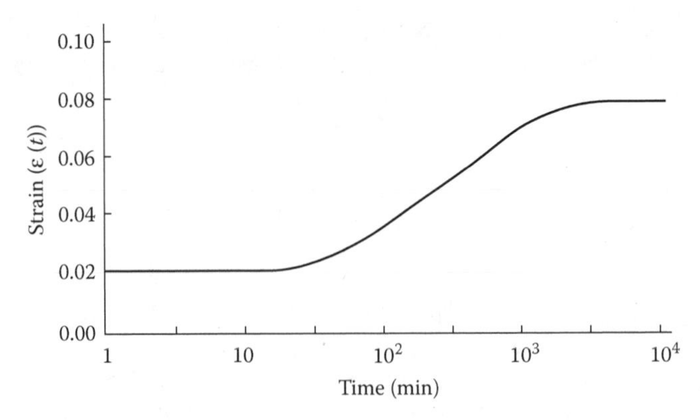

FIGURE 4.14 Strain (time).

HOMEWORK PROBLEMS

4.1. Find the creep buckling curve P_{cr} as a function of time for a pinned–pinned column made from a polymer material with data given in Figure 4.14 for a $\sigma = 2000$ psi creep test.

4.2. Consider two polymers A and B with creep rupture strengths given in Table 4.1 for $T = 70°F$. Choose polymer A or B for a load time history as shown in Figure 4.15. Consider tensile loading of a bar with a 0.5 in. diameter.

4.3. A cylindrical bar made of phenolic must withstand 10^6 cycles of load of P of 200 lb at 1800 cpm as shown in Figure 4.16. (a) Find the diameter of the bar required. (b) If the bar was subjected to cyclic strain rather than to cyclic stress, would a larger or smaller diameter be required? Explain.

4.4. Derive the relationship for energy loss per cycle for $\sigma = \sigma_0 \sin \omega t$ in terms of E', E'', and $\tan \delta$ only.

TABLE 4.1
Ultimate Strengths of Polymers A and B

	S_u (ksi)	
Time (min)	**A**	**B**
0	10	6
1	8	5.8
10	6	5
100	4	4
1000	3	3.7
10,000	2	3.4

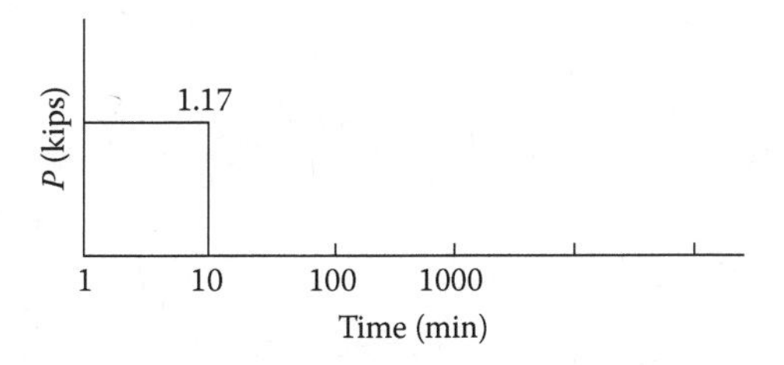

FIGURE 4.15 $P(t)$.

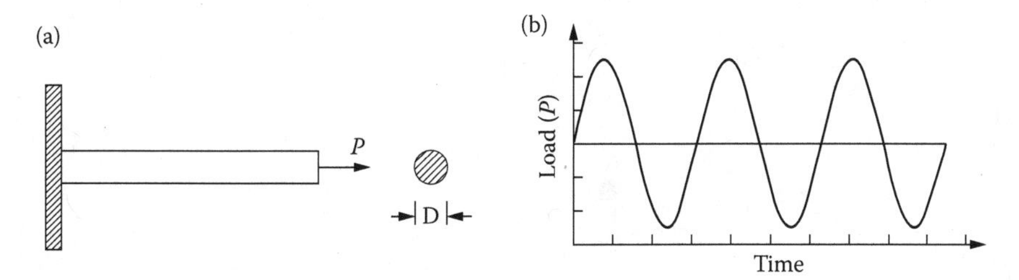

FIGURE 4.16 Bar with oscillating load.

4.5. Derive the relationship for energy loss per cycle for $\varepsilon = \varepsilon_0 \sin \omega t$ in terms of E', E'', and $\tan \delta$ only.

The reason that modulus is used instead of compliance for energy loss is that most cyclic applications utilize modulus values that are determined in cyclic testing.

REFERENCES

1. *Engineering Materials Handbook, Volume 2, Engineering Plastics*, ASM International, 1988, Figure 17, p. 666; Figure 25, p. 671.
2. Baer, E., *Engineering Design for Plastics*, Reinhold Book Corp., 1964, 283pp.
3. Riddle, M. N., Fatigue behavior of plastics, fatigue and impact resistance of plastics, *Proc. 4th Annual Plastics Conf.*, Eastern Michigan University, September, 1969, pp. 1–18.
4. Cessna, L. C. and J. A. Levens, Fatigue testing of reinforced thermoplastics, fatigue and impact resistance of plastics, *Proc. 4th Annual Plastics Conf.*, Eastern Michigan University, September, 1969, pp. 46–73.
5. Lockwood, P. A., Dynamic fatigue of glass fiber reinforced plastics, fatigue and impact resistance of plastics, *Proc. 4th Annual Plastics Conf.*, Eastern Michigan University, September, 1969, pp. 74–99.
6. Balazs, C. F., Fatigue tests and properties of plastics, Fatigue and impact resistance of plastics, *Proc. 4th Annual Plastics Conf.*, Eastern Michigan University, September, 1969, pp. 20–45.
7. Anonymous, *Effect of Creep and Other Time Related Factors on Plastics and Elastomers*, Plastics Design Library, William Andrews Publishing, www.knovel.com, 1991.
8. Anonymous, GE Plastics Library, www.ides.com
9. Gerdeen, J. C. and C. K. Sazawal, A review of creep instability in high-temperature piping and pressure vessels. *WRC Bulletin No.* 195, 33–56, 1974.

5 Impact Strength and Fracture Toughness

Other properties must be considered for design purposes. Two important mechanical properties that remain to be discussed are the impact strength and fracture toughness. These are considered in depth in this chapter.

5.1 IMPACT STRENGTH

The ability of a polymer to absorb energy under impact conditions is important in design applications, for both the obvious (e.g., the design of a plastic composite bumper for an automobile) and the not so obvious (e.g., the housing for an electric drill that might be accidentally dropped on the floor) reasons. Impact loading is a high strain rate type of loading that occurs in the crash of an automobile. In automobile design, this type of loading has opened a whole new field of study called "crashworthiness."

One measure of the impact strength is the area under the stress (strain) curve for a tensile test conducted at a high strain rate. This measure is called the modulus of toughness. Obviously, a thermoplastic-type polymer with a relatively high elongation (ductility) will have a higher modulus of toughness than does a more brittle thermoset.

The beam-bending tests, such as the Charpy and the Izod tests shown in Figure 5.1, have been commonly used for impact testing of metals as well as polymers. The student is familiar with these tests from courses in mechanics of materials. Recall that in these tests, a pendulum device with a hammer is used. The pendulum hammer of weight W is dropped from a height H_0. After striking the specimen, the hammer rises to a height H less than H_0. The difference in energy

$$U = W(H_0 - H) \tag{5.1}$$

is the energy absorbed, and this difference is used as a measure of the impact strength. This equation measures the loss of energy of the impact hammer and it is usually assumed that all of this energy is absorbed by the specimen as elastic strain energy prior to fracture. Any kinetic energy loss is neglected. This is really not the case as variations in clamping pressure causes friction losses as noted later in this section.

Notches are used in the Charpy and Izod specimens to create a stress concentration. The notch sensitivity is as important in impact as it is in fatigue. When the specimen is notched, most of the deformation takes place near the notch, so that the material near the notch experiences a much higher rate of straining than does the material away from the notch. The standard dimensions for a plastic Izod specimen are given by ASTM Standard D256,[1] and are shown in Figure 5.1. The V-notch

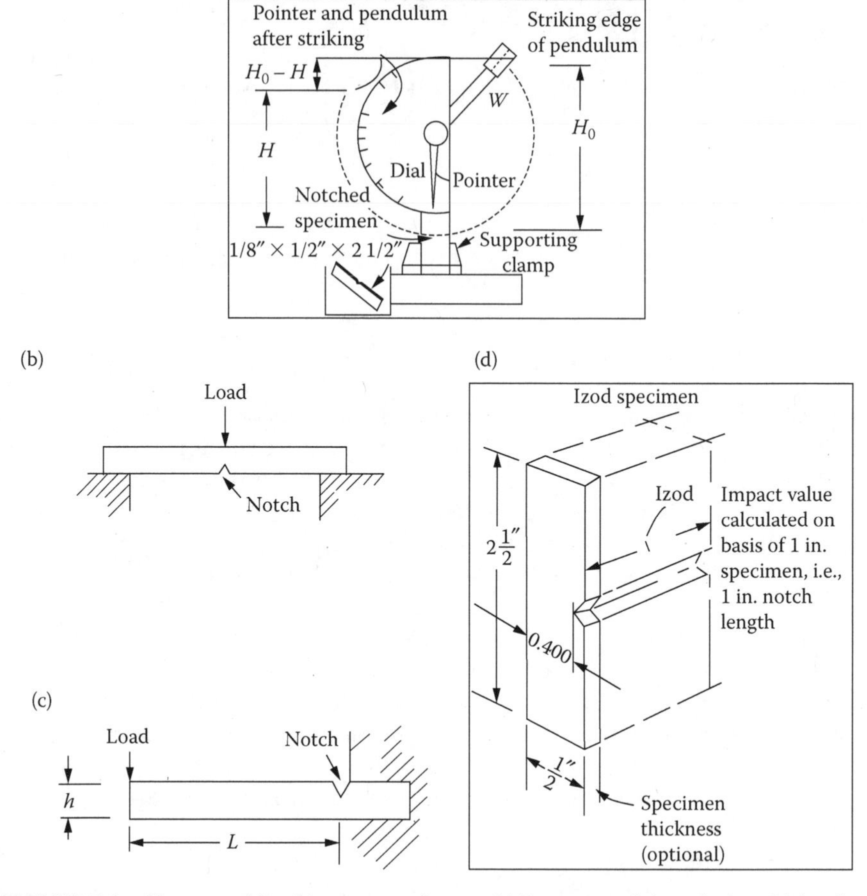

FIGURE 5.1 Charpy and Izod impact specimens. (a) Impact pendulum device, (b) loading of Charpy specimen, (c) loading of Izod specimen, and (d) dimensions of Izod specimen. (Adapted from *Tensile Impact Energy to Break Plastics and Electrical Insulating Materials*, ASTM Standard D1822, West Conshohocken, PA, 1999.)

should be milled and polished to a surface roughness of 10–100 μm and to a radius of 0.010 in. (0.254 mm). This should be done carefully as not to induce residual stresses from heating due to friction. The stress concentration K for a standard Izod test is approximately 3.0, and the theoretical value is assumed to be fully reached during an elastic stress condition at impact. However, flaws such as cracks or voids in the material will further enhance fracture. See the Fracture Toughness Analysis (Section 5.2) later in this chapter.

Table 5.1 shows some impact strength values for some polymers at 24°C. The units are energy (ft-lb) per length of notch (in.), the length parallel to the notch direction. The metric units are J/m where 1.0 J/m = 53.4 ft-lb/in. There is an appreciable range in some of the quoted values. This may be due to surface roughness variations

TABLE 5.1
Notched Izod Impact Strength Values of Some Polymers at 24°C

	Impact Strength	
Material	**ft-lb/in.**	**J/m**
PE (LDPE)	>16.0	>854.4
PE (HDPE)	0.5	26.7–1068.0
PP	0.5–2.0	26.7–106.8
PS	0.25–0.4	13.35–21.4
HIPS	0.5–8.0	26.7–427.2
ABS	1.0–10.0	53.4–534.0
PVC	0.4–3.0	21.4–160.2
PVC (polyblends)	3.0–20.0	160.2–1068.0
Polyacetal	1.1–2.3	58.7–122.8
Polymethyl methacrylate	0.4–0.5	21.4–26.7
Nylon 6	1.0–3.0	53.4–160.1
Nylon 6,6	1.0–3.0	53.4–160.2
Nylon 6,12	1.0–1.4	53.4–74.76
Nylon 11	1.8	96.12
PC	12.0–18.0	640.8–961.2
PTFE	2.0–4.0	106.8–213.6
Epoxy	0.20–5.0	10.7–267.0
Epoxy (glass filled)	10.0–30.0	534.0–1602.0
Polyphenylene oxide (PPO)	5.0	267
PPO (25% glass)	1.4–1.5	74.8–80.1
Polysulfone	1.3–5.0	69.42–267.0
Polyester	2.0–20.0	106.8–1068.0
Phenol-formaldehyde	0.25–0.35	13.0–19.0
Phenol-formaldehyde (cloth-filled)	1.0–3.0	53.0–160.0
Phenol-formaldehyde (glass fiber-filled)	10.0–30.0	534.0–1062.0

Sources: Adapted from *Guide to Plastics, Property and Specification Charts*, McGraw-Hill, New York, NY, 1985; Engineering Plastics, *Engineering Materials Handbook*, Vol. 2, ASM, Metals Park, OH, 1988.

and other variations in material. Also it has been found that uncontrolled application of a clamping pressure may lead to nonconsistent results.[4] The impact toughness values are useful to relatively compare materials under impact conditions, but they are not directly applicable to a particular design, because of size effects and because the loading conditions are generally different. If impact strength is critical in a design, the end user should perform their own tests or request separate data from the supplier for the particular material selected.

Polymers will exhibit a ductile/brittle transition temperature in impact tests similar to that observed in metals. In polymers, the transition temperature T_t, is close to the glass-transition temperature T_g for large-width specimens (plane strain condition),

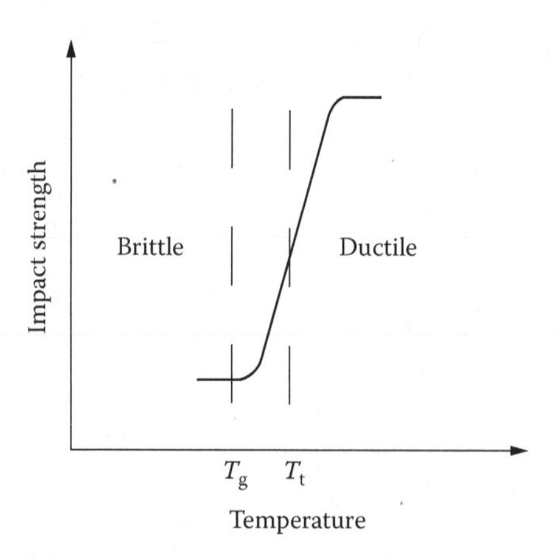

FIGURE 5.2 Ductile–brittle transition in polymers.

as shown in Figure 5.2, but can be much lower for thin Izod specimens as noted by Williams.[5]

An increase in impact strength is also found when the strain rate is increased. Impact modifiers, such as rubber particles, can be added to a polymer resin to increase its impact strength. For example, compare HIPS (high-impact modified PS) with PS in Table 5.1.

The stress condition that causes failure in the Izod test is primarily the bending stress and test values may not relate to other types of loading. Two other less common tests are the axial and the drop-weight impact test. The axial impact test (ASTM D1822) uses dogbone-shaped specimens (types S and L), as shown in Figure 5.3.[1] Obviously there is no stress concentration due to a notch as there is in the Izod test, and the stress is more uniform as it is in a tensile test. Failure may be initiated at an internal or surface flaw in the material. The specimens can be mounted in the pendulum device by use of special adapters. Some values for tensile impact for various polymers are shown in Table 5.2. Values are given in units of energy per area of the reduced cross section.

The drop-weight impact test (ASTM D 3029) is applied to thin-disk specimens. The test is designed primarily to study fracture initiation under biaxial loading. A weight with a hemispherical indenter (tup) is dropped onto a thin circular disk clamped on its periphery by a steel ring. In a constant-height test, a sequence of drops is made with increasing weights on successively fresh specimens until visible failure is observed. Some values from drop-weight tests are listed in Table 5.3. Values are given in units of energy (ft-lb), for a 50% statistical mean average of failure at that value of energy. The ring ID (inside diameter) is the clamped boundary of the disk specimen. The type of failure is a visible crack, except for the Nylon 6 failure, which was a puncture (P).

A drop-weight impact test standard (ASTM D7136)[6] has also been developed for fiber-reinforced polymer matrix composites. The specimen is a flat rectangular composite plate 4.000 in. × 6.000 in. (100.0 mm × 150.0 mm). The laminate consists of a number of plies to achieve a total cured thickness nearest to 5.0 mm (0.20 in.) with a stacking sequence of [(+45/–45)/(0/90)] (see Composites, Chapter 8). A

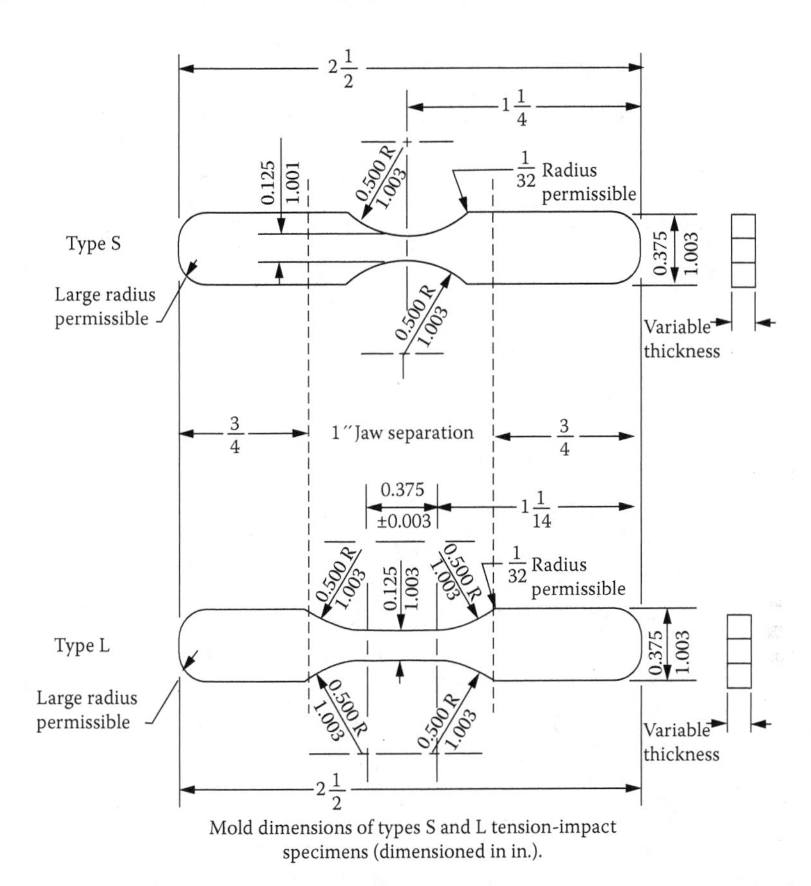

FIGURE 5.3 Types S and L tension-impact specimens. (Dimensions in inches.) (From *Tensile Impact Energy to Break Plastics and Electrical Insulating Materials*, ASTM Standard D1822, West Conshohocken, PA, 1999. Copyright ASTM International. Reprinted with permission.)

drop-weight (impactor) with a mass of 5.5 ± 0.25 kg (12 ± 0.5 lb_m), and with a smooth hemispherical striker tip of diameter of 16 ± 0.1 mm (0.625 ± 0.005 in.) is used. The weight is dropped from heights up to 300 mm (12 in.). If less impact energy is desired an impactor with a mass of 2.0 kg (4 lb_m) may be used. The damage resistance is quantified in terms of the resulting size and type of damage in the specimen. Various damage parameters can be specified including maximum diameter of dents developed, depth of dents, and measurements of crack lengths.

Circular plates of HDPE have also been tested under impact by Clausen et al.[7] The results agreed with a hyperelastic-viscoplastic constitutive model.

Stenzler[8] conducted instrumented impact tests of laminates with soft polymer interlayers have been conducted on three-layered samples consisting of a PMMA front, a polymer interlayer or adhesive, and a PC layer. The inner layers were either polyurethane (PUR) or polyacrylate. Strain rates on the order of 100 s^{-1} at impact velocities 12 and 22 m/s were achieved using a gas gun. The interlayers increased impact resistance by increasing energy dissipation.

TABLE 5.2
Tensile Impact Strength Values of Some Polymers at 23°C

Name	Type of Specimen	Impact Strength (ft-lb/in.^{2})
Acrylic—cast sheet (Plexiglas)	L (S)	25 (15)
PE (HDPE)	S	18–60
PP	S	25–40
ABS—sheet extrusion compound	L	69–80
ABS—5% glass fiber reinforced	S	19–37
PVC (rigid)	S	50–70
Polyacetal—injection molding resin	L	60–90
Nylon 6, molding resin	L, Dry	250
Nylon 6/6, molding resin	L, Dry (S, Dry)	240 (75)
PC, injection molded	S	225–300
Polyester PET, injection molded	L	23

Sources: Adapted from Brostow, W. and R. D. Corneliussen, *Failure of Plastics*, Hanser Publishers, Munich, West Germany, 1986; *Guide to Plastics, Property and Specification Charts*, McGraw-Hill, New York, NY, 1985.

5.1.1 Thickness Effects

The thickness of the Izod specimens is generally very thin, typically 0.10 in. (2.54 mm). This corresponds to a plane stress state. The standard Charpy specimen is 0.394 in. (10.0 mm), and results in a more nearly plane strain state. This difference causes the failure plane under crack propagation to be different in each case. It also means that the

TABLE 5.3
Drop-Test Impact Strength Values of Some Polymers at 23°C

Name	Specimen Thickness (in.)	Ring ID (in.)	Tup Dia. (in.)	Impact Strength (ft-lb)
PP—homopolymer	0.090	3.00	0.5	8
—copolymer	0.090	3.00	0.5	>32
ABS—sheet extrusion compound		2.75	0.5	19
PVC (rigid)	0.060	0.625	0.5	14
Nylon 6, injection molding resin, dry	0.125	3.00	1.0	125 P
Nylon 6/6, molding resin, mineral reinforced, dry	0.100	2.25	1.5	2.8
PC, injection molding resin	0.125	3	1	>125

Source: Adapted from *Guide to Plastics, Property and Specification Charts*, McGraw-Hill, New York, NY, 1985.

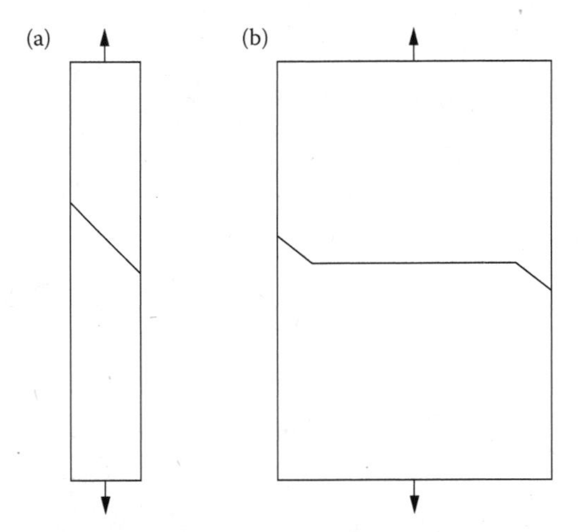

FIGURE 5.4 Failure planes under plane stress or plane strain depending on the thickness. (a) 45 degree shear plane in thin specimen and (b) zero degree normal fracture plane in thick specimen (small shear lips at surface).

Izod results are not directly applicable to thicker parts. Ductile failure will generally occur on a maximum-shear plane at 45° to the surface. Figure 5.4 shows the difference in failure planes for different thicknesses.

5.1.2 Rate Effects

Adams and Wu conducted experiments[10] to measure the load deflection behavior under different rates of deflection. It has been found that the fracture energy absorbed by the specimen is less under high rates than under the quasi-static case. Figure 5.5 shows the general effect. Under high rates the fracture exhibits more brittle behavior. Crack initiation occurs at the peak of the loading curve and crack propagation occurs

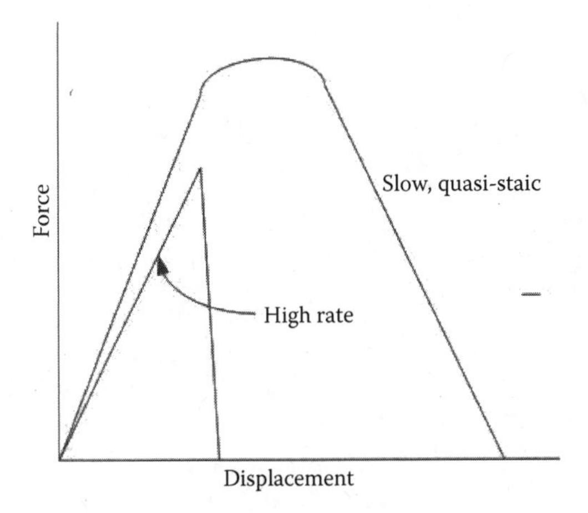

FIGURE 5.5 Force (displacement) curves for different rates of loading.

TABLE 5.4
Rate Effects for Charpy Impact Specimens

Material	Velocity (m/s)	Energy Absorbed (J)
Nylon 6,6	2.0×10^{-5}	1.43
	2.0	0.43
Rubber-toughened nylon	2.0×10^{-5}	5.78
	2.0	4.97
HIPS	2.0×10^{-5}	0.63
	2.0	0.31

during the unloading portion. Table 5.4 lists some example data. Many more such data and extensive analyses may be found in Williams[5] and Adams and Wu.[10]

5.1.3 Combined Stiffness and Impact Properties

A company advertised a new polymer material that was supposedly better than any material known when rated on the basis of combined stiffness and impact strength, as shown in Figure 5.6, where the new material falls above the envelope of the data for all the others. This result raised the question of whether this way of plotting the combined properties is applicable to any real situation.

First, the known data for many materials were plotted as shown in Figure 5.7, and the code numbers of the materials are defined in Table 5.5. This result confirmed the envelope of the relation that showed how the Izod impact strength decreases with stiffness for most polymers.

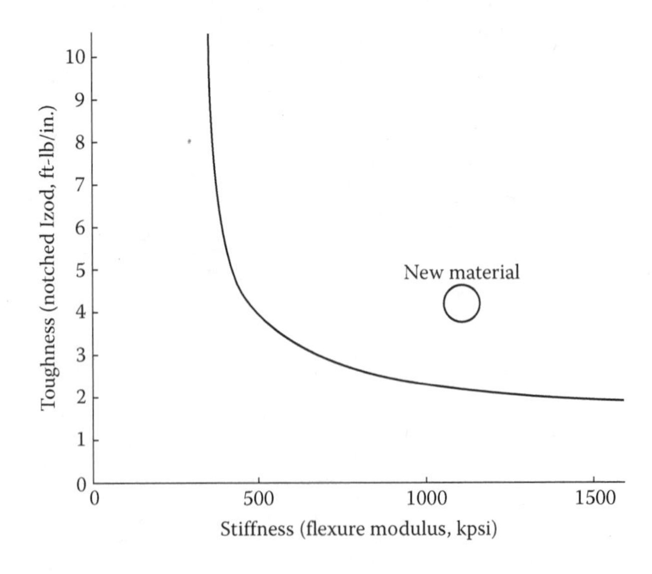

FIGURE 5.6 "Breakthrough" in toughness compared to stiffness.

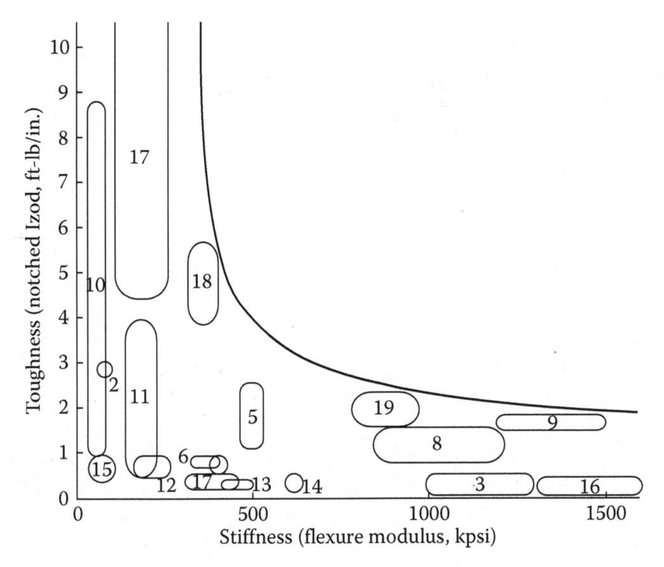

FIGURE 5.7 Toughness of various polymers (material number code in Table 5.5).

TABLE 5.5
Code Numbers for Materials in Figure 5.7

Number	Material Name
1	PMMA (Acrylic)
2	PTFE (Polytetraflourethylene)
3	PF (Phenolic, wood flour filled)
4	Nylon 6 (PA, polyamide)
5	PC (Polycarbonate, 10% glass)
6	PBT (Polyester, polybutylene terephthalate)
7	PET (Polyester, polyethylene terephthalate)
8	PBT (Glass reinforced)
9	PET (Glass reinforced)
10	LDPE (Low-density polyethylene)
11	HDPE (High-density polyethylene)
12	PP (Polypropylene)
13	PS (Polystyrene)
14	PUR (Polyurethane, unsaturated)
15	PVC (Polyvinylchloride)
16	UREA
17	Impact modified thermoplastics
18	PPO (Polyphenylene oxide)
19	Nylon (Glass reinforced)

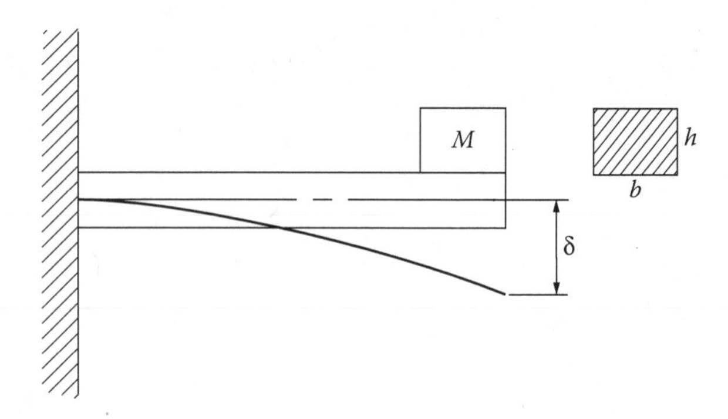

FIGURE 5.8 Impact loading of a cantilever beam.

Next, a problem was conceived in which the combined properties would be important in the design. Consider the cantilever beam problem in Figure 5.8. Assume at time $t = 0$ that the beam is unstressed and the mass M is held immediately above the beam by an external agency. At time $t = 0^+$, the mass is suddenly released. This application is a dynamic load application, which is typical of loading on leaf springs in vehicles.

Let δ and P be the maximum deflection and load reached. The input energy is the potential energy of the mass before deflection given by

$$V = Mg\delta \tag{5.2}$$

The impact energy per width then is

$$V^* = \frac{V}{b} \tag{5.3}$$

The factor of safety on impact strength then is

$$\eta = \frac{T}{V^*} \tag{5.4}$$

Next, assume that all of the input energy goes into elastic-strain energy to deflect the beam, such that

$$V = Mg\delta = \frac{1}{2}P\delta \tag{5.5}$$

where T is the Izod toughness of the material.

The dynamic load then is twice the static load and is given by

$$P = 2Mg \tag{5.6}$$

The deflection of a cantilever beam is given by

$$\delta = \frac{PL^3}{3EI} = \frac{8MgL^3}{Ebh^3}. \tag{5.7}$$

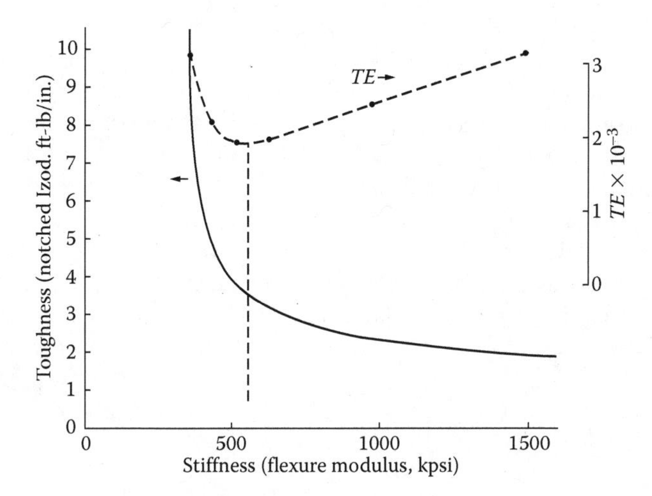

FIGURE 5.9 Plot showing product of impact strength and stiffness that is important in dynamic loading of a cantilever beam.

Thus,

$$\frac{V}{b} = \frac{8(Mg)^2L^3}{Eb^2h^3}. \tag{5.8}$$

Example 5.1

Assume the length L and depth h are constant, but the width b needs to be selected to satisfy a certain factor of safety given by Equation 5.4. Thus, select b according to

$$b^2 = FS\frac{8(Mg)^2L^3}{TE\,h^3}, \tag{5.9}$$

that is,

$$b^2 \approx \frac{1}{TE}. \tag{5.10}$$

Therefore, the product of impact strength and stiffness is important in this problem to minimize the size of beam required. Figure 5.7 is replotted in Figure 5.9, and the product (TE) is also plotted as shown. The TE curve shows that materials on both ends of the spectrum with either high impact strength or high stiffness are candidate materials and that the new material (Figure 5.6) has the best rating in the intermediate range.

5.2 FRACTURE TOUGHNESS

Whereas the impact strength is the measure of a ductile polymer to absorb energy, the fracture toughness is a measure of a more brittle polymer to resist crack propagation. Polymers can contain flaws from processing, such as cracks, voids, and inclusions. These flaws act as stress concentrations. Often these flaws are microscopic.

5.2.1 Brittle Fracture

Historically the theory of brittle fracture is known as the Griffith theory after A. A. Griffith,[11] who developed a theory for the brittle fracture of glass. He calculated that the theoretical tensile strength of glass should be about 1.6 (10^6) psi, but actual tests showed strength values 100 to 1000 times lower. He also found a size effect: Smaller-diameter filaments had higher strengths, as shown in Table 5.6.

Anderegg[12] found a similar effect when testing even smaller-diameter filaments, as shown in Table 5.7. Griffith postulated and showed that the reason for the size effect was flaws or cracks in the material. The smaller-diameter filaments must have smaller cracks and, thus, must have a higher strength. Griffith used an energy approach to prove his postulate.

First consider the elastic energy stored in a uniform bar in tension, as shown in Figure 5.10. Assume for now that the bar has no cracks. The external work done on the bar is

$$W = \frac{1}{2} Pu \tag{5.11}$$

where P is the externally applied load and u is the deflection. This work can be related to the internal elastic energy stored in the bar in terms of stress and strain in the bar. The stress and strain are

$$\sigma = \frac{P}{A}, \quad \varepsilon = \frac{u}{L}. \tag{5.12}$$

For an elastic bar,

$$\varepsilon = \frac{\sigma}{E} \tag{5.13}$$

TABLE 5.6
Tensile Strength of Glass Filaments

Diameter (mm)	1.02	0.0508	0.0178	0.0033
S_u (kg/cm^2)	1750	5610	11,500	34,600
S_u (psi)	2490	7980	16,350	49,200

Source: Adapted from Griffith, A. A., *Philos. Trans. R. Soc.*, A221, 163–198, 1921.

TABLE 5.7
Tensile Strength of Glass Filaments

Diameter (10^{-6} m)	3.1	4.1	6.6	8.4
S_u (kg/cm^2)	364,000	174,000	121,000	97,000
S_u (psi)	517,600	247,400	172,000	138,000

Source: Adapted from Anderegg, F. O., *Ind. Eng. Chem.*, 31, 290–298, 1939.

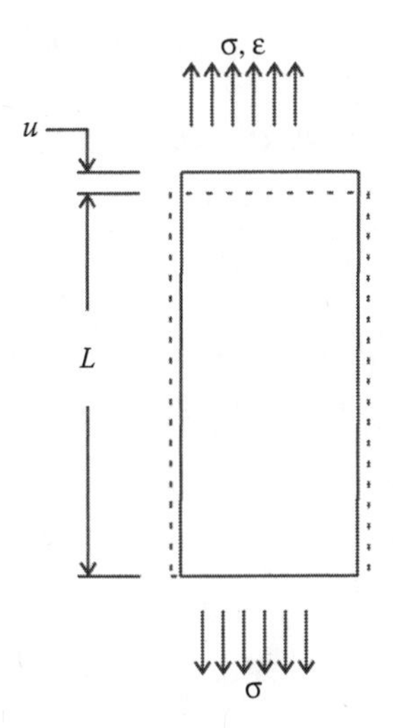

FIGURE 5.10 Work done on uniform bar in tension.

Thus,

$$W = \frac{1}{2}\sigma A\left(\frac{\sigma}{E}\right)L = \frac{1}{2}\left(\frac{\sigma^2}{E}\right)AL = UV \tag{5.14}$$

where V = volume, and the strain energy per unit volume in tension is

$$U = \frac{1}{2}\frac{\sigma^2}{E} \tag{5.15}$$

Next, consider the energy stored in a bar with a crack of length $2a$, as shown in Figure 5.11. (The length is taken as $2a$ so that from symmetry, the same solution will

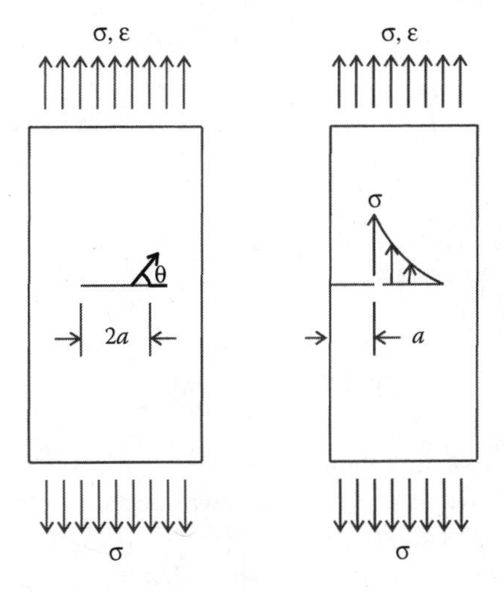

FIGURE 5.11 Crack in bar under simple tension.

apply to a side crack of length α in a bar half as wide as shown.) From the theory of elasticity,[13] a complex variable solution is used to show that the elastic energy stored in the bar with a crack is less by an amount of

$$\delta U_E = \pi a^2 \frac{\sigma^2}{E}, \quad \text{per unit thickness} \tag{5.16}$$

where, for now, we assume that the width of the bar is relatively wide and long compared with the size of the crack, and the release (difference) in strain energy all goes into forming the crack.

Let γ be the surface energy per surface area required to create a new free surface in the material (e.g., $\gamma = 3.12\ (10^{-3})$ in.-lb/in.^{2} in glass). Then, the change in surface energy required to produce a crack of length $2a$ is

$$\delta U_S = 4\gamma a \text{ per unit thickness} \tag{5.17}$$

Consider crack extension (assume a preexisting crack), and postulate that the rate of decrease of elastic energy is equal to the rate of surface energy:

$$\frac{d(\delta U_E)}{da} = \frac{d(\delta U_S)}{da}, \tag{5.18}$$

which gives

$$2\pi a \frac{\sigma^2}{E} = 4\gamma$$

or

$$\sigma_{cr} = S_u = \sqrt{\frac{2E\gamma}{\pi a}}. \tag{5.19}$$

The Griffith factor is

$$G_C = 2\gamma, \tag{5.20}$$

and the product $G_C E$ (a constant) is related to a material property known as the fracture toughness K_{IC}, where

$$K_{IC} = \sigma_{cr}\sqrt{\pi a}. \tag{5.21}$$

Thus, we can specify a maximum permissible stress for a given crack (flaw) size, or we can specify a maximum permissible flaw size for a given ultimate stress. This feature is important for nondestructive evaluation (NDE) and quality control. (The subscript *IC* in Equation 5.21 refers to mode *I* fracture in simple tension. Other modes of fracture, modes II and III, are shown in Figure 5.12.)

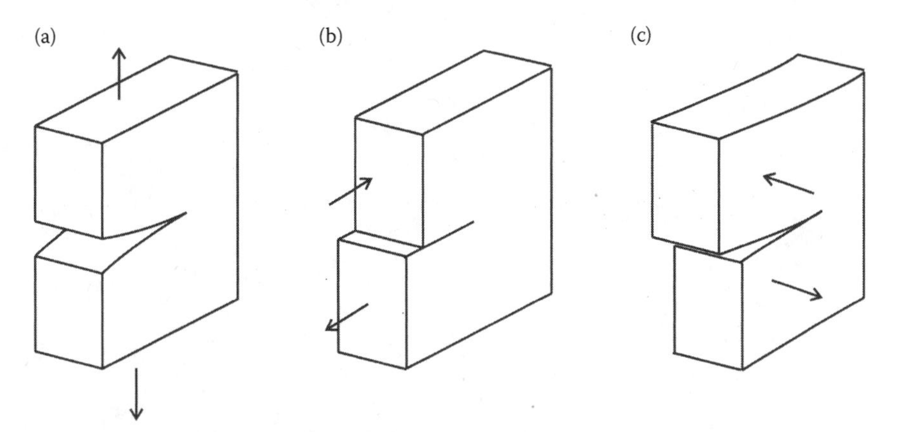

FIGURE 5.12 Different modes of fracture: (a) mode I opening mode, (b) mode II sliding mode, and (c) mode III tearing mode.

If the stress is less than the critical value to cause failure, then stress intensity K_I under tension can be defined as

$$K_I = \sigma\sqrt{\pi a}, \tag{5.22}$$

where $K_I < K_{IC}$ represents a factor of safety against fracture. From the theory of elasticity applied to linear elastic fracture mechanics (LEFM), the stresses in the vicinity of a crack tip can be shown to be represented by

$$\sigma_{ij} = \frac{K_I f_{ij}(\theta)}{\sqrt{2\pi r}}, \tag{5.23}$$

where r and θ are the coordinates shown in Figure 5.11.

The stress-intensity factor K_I is like a stress concentration factor with the singularity $r^{1/2}$ separated out. As r approaches 0, the stress concentration goes to infinity.

5.2.2 Ductile Fracture

In ductile materials some blunting of the crack occurs, as shown in Figure 5.13, where the radius of the plastic zone is given by[5]

$$r_p = \frac{(K_{IC}/S_y)^2}{2\pi}. \tag{5.24}$$

Thus, the ductility factor $(K_{IC}/S_y)^2$ is a measure of the degree of crack blunting.

Table 5.8 gives some fracture-toughness parameters and ductility factors for a number of materials.[9] Figure 5.14 shows the variation of fracture toughness value K_{IC} for finite-width plates. Shigley and Mischke[14] and Sih[15] give values for other modes of fracture.

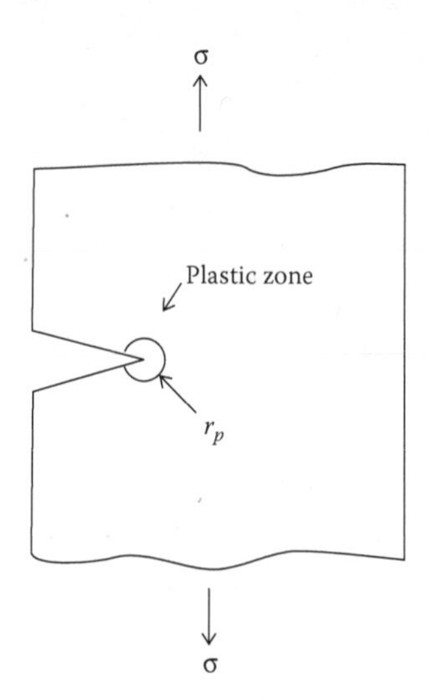

FIGURE 5.13 Blunting of crack tip due to yielding of material.

5.2.3 General Theory of Fracture Instability

To develop a general theory and investigate the stability or instability of existing cracks in some known geometries, a connection with Griffith's postulate is made. Consider an energy balance:

$$W = U_e + U_k + U_d, \tag{5.25}$$

TABLE 5.8
Fracture Toughness Values for a Range of Materials at 20°C

Material	G_{IC} (kJ/m²)	K_{Ic} (MN/m$^{3/2}$)	(K_{Ic}/S_y)	$(K_{Ic}/S_y)^{1/2} \times (10^3)$
ABS	5	2–4	0.13	17
Acetal	1.2–2	4	0.08	6
Acrylic	0.35–1.6	0.9–1.0	0.014–0.023	0.2–0.5
Epoxy	0.1–0.3	0.3–0.5	0.005–0.008	0.02–0.06
Glass-reinforced polyester	5–7	5–7	0.12	14
LDPE	6.5	1	0.125	16
Medium density polyethylene (MDPE)/HDPE	3.5–6.5	0.5–5	0.025–0.25	5–100
Nylon 66	0.25–4	3	0.06	3.6
PC	0.4–5	1–2.6	0.02–0.5	0.4–2.7
PP copolymer	8	3–4.5	0.15–0.2	22–40
PS	0.3–0.8	0.7–1.1	0.02	0.4
uPVC	1.3–1.4	1–4	0.03–0.13	1.1–18
Glass	0.01–0.02	0.75	0.01	0.1
Mild steel	100	140	0.5	250

Sources: Adapted from Brostow, W. and R. D. Corneliussen, *Failure of Plastics*, Hanser Publishers, Munich, West Germany, 1986; Engineering Plastics, *Engineering Materials Handbook*, Vol. 2, ASM, Metals Park, OH, 1988.

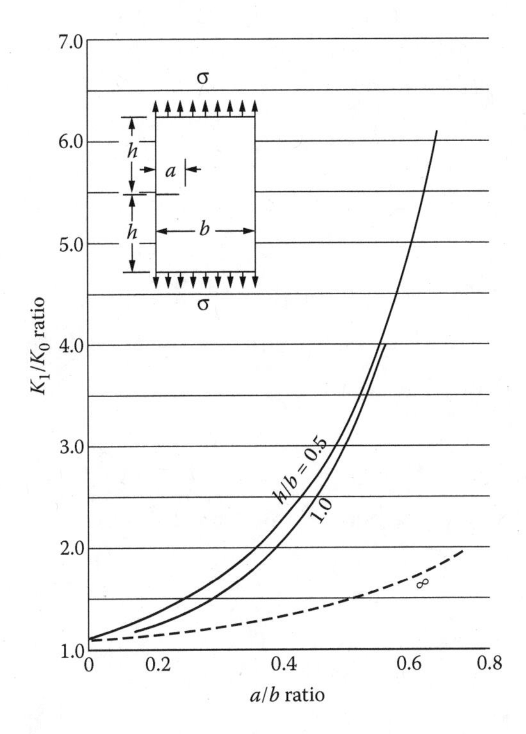

FIGURE 5.14 Values of fracture toughness for finite-width plates. (Adapted from Shigley, J. E. and C. R. Mischke, *Mechanical Engineering Design*, 5th ed., McGraw-Hill, New York, 1989, 5.22, p. 224.).

where W is the external work done by the loads on the body, U_e is the increase in internal elastic-strain energy, U_k is the change in kinetic energy, and U_d is the energy dissipated in the body by heat or through fracture.

Define energy release rate G as[5]

$$G = \frac{\partial W}{\partial A} - \frac{\partial U_e}{\partial A}, \tag{5.26}$$

and define fracture resistance by

$$R = \frac{\partial U_r}{\partial A}, \tag{5.27}$$

where A is the area of the fracture surface and ∂A is the increase in fracture surface area. Thus, the energy balance in terms of G and R becomes

$$G - R = \frac{\partial U_k}{\partial A}. \tag{5.28}$$

For fracture initiation, the kinetic energy U_k is 0 and $(\partial U_k / \partial A) > 0$ for initiation to occur

$$G - R \geq 0. \tag{5.29}$$

For monotonic fracture, $G = R$, and the fracture is unstable if

$$\frac{\partial G}{\partial A} - \frac{\partial R}{\partial A} > 0. \tag{5.30}$$

For constant fracture resistance R (Griffith theory for a brittle elastic material),

$$\frac{\partial G}{\partial A} > 0 \tag{5.31}$$

for unstable crack growth.

Next, G is formulated for a linear elastic body. From the definition of work,

$$\partial W = P\, \partial u. \tag{5.32}$$

For a linear elastic body,

$$U_e = \frac{1}{2} Pu \tag{5.33}$$

such that

$$\frac{\partial U_e}{\partial A} = \frac{1}{2}\left(P \frac{\partial u}{\partial A} + u \frac{\partial P}{\partial A} \right) \tag{5.34}$$

Substitution of Equations 5.33 and 5.34 into Equation 5.26 gives

$$G = \frac{1}{2}\left(P \frac{\partial u}{\partial A} - u \frac{\partial P}{\partial A} \right) \tag{5.35}$$

for a linear elastic body.

If

$$P = ku, \tag{5.36}$$

where k is the stiffness (spring constant) of the elastic body, then

$$\frac{\partial P}{\partial A} = k \frac{\partial u}{\partial A} + u \frac{\partial k}{\partial A}$$

and thus,

$$G = -\frac{1}{2}\left(\frac{P}{k} \right)^2 \frac{\partial k}{\partial A}, \tag{5.37}$$

or equivalently

$$G = -\frac{1}{2}u^2\frac{\partial k}{\partial A}, \tag{5.38}$$

which shows that the energy release rate is represented by a decrease in stiffness for larger cracks.

From Equations 5.33 and 5.36, G can be written as

$$G = -\frac{U_e}{k}\frac{\partial k}{\partial A}. \tag{5.39}$$

For the plane case of a sheet of thickness b, $\partial A = 2b\,\partial a$ where a is the crack length,

$$G = -\frac{U_e}{2bk}\frac{\partial k}{\partial a}. \tag{5.40}$$

Alternatively, from Equations 5.48 and 5.46:

$$G = -\frac{1}{4b}u^2\frac{\partial k}{\partial a} \tag{5.41}$$

or

$$G = -\frac{1}{4b}\left(\frac{P}{k}\right)^2\frac{\partial k}{\partial a}. \tag{5.42}$$

Example 5.2

As the first example, consider the double-cantilever beam in Figure 5.15. For this case, the deflection u is twice the deflection of a cantilever beam of length a; that is,

$$u = 2\frac{Pa^3}{3EI}. \tag{5.43}$$

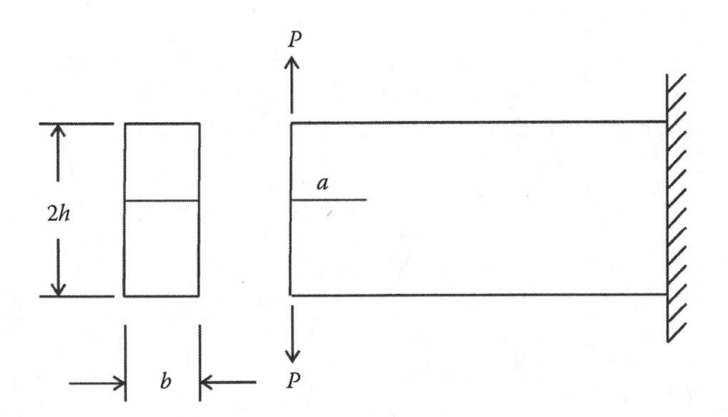

FIGURE 5.15 Double-cantilever beam.

For the second moment of area $I = bh^3/12$,

$$u = 8\frac{pa^3}{Ebh^3}. \tag{5.44}$$

$$k = \frac{Ebh^3}{8a^3}. \tag{5.45}$$

Thus,

$$G = \frac{3}{32}u^2\frac{Eh^3}{a^4}, \tag{5.46}$$

or

$$G = \frac{6P^2a^2}{Eb^2h^3}. \tag{5.47}$$

Next, the stability is checked. Assume a constant fracture resistance R, as in the Griffith theory. Then check Equation 5.31. Instability can be checked either under constant load P or under constant deflection u. First under constant deflection u, from Equation 5.46,

$$\frac{\partial G}{\partial a} = -\frac{12}{32}u^2\frac{Eh^3}{a^5} < 0, \tag{5.48}$$

which represents stability (i.e., crack will not grow).

Second, for constant force P (and constant width b),

$$\frac{\partial G}{\partial a} = \frac{12P^2a}{Eb^2h^3} > 0, \tag{5.49}$$

which indicates unstable crack growth. Thus, if G is at a critical value G_C, and the load is held constant, unstable crack growth is indicated.

Example 5.3

As the second example, consider the adhesion-peel test in Figure 5.16. This test is used to check the strength of adhesive joints. If the flexible strip is unpeeled, an additional amount ∂L is accounted for and then the change in external work is

$$\partial W = P(1+\varepsilon)\partial L, \tag{5.50}$$

where

$$\varepsilon = \frac{\sigma}{E} = \frac{P}{bhE} \tag{5.51}$$

Note that the stiffness from Equation 5.36 cannot be used, because ∂L itself is a displacement parallel to the load, in addition to the displacement dependent

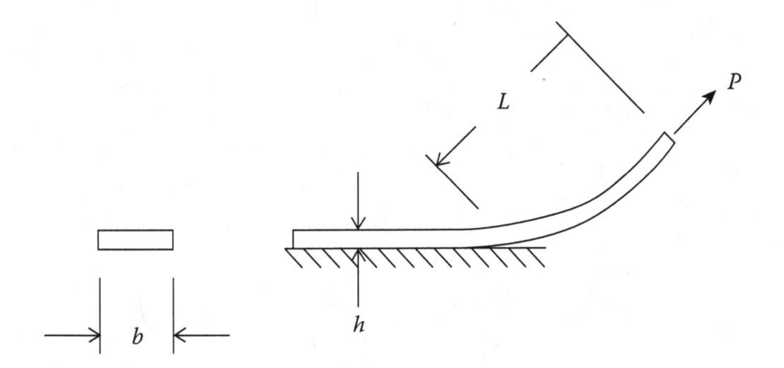

FIGURE 5.16 Adhesion-peel test.

on the stiffness of the strip. The strain energy is $(1/2)\sigma\varepsilon(bhL)$, and the change in strain energy is

$$\partial U_e = \frac{P^2}{2Ebh}\partial L = \frac{P}{2}\varepsilon\,\partial L. \tag{5.52}$$

For $\partial A = 2b\partial L$ Equation 5.26 yields

$$G = \frac{P}{2b}\left(1+\frac{\varepsilon}{2}\right) = \frac{P}{2b}\left(1+\frac{u}{2L}\right), \tag{5.53a}$$

$$G = \frac{P}{2b}\left(1+\frac{P}{2bhE}\right), \tag{5.53b}$$

$$G = \frac{bhEu}{L}\left(1+\frac{u}{2L}\right) \tag{5.53c}$$

If $\varepsilon \ll 1$, G is independent of material properties, which does not make sense. The result can be considered practical for elastomer materials with large strain. Checking the stability for P = constant gives

$$\frac{\partial G}{\partial L} = 0, \tag{5.54}$$

which indicates a neutral stability for crack growth. Thus, P has to increase for unstable crack growth. (The reader can experimentally check this with a Band-Aid. A Band-Aid applied over hair verifies this result.) Checking the stability for u = constant gives

$$\frac{\partial G}{\partial L} < 0, \tag{5.55}$$

which indicates a stable crack.

5.3 ANALYSIS OF THE CHARPY AND IZOD IMPACT TESTS USING FRACTURE MECHANICS

The analysis of fracture mechanics has been applied to the Charpy impact test by Williams,[5] and Adams and Wu.[10] They use the compliance C in their analysis. We use the stiffness k, where $k = 1/C$.

Use Equation 5.40 for the elastic energy,

$$G = -\frac{U_e}{2bk}\frac{\partial k}{\partial a},$$

where G is the energy release rate for brittle fracture.

Solve for U_e,

$$U_e = -\frac{(2bk\ G)}{(\partial k/\partial a)} = \frac{(2bGC)}{(\partial C/\partial a)} = G(2b\theta) \tag{5.56}$$

where

$$\theta = \frac{(C)}{(\partial C/\partial \mathrm{a})}.$$

Note that θ is a function of stiffness (or compliance) which in turn is a function of the material and geometry. Experimental results[5] show that when U_e for the Charpy test and for the Izod test is plotted against $2b\theta$ for ductile polymers like PE and PP a linear relation is found where the slope is G. This shows that the fracture mechanics relations apply to these tests even for polymers exhibiting some ductile yielding. Why is this? It is because the stiffness k of the specimens is largely dependent on the elastic portion of the beam specimen and much less dependent on the small yield region near the crack.

Results[5] also show for the Charpy test that unstable brittle fracture ($\mathrm{d}G/\mathrm{d}a > 0$) occurs for crack lengths from initiation to $a/h = 0.13$ where h is the depth of the beam specimen. Unstable fracture with arrest at $a/h = 0.36$ occurs for $0.13 < a/h < 0.36$ and stable fracture occurs for $a/h > 0.36$. (This is an important result for small a/h. Recall that for NDE and quality control, inspection for flaws and voids is important. From Equation 5.22, $K_I = \sigma\sqrt{\pi a}$ so that more stress can be applied for smaller cracks, but that the results here show that the failure will be more brittle and catastrophic which is bad for pressure vessels or piping under pressure.)

Impact fatigue tests[4] have also been conducted on some polymers namely Nylon 66, HIPS acrylonitrile butadiene styrene (ABS), and rubber-toughened Nylon. The impact energy was 98% of the crack initiation energy. Specimens withstood 100–200 cycles of loading before failure.

5.4 ANALYSIS OF IMPACT SPECIMENS AT THE NANOSCALE

In recent years there has been much interest in the application of polymer materials at the micro- and nanoscale as microelectronic devices are made smaller and smaller. Constantinides[16] describes an analysis and experiments of materials at the nanoscale. An instrumented pendulum device with a diamond Berkovich[17] indenter was used to indent polymer specimens at a rate of 0.7–1.5 mm/s. The highest impact velocity (1.5 mm/s) corresponded to an impulse energy of 250 nJ. (The Berkovich nanoindenter similar to the Vickers type is normally used for testing the hardness of a material. It has a three-sided pyramid shape. It has also been used to measure bulk materials and films greater than 100 nm (3.9 (10^{-6}) in.) thick.)

Thin films of materials tested[16] included two semi-crystalline polymers PE and PP above their corresponding glass transition temperatures T_g, and three fully amorphous polymers PS, PC, and PMMA well below T_g. The authors concluded: "The maximum penetration depth x_{max} was found to scale linearly with the residual depth x_r for all six polymers, irrespective of material strain rate sensitivity. The ratio x_r/x_{max} is related directly to the ratio of total impact energy W_p/W'' dissipated via an elastic deformation of the material. .

HOMEWORK PROBLEMS

5.1. Consider a polybutylene terephthalate (PBT) polymer for an impact situation in which both toughness and stiffness are important. Assume a cantilever beam loaded by a mass M at the end under suddenly applied loads. How much can the size (width b) be reduced by adding glass reinforcement? How much can the weight of the beam be reduced with the reinforced PBT? Use the data in Table 5.9.

5.2. Find the energy release rate G for the cantilever-beam specimen loaded by the moments as shown.

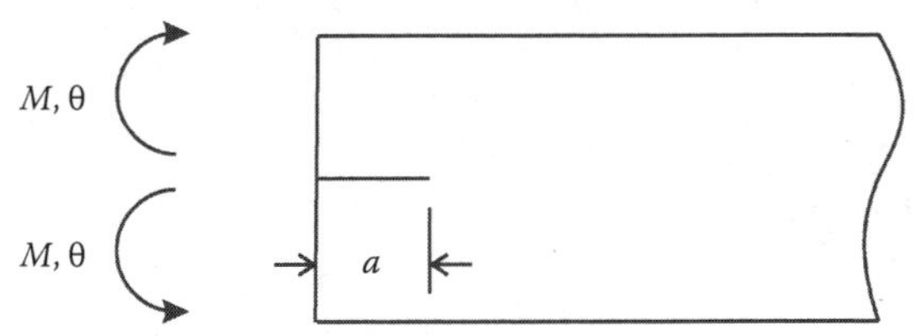

5.3. Find the energy release rate G for the split-bar specimen loaded in double torsion as shown.

T, θ

T, θ

a

5.4. Find the energy release rate G for the uniformly loaded beam specimen by the pressure P on each side as shown.

TABLE 5.9
Properties of Polymer PBT

	Unfilled	30% Glass Fiber Reinforced
Tensile modulus (ksi)	280	1300
Flexural modulus (ksi)	330	850
Izod impact (ft-lb/in.)	0.7	0.9
Specific gravity	1.30	1.48

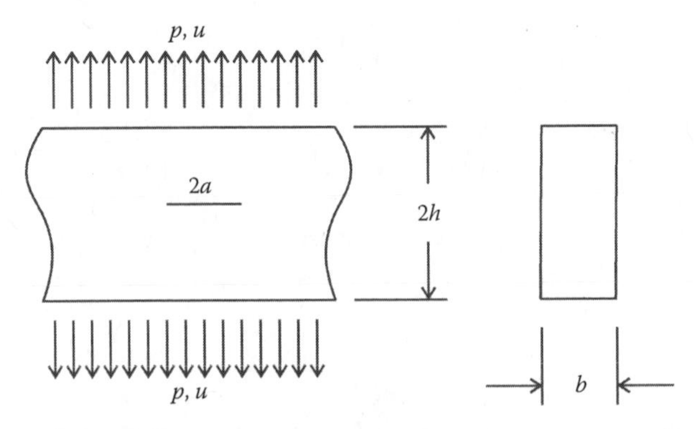

5.5. Find the energy release rate G for the doubly loaded cantilever beam (Example 5.1) but where the material behavior is nonlinear and given by

$$\sigma = E\varepsilon^{n} \tag{5.57}$$

5.6. Find the energy release rate G for the uniformly loaded beam specimen loaded by pressure G on each side as shown, but where the boundaries translate vertically as a rigid body and where the material behavior is given by Equation 5.57.

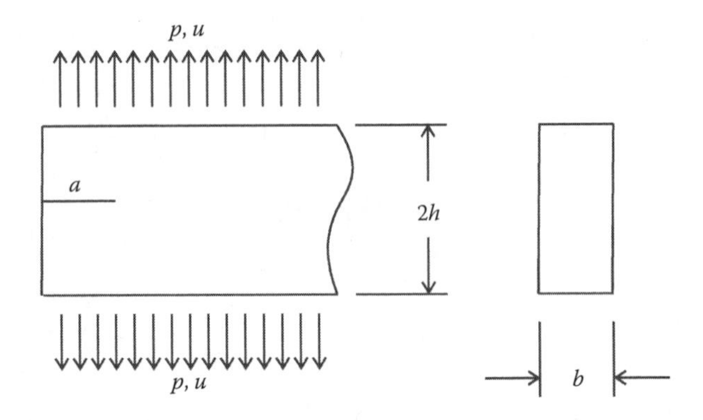

5.7. Consider a simply supported beam loaded transversely by a uniformly distributed load w (force/length). If the temperature of the beam is suddenly decreased by δT, what is the energy release rate G? Neglect any heat energy. Relate δT to δL, and express in terms of L/R, the coefficient of thermal expansion.

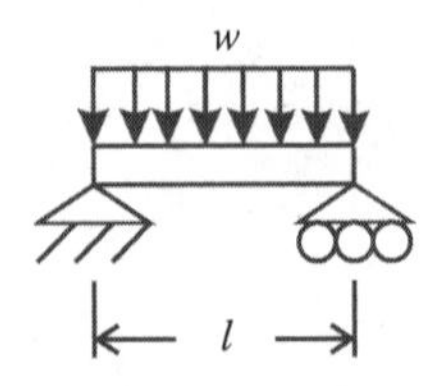

REFERENCES

1. *Tensile Impact Energy to Break Plastics and Electrical Insulating Materials*, ASTM Standard D1822, American Society for Testing Materials, West Conshohocken, PA, 1999.
2. *Guide to Plastics, Property and Specification Charts*, McGraw-Hill, New York, NY, 1985.
3. Engineering Plastics, *Engineering Materials Handbook*, Vol. 2, ASM, Metals Park, OH, 1988.
4. Domíngueza, C., M. Arocab, and J. Rodríguezc, Izod impact tests of polypropylenes: The clamping pressure influence, *Polym. Test.*, 25(1), 49–55, 2006.
5. Williams, J. G., *Fracture Mechanics of Polymers*, Ellis Horwood Ltd., Chichester, West Sussex, England, pp. 23–25, 100, 1984.
6. *Standard Test Method for Measuring the Damage Resistance of a Fiber-Reinforced Polymer Matrix Composite to a Drop-Weight Impact Event*, ASTM D7136/D7136M – 07, American Society for Testing Materials, West Conshohocken, 2007.
7. Clausen, A. H. et al., Polymer plates subjected to impact: Experimental tests and numerical simulations, *DYMAT 2009–9th International Conference on the Mechanical and Physical Behaviour of Materials under Dynamic Loading*, 2, 1537–1543, 2009.
8. Saul Stenzler, J. S., *Impact Mechanics of PMMA/PC Multi-Laminates with Soft Polymer Interlayers*, Master's Thesis, Virginia Polytechnic Institute, Blacksburg, VA, November 30, 2009.
9. Brostow, W. and R. D. Corneliussen, *Failure of Plastics*, Hanser Publishers, Munich, West Germany, 1986.
10. Adams, G. C. and T. K. Wu, Materials characteristics by instrumented impact testing, Chapter 8, Brostow, W. and Corneliussen, R.D., eds., *Failure of Plastics*, Hanser Publishers, New York, NY, 1986.
11. Griffith, A. A., The phenomena of rupture and flow in solids, *Philos. Trans. R. Soc.*, A221, 163–198, 1921.
12. Anderegg, F. O., Strength of glass fibers, *Ind. Eng. Chem.*, 31, 290–298, 1939.
13. Timoshenko, S. P. and J. N. Goodier, *Theory of Elasticity*, 3rd ed., McGraw-Hill, New York, NY, 1987.
14. Shigley, J. E. and C. R. Mischke, *Mechanical Engineering Design*, 5th ed., McGraw-Hill, New York, 1989.
15. Sih, G. C. M., *Handbook of Stress Intensity Factors*, Lehigh University, Bethlehem, PA, 1973.
16. Constantinides, G. et al., Quantifying deformation and energy dissipation of polymeric surfaces under localized impact, *Mater. Sci. Eng., A*, 489, 403–412, 2008.
17. Khrushchov, M. M. and E. S. Berkovich, Methods of determining the hardness of very hard materials: The hardness of diamond, *Ind. Diamond Rev.*, 11, 42–49, 1951.

6 Selection of Polymers for Design Applications

6.1 INTRODUCTION

The reader now has learned something about the mechanical properties of polymers, and hopefully has gained an appreciation of the relative complexity of polymer behavior compared to that of metals commonly used for engineering design applications. Because of this complexity, it has been necessary to build a sufficient knowledge base so that the student can more intelligently design with polymers. A systematic design procedure would be helpful so that one could chart some kind of course or direction through the maze of choices one has in selecting polymers for design. Such a systematic procedure is now presented. With it one can select materials based on criteria such as cost-to-strength ratio, weight-to-stiffness ratio, and so on.

To establish a procedure, it is helpful to sort the data, which need to be considered, into four categories:

1. What are the important material properties?
2. What are the performance parameters?
3. What are the loading conditions and geometrical configurations?
4. What materials are available?

The design selection procedure suggested here is to list candidate materials in material comparison tables, and list relative size ratios, relative cost ratios, relative weight ratios, and so on, in these tables, and then to select the materials based on these one of the relative ratios, typically cost.

6.2 BASIC MATERIAL PROPERTIES

The material properties that need to be considered will vary depending on the particular design application. One of the basic material properties may be a strength value, namely:

Ultimate strength, S_u
Yield strength, S_y
Endurance limit in fatigue, S_e
Creep rupture strength and critical time, (S_c, t_c)
Ultimate strain, ε_u
Creep strain limit, ε_c

On the other hand, stiffness may be the more important material property. The stiffness value may be one of the following:

Instantaneous elastic modulus, E
Creep compliance, $J(t)$
Equivalent modulus for a composite, E_c
Complex modulus, E^*

In some applications involving temperature change, thermal properties may be important such as the thermal conductivity, k, or the coefficient of thermal expansion, α. The weight of a component is always important for performance as well as cost. Therefore, we need to know the density, ρ, or the specific gravity, γ. Cost is always important. Assume the cost is measured by the price per unit weight, P.

6.3 PERFORMANCE PARAMETERS

Next, there must be some way of measuring the performance of a design. Consider the following performance parameters:

Relative depth (h) or diameter (d) of a section for equal strength
Relative depth (h) or diameter (d) of a section for equal stiffness
Relative volume (V) for equal strength
Relative volume (V) for equal stiffness
Relative weight (W) for equal strength
Relative weight (W) for equal stiffness
Relative cost (C) for equal strength
Relative cost (C) for equal stiffness

6.4 LOADING CONDITIONS AND GEOMETRICAL CONFIGURATIONS

The requirements of the design must be defined. There may be some basic loading conditions such as tension, bending, torsion, or internal pressure. The general configuration must be classified. For example, it may be rectangular, round, solid, or hollow.

6.5 AVAILABILITY OF MATERIALS

The material availability must be established. Of course, an infinite number of polymers available when one considers various blends and composites. Here, some GO/NO-GO criteria need to be used for preliminary screening, such as

Fabricability
Heat-distortion temperature
Compatibility with environment
Glass transition temperature
Toxic or other adverse health effects

6.6 A RECTANGULAR BEAM IN BENDING

Example 6.1

As a first example of the design selection procedure, a rectangular beam in bending is considered as shown in Figure 6.1. Some assumptions need to be made to define the problem. Assume the following parameters to be held constant:

1. Loading, $w(x)$
2. Length, L
3. Maximum bending moment, M (as a result of (a) and (b))
4. Width of the cross section, b
5. Factor of safety, η

Problem definition

Vary the depth of the section h to meet the static strength requirement S_u.

Procedure

Calculate the maximum stress.

$$\sigma = \frac{Mc}{I} = \frac{6M}{bh^2} \tag{6.1}$$

Define the factor of safety.

$$\eta \equiv \frac{S_u}{\sigma} \tag{6.2}$$

Compare any two materials, 1 and 2, on a strength basis with the same factor of safety.

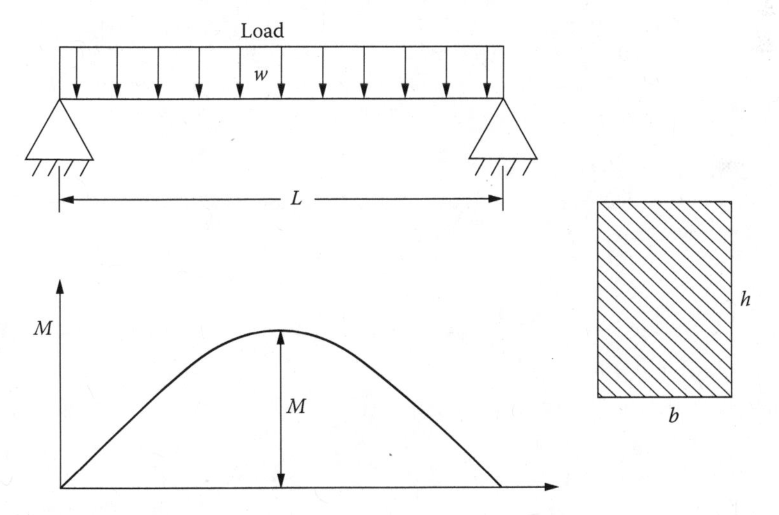

FIGURE 6.1 Rectangular beam in bending.

$$\eta_1 = \eta_2$$

$$\frac{S_{u1}}{\sigma} = \frac{S_{u2}}{\sigma}$$

$$\frac{bh_1^2 S_{u1}}{6M} = \frac{bh_2^2 S_{u2}}{6M}$$

Thus, the relative depth ratio is

$$\frac{h_2}{h_1} = \sqrt{\frac{S_{u1}}{S_{u2}}} \tag{6.3}$$

Then, calculate the relative volume ratio.
From $V_1 = bh_1L$ and $V_2 = bh_2L$,

$$\frac{V_2}{V_1} = \frac{h_2}{h_1} \tag{6.4}$$

Calculate the relative weight ratio.
From $W_1 = V_1\rho_1$ and $W_2 = V_2\rho_2$,

$$\frac{W_2}{W_1} = \frac{\rho_2}{\rho_1}\frac{h_2}{h_1} \tag{6.5}$$

Finally, calculate the relative cost ratio. From $C_1 = W_1P_1$ and $C_2 = W_2P_2$,

$$\frac{C_2}{C_1} = \frac{\rho_2}{\rho_1}\frac{h_2}{h_1}\frac{P_2}{P_1} \tag{6.6}$$

Equations 6.3 and 6.6 are used to prepare a material comparison table (Table 6.1) for rating three PF's. The first PF GP where GP refers to general purpose is used as the basis (material number 1 in the above equations) and the second and third PF's in the table are used alternatively as material number 2 for rating purposes. As evident the third material is better on a relative weight basis but the first is better on a relative cost basis.

Next, consider a new problem definition—compare the materials on a stiffness basis rather than the strength basis keeping the same assumed constants.

Problem Definition

Vary the depth of the section h, to keep the deflection, δ the same.

TABLE 6.1
Material Comparison Table for Rectangles in Bending for PF Phenolics on an Equal Strength Basis

Material	S_{ut} (ksi)	ρ (lb/in.^{3})	P (\$/lb)(h)	h_2/h_1	V_2/V_1	W_2/W_1	C_2/C_1
PF-GP	3	0.0486	0.75	1.000	1.000	1.000	1.000
PF-MI	4	0.0485	1.05	0.866	0.866	0.866	1.212
PF-HI	6	0.0522	1.60	0.707	0.707	0.759	1.619

TABLE 6.2
Material Comparison Table for Rectangles in Bending for PF Phenolics on an Equal Stiffness Basis

Material	E (ksi)	ρ (lb/in.³)	P ($/lb)	h_2/h_1	V_2/V_1	W_2/W_1	C_2/C_1
PF-GP	0.4	0.0486	0.75	1.000	1.000	1.000	1.000
PF-MI	0.6	0.0485	1.05	0.874	0.874	0.874	1.224
PF-HI	0.9	0.0522	1.60	0.763	0.763	0.820	1.749

Procedure

Calculate the maximum deflection:

$$\delta = \frac{5wL^4}{384EI} = \frac{160wL^4}{Ebh^3} \tag{6.7}$$

For the same deflection

$$\delta_1 = \delta_2$$
$$\frac{0.156wL^4}{E_1bh_1^3} = \frac{0.156wL^4}{E_2bh_2^3} \tag{6.8}$$

Thus, for this case the relative depth ratio is (in place of Equation 6.3):

$$\frac{h_2}{h_1} = \left(\frac{E_1}{E_2}\right)^{1/3}. \tag{6.9}$$

Now using Equations 6.9 and 6.4 through Equation 6.6, a material comparison table (Table 6.2), can be prepared to rate the materials on a stiffness basis.

Again, the third material is better on a relative weight basis, but the margin over the second is reduced. The first is still best on the cost basis because of the lower price *P*.

The reader may ask what to do if both weight and cost are both important. The procedure can be modified using weighting factors to address this issue.

6.7 WEIGHTING-FACTOR ANALYSIS

Let the weighting factors w_i be defined as follows:

$$\sum_{i=1}^{N} w_i = 1.0 \quad \text{where each } w_i \leq 1.0 \tag{6.10}$$

For example, consider two weighting factors: w_i for the cost basis and w_j for the weight basis, such that $w_i + w_j = 1$. Let *R* be any rating factor (depth *h*, volume *V*, weight *W*, or cost *C*).

For weight and cost to be rated together as both important,

$$\frac{R_2}{R_1} = w_i\left(\frac{W_2}{W_1}\right) + w_j\left(\frac{C_2}{C_1}\right). \tag{6.11}$$

TABLE 6.3
Material Comparison Table for Rectangles in Bending for PF Phenolics Rated 60% on Weight Basis and 40% on Cost Basis

Material	W_2/W_1	C_2/C_1	R_2/R_1
PF-GP	1.000	1.000	1.000
PF-MI	0.866	1.212	1.004
PF-HI	0.759	1.619	1.103

Also, a material comparison table can be prepared using a combined strength and deflection basis, with weighting factors deflection basis, with weighting factors w_i and w_j, respectively, as follows:

$$\frac{h_2}{h_1} = w_i\left(\frac{h_2}{h_1}\right)_{\text{strength}} + w_j\left(\frac{h_2}{h_1}\right)_{\text{deflection}} \tag{6.12}$$

$$\frac{W_2}{W_1} = w_i\left(\frac{W_2}{W_1}\right)_{\text{strength}} + w_j\left(\frac{W_2}{W_1}\right)_{\text{deflection}} \tag{6.13}$$

$$\frac{C_2}{C_1} = w_i\left(\frac{C_2}{C_1}\right)_{\text{strength}} + w_j\left(\frac{C_2}{C_1}\right)_{\text{deflection}} \tag{6.14}$$

For example, using Equation 6.11 with $w_i = 0.6$ and $w_j = 0.4$, comparison of Table 6.3 is obtained for an equal strength basis (from Table 6.1). On the combined weight and cost basis, material 2 is nearly equal to material 1 as the best choice.

6.8 THERMAL GRADIENT THROUGH A BEAM

Example 6.2

Consider a thermal stress problem caused by a temperature differential across a beam as shown in Figure 6.2. This type of problem can result in the housing of

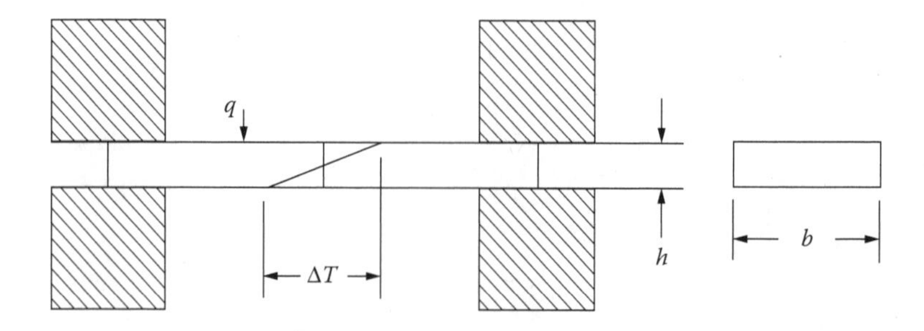

FIGURE 6.2 Example 6.2—Thermal gradient through a beam.

an electrical drill, for example, where thin plate or shell theory should be used rather than beam theory. However, the beam example illustrates the problem well enough and the student can wait for graduate school to learn the theory of plates and shells. Again some assumptions need to be made to define the problem. Assume the following parameters to be held constant:

1. Heat flow q(BTU/h)
2. Surface area $A = b1$
3. Factor of safety η

Problem definition

Vary depth h, thermal conductivity k, elastic modulus E, and coefficient of thermal expansion α to meet the static strength requirement S_u.

Procedure

Calculate the maximum stress.

$$\sigma = \frac{E\alpha\Delta T}{2}. \tag{6.15}$$

Calculate the thermal gradient.

$$\Delta T = \frac{qh}{kA}. \tag{6.16}$$

Combine Equations 6.15 and 6.16:

$$\sigma = \frac{E\alpha qh}{2kA}. \tag{6.17}$$

Compare any two candidate materials, 1 and 2, on a strength basis using the same factor of safety:

$$\begin{aligned} \eta_1 &= \eta_2, \\ \frac{S_{u1}}{\sigma_1} &= \frac{S_{u2}}{\sigma_2}, \\ \frac{2S_{u1}k_1A}{E_1\alpha_1qh_1} &= \frac{2S_{u2}k_2A}{E_2\alpha_2qh_2}. \end{aligned} \tag{6.18}$$

From Equation 6.18 the relative depth ratio is established for this problem:

$$\frac{h_2}{h_1} = \frac{E_1\alpha_1k_2S_{u2}}{E_2\alpha_2k_1S_{u1}}. \tag{6.19}$$

Thus, it is found that the size ratio (h_2/h_1) in this case is *directly* proportional to the strength ratio, whereas in Example 6.1, Equation 6.3, it was *inversely* proportional to the square root of the strength ratio. Why is this? Examine Equations 6.15 and 6.16: σ is proportional to ΔT but ΔT is proportional to h. However, in the beam bending problem σ was inversely proportional to h^2 in Equation 6.1. The reader should keep in mind this difference in stresses caused by external loads and internal temperature gradients. (As a mental challenge, consider combined external loads with internal temperature gradients.)

TABLE 6.4
Material Property Table for PF Phenolics

Material	E (ksi)	α (10^{-5}/°F)	k (btu/h/ft²/°F/ft)	S_u (ksi)	ρ (lb/in.³)	P ($/lb)
PF-GP	0.8–1.3	1.6–2.5	0.097–0.3	5–8.5	0.0558	0.75
PF-MI	0.9–1.4	1.6–2.2	0.097–0.17	5–9	0.0515	1.05
PF-HI	3.0–3.3	0.88	0.20	5–10	0.0684	1.60

TABLE 6.5
Material Comparison Table for Thermal Gradient through a Beam for PF Phenolics on an Equal Strength Basis

Material	E_1/E_2	α_1/α_2	k_2/k_1	S_{u2}/S_{u1}	h_2/h_1	W_2/W_1	C_2/C_1
PF-GP	1.000	1.000	1.000	1.000	1.000	1.000	1.000
PF-MI	0.929	1.136	0.567	1.059	0.634	0.585	0.819
PF-HI	0.394	2.840	0.667	1.176	0.878	1.076	2.295

Before preparing a material comparison table for the thermal gradient problem, at least one no/no-go criterion should be applied—namely, the heat distortion temperature. This leads to the selection of thermosetting-type polymers rather than thermoplastic type. Therefore, three types of PF's (see Table 6.4) with heat distortion temperatures of 220°F and greater are chosen and these are compared in Table 6.5.

Note that ranges of values of E, α, k, and S_u are available. In the calculations extreme values from the ranges are chosen here (indicated by underlining) in order to minimize h and present each material in its best available condition. However, this choice is not arbitrary, that is, one cannot expect to choose maximum E from the range and minimum S_u from its range since the reinforcement that maximized E will also maximize S_u. Therefore, choosing the maximum value for E forces one to choose maximum values for all the other properties. It is assumed that these chosen extreme values can be specified and provided by a supplier with the appropriate composition of the fillers and reinforcements. From Table 6.5, the second material (the medium impact (PF-MI)) is rated best on both a weight and cost basis.

First, the choice of material is made, then the calculation has to be made absolute rather than relative, that is, the depth h required has to be calculated for a specific q, A, and $\eta > 1$.

6.9 RATING FACTORS FOR VARIOUS LOADING REQUIREMENTS

Many other loading conditions and geometrical configurations can be considered. Rating factors, relative sizes for equal strength and for equal stiffness, are listed in Table 6.5 for rectangles in bending, solid cylinders in bending, solid cylinders in torsion, solid cylinders in tension, solid cylinders as columns, and cylindrical pressure vessels. The student should be able to verify the ratios given in Table 6.6.

TABLE 6.6
Rating Factors for Various Loading Requirements

Loading and Cross Section	Variable Dimension	Equal Strength h_2/h_1 or d_2/d_1	Equal Stiffness h_2/h_1 or d_2/d_1
Rectangles in bending	Depth h	$(S_{u1}/S_{u2})^{1/2}$	$(E_1/E_2)^{1/3}$
Solid cylinders in bending	Diameter d	$(S_{u1}/S_{u2})^{1/3}$	$(E_1/E_2)^{1/4}$
Solid cylinders in torsion	Diameter d	$(S_{us1}/S_{us2})^{1/3}$	$(G_1/G_2)^{1/4}$
Solid cylinders in tension	Diameter d	$(S_{u1}/S_{u2})^{1/2}$	$(E_1/E_2)^{1/2}$
Solid cylinders as columns	Diameter d	$(S_{u1}/S_{u2})^{1/2}$	$(E_1/E_2)^{1/4}$
Cylindrical pressure	Thickness h	S_{u1}/S_{u2}	E_1/E_2

6.10 DESIGN OPTIMIZATION

The material selection procedure for design described above is a method of rating materials for a particular design application that is based on a single design variable such as varying the depth h or diameter d in Table 6.5. This allows use of the design variable in a linear form.

In this section, design optimization theory is presented, where multiple design variables are allowed and where nonlinear constraints can be imposed.[1,2] The following is the standard format for posing a design optimization problem.

Let the set of design variables be defined as

$$X_i, \quad i = 1,2,3,\ldots,N \tag{6.20}$$

Let the objective function to minimized (or maximized) be defined as

$$F = F(X_i) \tag{6.21}$$

Assume that there are some inequality constraints given by

$$g_j(X_i) \leq 0, \quad j = 1,2,3,\ldots,J \tag{6.22}$$

Assume that there are some equality constraints given by

$$h_k(X_i) = 0, \quad k = 1,2,3,\ldots,K \tag{6.23}$$

Assume that there are side constraints given by

$$X_i^{\min} \leq X_i \leq X_i^{\max}, \quad i = 1,2,3,\ldots,I \tag{6.24}$$

where the superscripts denote minimum and maximum limits. The functions $F(X_i)$ and $g_j(X_i)$ may be linear or nonlinear, and explicit or implicit functions of the X_i. However, these functions should be continuous with continuous first derivatives.

Design codes incorporate computer software to automatically solve the optimization problem as posed above. Before describing these, a graphical method of solution will be illustrated by an example problem.

6.10.1 Graphical Solution

Example 6.3

A thin-walled tube under a compressive load P and with a fixed support at the bottom is considered, as shown in Figure 6.3. (This problem is similar to the one from Ref. 1, p. 55, except a polymer nylon is considered instead of steel.) The problem is to design a minimum weight tubular column subject to the loading and support conditions. Assume that the length L and the load P are given as constants. The design variables are the mean radius R and the thickness t. Thus,

$$X_1 = R \quad \text{and} \quad X_2 = t. \tag{6.25}$$

For a thin-walled tube with $R \gg t$, the area and the second area moment of inertia are given by

$$A = 2\pi Rt \quad \text{and} \quad I = \pi R^3 t. \tag{6.26}$$

The objective function to be minimized is the weight given by

$$F(X_1, X_2) = \rho AL = 2\rho\pi RtL. \tag{6.27}$$

where ρ is the density of the material.

For a column in axial compression the beam must be checked for buckling failure. As discussed in Chapter 2, a column will buckle with a critical buckling load that is either described by Euler buckling or Johnson buckling failure criterion dependent on the ratio of length to the radius of gyration. The buckling analysis implicitly checks for yield failure for this simple loading case as it is in the Johnson criteria. Furthermore, let us add geometric constraints that can be applied from manufacturing, such that the ratio t/R is bounded. It is also required that t and R must be greater than zero.

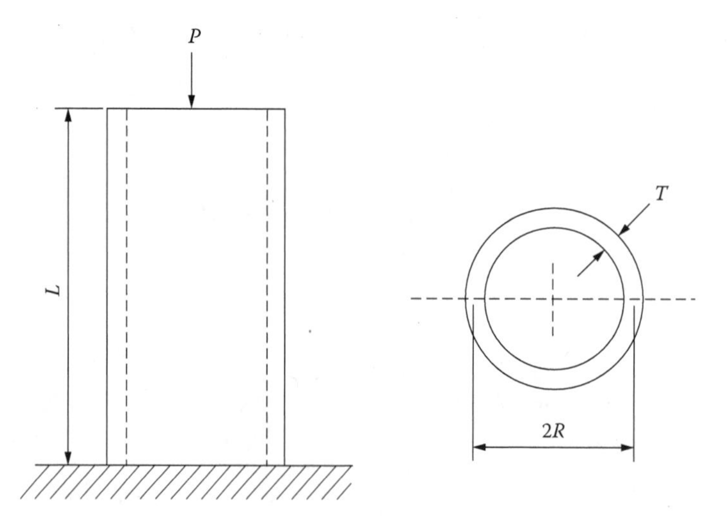

FIGURE 6.3 Tubular column under compression.

The inequality constraints come from requiring the load to be less than the critical buckling load for the column. If

$$\left(\frac{l}{k}\right) \geq \left(\frac{l}{k}\right)_{cr} \tag{6.28}$$

then

$$\frac{P_{cr}}{A} = \frac{n\pi^2 E}{(l/k)^2} \tag{6.29}$$

else when

$$\left(\frac{l}{k}\right) \leq \left(\frac{l}{k}\right)_{cr} \tag{6.30}$$

$$\frac{P_{cr}}{A} = S_y - \left(\frac{S_y}{2\pi}\right)^2 \frac{1}{nE}\left(\frac{l}{k}\right)^2 \tag{6.31}$$

Regardless of which failure criteria is applicable, the applied compressive load has to be less than the buckling load,

$$P \leq P_{cr} \quad \text{or} \quad \text{equivalently} \; \frac{P}{A} \leq \frac{P_{cr}}{A} \tag{6.32}$$

For the problem the side constraints would normally be given by

$$-R \leq 0 \tag{6.33}$$

and

$$-t \leq 0. \tag{6.34}$$

Often manufacturing restrictions impose additional geometric constraints on the t/R ratio. In general,

$$(t/R)_{min} \leq t/R \leq (t/R)_{max}. \tag{6.35}$$

However, it will become obvious when we run the optimization solution that the maximum t/R ratio, while a manufacturing limitation, does not minimize weight. However, a minimum t/R ratio, should be considered

$$(t/R)_{min} \leq t/R \tag{6.36}$$

or

$$(t/R)_{min} - t/R \leq 0.$$

To illustrate a graphical solution, the following loading and data for nylon from the polymer data base are used:

$$P = 500\text{ N},\ E = 1.59(10^9)\text{ Pa},\ \rho = 1130\text{ kg/m}^3,\ L = 0.5\text{ m},$$
$$S_y = 4.48(10^7)\text{ Pa, and } \mu = 0.35. \tag{6.37}$$

Note that a factor of safety can be applied to the load constraints by

$$P \le \frac{P_{cr}}{\eta}. \tag{6.38}$$

However, let us first consider the situation where $\eta = 1$. Thus, the design optimization problem is presented numerically as

$$F(R,t) = (1130)\, 2\pi Rt\, (0.5) \tag{6.39}$$

When the Euler buckling criteria is applicable

$$\begin{aligned} g_1(R,t) &= P - \frac{P_{cr}}{A} \le 0 \\ &= P - \frac{n\pi^2 E}{(l/k)^2} \le 0 \end{aligned} \tag{6.40}$$

where for a fixed–free column, the end condition constant $n = 1/4$.

If Johnson buckling is the controlling constraint

$$g_2(R,t) = P - \left(S_y - \left(\frac{S_y}{2\pi}\right)^2 \frac{1}{nE}\left(\frac{l}{k}\right)^2 \right) \le 0. \tag{6.41}$$

Column buckling transitions between the Johnson and Euler criteria at the critical length over radius of gyration ratio as defined in Chapter 2.

From Flugge[4] it is noted that shell buckling will occur if $L/R < 25$, with a buckling load given by

$$P_{cr} = \frac{8\pi ERt\,(10^{-3})}{1-\mu^2} \tag{6.42}$$

Thus, an additional load constraint is added

$$\begin{aligned} g_3(R,\overline{T}) &= P - P_{cr} \le 0 \\ &= P - \frac{8\pi ERt\,(10^{-3})}{1-\mu^2}. \end{aligned} \tag{6.43}$$

Instead of adding the $L/R < 25$ as a constraint we will only apply the shell buckling criteria when this condition holds.

The geometric constraints are

$$g_4(R,t) = -R \le 0, \tag{6.44}$$

$$g_5(R,t) = -t \le 0, \tag{6.45}$$

and

$$g_6(R,t) = (t/R)_{\min} - t/R \le 0. \tag{6.46}$$

We will first present a graphical solution to this two-variable optimization problem. A graphical solution has various advantages. One advantage is that it aids the designer or analyst in explain the optimization procedure to those not familiar with optimization techniques. It is possible to create a three-dimensional graphical solution, for a three-variable optimization problem. It can be advantageous for problems with more than two variables to create graphical solutions for explanation purposes by holding all but two variables constant. For this specific problem a graphical solution is found by plotting the prior constraints g_1 to g_6 in the design variable space $R(t)$ as shown in Figure 6.4. The cross-hatched side of the lines is the forbidden (infeasible) side of the inequalities. There are a few points of interest in Figure 6.4. All of the load constraints are only shown over the range that they are applicable. While the equations for Johnson and Euler buckling could be plotted beyond the transition point shown in Figure 6.4, it is not strictly valid to do so. In addition, shell buckling is a complex topic that has received much attention in the last 20–30 years and the reader would benefit by ensuring that there is not a more applicable shell buckling theory or experimental results that are more applicable to his specific problem. It can be seen from Figure 6.4 that there is a limited range of acceptable values of R and t. The optimal values of (t, R), where the weight is minimized are (0.0011, 0.0212).

The solution as presented in Figure 6.4 can be improved by including a factor of safety. In addition to the factor of safety we will change the t/R ratio. The factor of safety will be incorporated for all three load constraints should really be applied as well on the buckling mode in Equation 6.37. Buckling is very sensitive to imperfections in geometry and a factor of safety must be included. For all three load constraints, the application of the factor of safety will be to shift the g_1, g_2, and g_3 curves upward relative to R. Specifically, let us consider the new conditions $\eta = 2$ and $0.0025 \le t/R$.

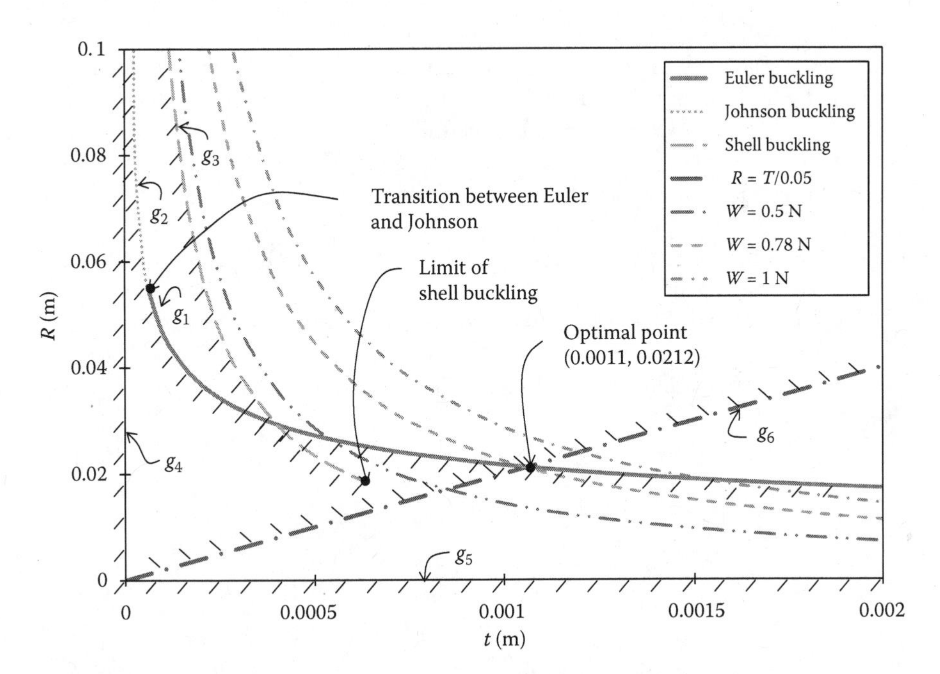

FIGURE 6.4 Graphical solution for a minimum-weight nylon tubular column.

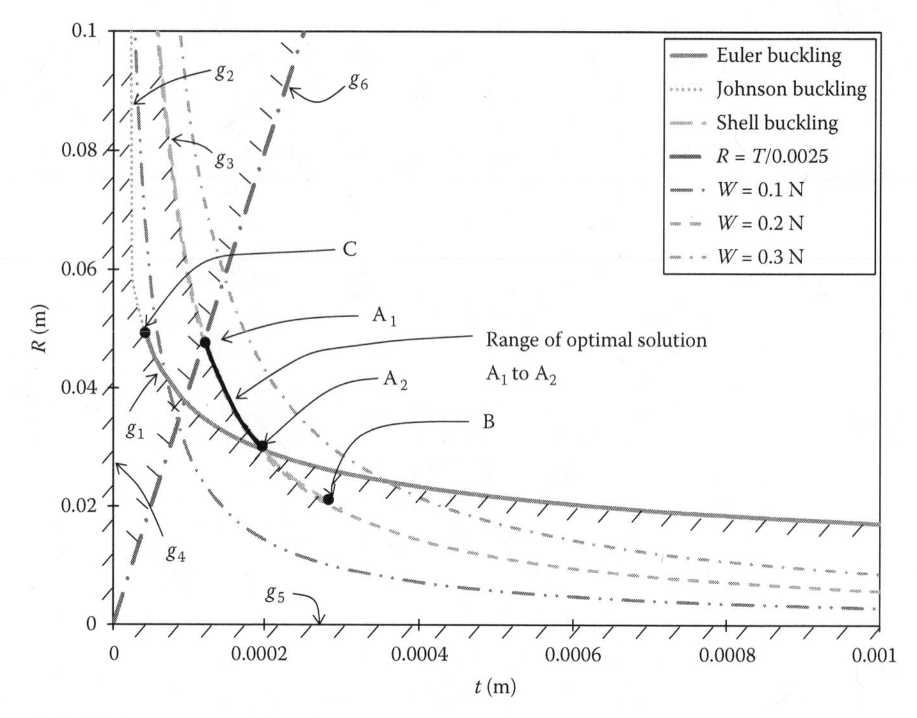

FIGURE 6.5 Graphical solution for a minimum-weight nylon tubular column with a factor of safety of 2.

The new graphical solution is shown in Figure 6.5. The solution is now controlled not by column buckling, g_1 but shell buckling g_3 and the t/R ratio. It is noted that the curves of constant weight are parallel to the shell buckling, g_3 constraint as shown in both Figures 6.4 and 6.5. The t/R ratio was chosen to illustrate that now instead of a single optimal point A as from the previous solution, there is now a range of optimal solutions from A_1 to A_2 as shown in Figure 6.5. Since all solutions on this curve have the same minimal weight, any solution on this curve is equally valid relative to the objective function.

Now consider that since the column is a polymer material, that through the viscoelastic correspondence principle, E in Equations 6.29, 6.31, and 6.42 is replaced by $1/J(t)$. Thus, the constraints g_1, g_2, and g_3 in Equations 6.4, 6.41, and 6.43 are all functions of time

$$g_1 = P - \frac{n\pi^2 E}{(l/k)^2} \leq 0 \tag{6.47}$$

$$g_2(R,t) = P - \left(S_y - \left(\frac{S_y}{2\pi}\right)^2 \frac{1}{nE}\left(\frac{l}{k}\right)^2 \right) \leq 0 \tag{6.48}$$

$$g_3(R,T) = P - \frac{8\pi ERt\,(10^{-3})}{1-\mu^2} \tag{6.49}$$

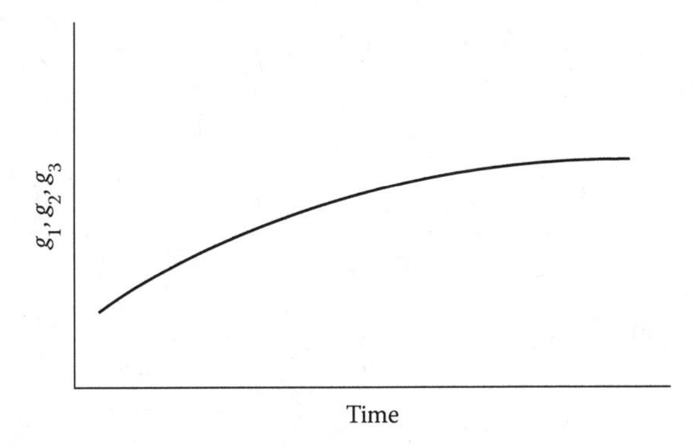

FIGURE 6.6 Normalized increases in functions g_1, g_2, and g_3 with time.

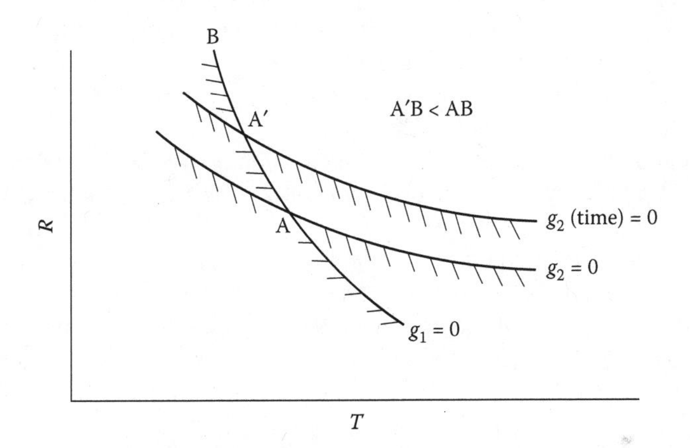

FIGURE 6.7 Decrease of solution region A′B for polymer with creep behavior.

The creep compliance $J(t)$ increases with time. Therefore, the functions g_1, g_2, and g_3 will increase with time as illustrated in Figure 6.6. As a consequence the range of optimum solutions for R and T has been decreased where A′B < AB. The Euler buckling curve, g_1 will shift up with $J(t)$ as shown in Figure 6.7. (An example of creep buckling of a cylindrical shell has been given earlier in Chapter 4, Example 4.3.)

6.10.2 Computer Solution

It is not the intent here to review all the numerical methods used in optimization theory. There are other references one can consult for a thorough review, for example.[1,2] However, a brief outline of the computer search strategy of modern approaches is given.

Let the vector

$$X = X(X_i) \tag{6.50}$$

represent the design variables. Then let an initial trial solution be X^0. Then the solution is updated in steps given by

$$X = X^0 + \delta X. \tag{6.51}$$

The best search direction S will be the one that will minimize $F(x)S$, subject to:

$$g_j(X)S \le 0, \quad j = 1,2,3,\ldots,J, \tag{6.52}$$

where J is the number of active or critical constraints.

The gradient in design space is given by

$$\bar{F} = F^0 + \nabla F^T \delta X + \frac{1}{2}\delta X^T H \delta X. \tag{6.53}$$

Thus, the objective function needs to be differentiable. The gradient expression 6.52 determines the feasible and infeasible regions of the solution as shown in Figure 6.8.

A very important consideration in the use of optimization routines needs to be noted at this juncture. When we first had students work on optimal designs in the classroom, a traditional optimization code (DOT, Design Optimization Tools[3]) which was written in Fortran was used. This particular code required a 20–40-line main Fortran program to call the DOT subroutine. Currently, we use the Solver routine in Microsoft Excel to perform the optimization or more generally the minimization of weight for the design comparisons. While use of the Solver routine is not as robust as other optimization methodologies, for simple problems it will typically find the same optimal point. In addition, it is universally available. However, the central issue here is that regardless of which software tool the designer uses to perform optimization, it is necessary to run the optimization program multiple times with various initial solution

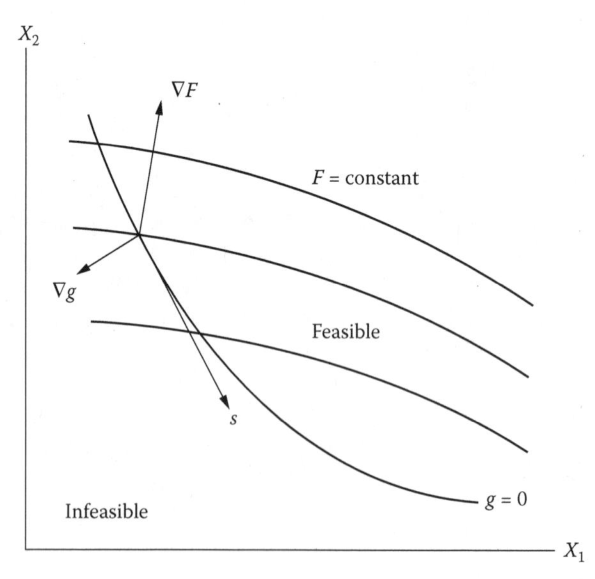

FIGURE 6.8 Feasible and infeasible regions of design space as determined by the gradient of the constraints.

guesses. This is particularly true of the buckling example used previously in this chapter. The reason for this is that it is possible to find local minima or optimal points that are not the global minima or optimal point. If the optimization routine finds a vector space that is a local minimum the routine will make step the variables to check that this is the global minimum. However, if the global minimum is either sufficiently far away or if there is not gradient that drives the routine to this solution, the search routine will stop. One can visualize this by imagining a blanket that has most of the surface at the same height. However, close to the four corners there are four independent indentations. One of them is deeper than the other, forming the global minimum of the 2-D vector space. If the initial guess in one of the other indentations, the search routine will converge on this location as the minimum. From this simple example, one can see that there should be perhaps 4, if not 5 initial guesses to start the search routine. The 5th guess is starting in the middle. As the number of design variables get large, it may seem an insurmountable task to select initial guesses. However, often there are some variables that are obviously more dominant than others. In addition, the speed of modern optimization routines is so fast that multiple runs of most problems occurs so quickly that the time required to do so is insignificant.

6.10.3 Microsoft Excel Solver Routine

While it is undoubtedly worthwhile for every design engineer to take a design optimization class, our intent here is to present the concept of optimization and have everyone who reads this text be able to run simple optimal design problems without the purchase of specialized and costly software. Admittedly, there are far more powerful and sophisticated optimization routines that can be run compared to utilizing Microsoft Excel. However, the latest versions of Solver utilize a nonlinear optimization routine that has been proven by students and their work colleagues to handle a variety of problems. The use of Solver for the class had come about by utilizing it for optimal nonlinear least squares fits in industry. Initially the Solver routine would handle a design problem with a few variables where the problem was highly nonlinear. However, at the end of the class summary one semester, after the instructor had stated that use of Solver was limited in the number of variables that it could successfully handle, one of the graduate students stated that after his coworker at an aerospace company had optimized a design with 50 variables. The caveat was that the problem was only slightly nonlinear. You can read more about the Solver routine in the Excel help files.

The first thing to note is that the Solver routine is not automatically installed when Excel is installed. In order to install the Solver routine a you must perform custom installation or install the routine specifically after the standard installation. We suspect you are now wondering where you put your Microsoft installation disks. The Solver routine is found on the Excel main menu toolbar under Tools.

Let us now reconsider the example in Section 6.10.1, and use the Solver routine to find the solution. The problem will initially be set up in the same manner that would traditionally be used in optimization; that is the constraints will be written as inequality constraints as shown in Equation 6.22. Table 6.7 shows the initial Excel file under the columns headed by "Results with Initial Guesses" and it is also available on the website www.taylorandfrancis.com under the file name optimization.xls. Note that

TABLE 6.7
Optimization Results for Example Problem 6.3

	Results with Initial Guesses				Optimization Results $\eta = 2$ and $0.05 \le t/R$	Optimization Results $\eta = 2$ and $0.0025 \le t/R$
	A	B	C	D	E	F
1	*Engineering Design with Polymers and Composites*, 2nd Edition					
2	Ron Rorrer					
3	Example 6.3 from text					
4						
5	R	m	X1	0.2000	0.0212	0.0294
6	T	m	X2	0.0100	0.0011	0.0002
7	L	m		0.5	0.5	0.5
8	D_i	m		0.3900	0.0414	0.0586
9	D_o	m		0.4100	0.0435	0.0590
10	ρ = density	kg/m^3		1130	1130	1130
11	P	N		500	500	500
12	E	Pa		1.59E+09	1.59E+09	1.59E+09
13	S_y	Pa		4.48E+07	4.48E+07	4.48E+07
14	mu			0.25	0.25	0.25
15	N = FS			1	1	2
16	Pall = P/FS	N		500	500	250
17	$A = 2\pi Rt$	m^2		1.26E–02	1.42E–04	3.69E–05
18	$I = \pi tR$^3	m^4		2.51E–04	3.19E–08	1.60E–08
19	$k = (I/A)$^0.5	m		0.1415	0.0150	0.0208
20	n-fixed free			0.25	0.25	0.25
21	g	m/s^2		9.81	9.81	9.81
22						
23	Objective function					
24	Weight = ρAL	N		69.65	0.78	0.78
25						
26	Constraints					
27						
28	Check Euler and Johnson Buckling (stress failure is part of Johnson!)					
29	(l/k)critical = ((2(π^2)nE)/S_y)^0.5			13.2	13.2	13.2
30	(l/k)actual			3.53	33.29	24.05
31	If (l/k)actual <= (l/k)critical use Johnson Failure Criterion					
32	(P_{cr}/A) Johnson			4.32E+07	–9.74E+07	–2.94E+07
33	If (l/k)actual > (l/k)critical use Euler Failure Criterion					
34	(P_{cr}/A)			3.13E+08	3.53E+06	6.77E+06
35	Logic test			4.32E+07	3.53E+06	6.77E+06
36	(P/A)			3.98E+04	3.53E+06	1.35E+07
37	(P/A)-$(P_{cr}/A) \le 0$			–4.32E+07	0.00E+0	–2.11E–07
38						
39	Check shell buckling (L/R > 25)					

TABLE 6.7 (continued)
Optimization Results for Example Problem 6.3

	Results with Initial Guesses				Optimization Results $\eta = 2$ and $0.05 \leq t/R$	Optimization Results $\eta = 2$ and $0.0025 \leq t/R$
	A	B	C	D	E	F
40	$25 - L/R \leq 0$			−22.50	−1.45	−8.00
41	P_{cr} Shell = $(8\pi ERT(10^{-3})/(1-\mu^2)$			8.50E+04	9.58E+02	250.00
42	$P - P_{cr} \leq 0$			−8.45E+04	−4.58E+02	−1.56E−12
43						
44	Geometric constraints					
45						
46	$-R \leq 0$			−0.2000	−0.0212	−0.0294
47	$-t \leq 0$			−0.0100	−0.011	−0.0002
48	t/R	(t/R)min $*R - t \leq 0$		0	0	−0.0001

there are only 32 lines of input needed to solve this problem! Formulas are shown in the body of the spreadsheet for clarity only. The initial guesses to this problem are $R = 0.2$ m and $T = 0.010$ m in cells D-5 and D-6, respectively. Cells D-7, D-10 to D-14 are specific geometric and material property data for the problem. Note that accommodation has been provided for a factor of safety in cell D-15. Cells D-17 and D-18 are intermediary calculations of area and inertia, respectively. The objective function is the weight and is cell D-2418. The loading constraints are shown in cells D-36 for Euler or Johnson buckling and D-42 for shell buckling. Geometric constraints are shown in cells D-46 and D-47 to require that both the radius and wall thickness are kept greater than zero. Additionally, the thickness to radius ratio constraint is shown in D-48.

When run the Solver routine returns the results shown in the column E "Optimization Results with $\eta = 1$ and $0.05 \leq t/R$." Actually, this is not correct, when the initial guess of $R = 0.2$ and $t = 0.01$ are chosen the optimization routine will not converge or successfully run. Another guess will have to occur closer to the optimal result. The reader can verify this by running the program with a result close to the optimal values. Note that when this runs in the Excel program a new column is not created as shown, column D is changed instead. Note that the weight is at the optimal value of 0.78 N as shown previously in the graphical solution. Note that the column buckling and thickness over radius ratios are zero ($1.75(10^{-7})$ is essentially zero).

Analogous to the second graphical solution shown in Figure 6.5, the optimization routine is modified to accommodate $\eta = 2$ and $0.025 \leq t/R$. Note that the results of $R = 0.0294$ and $t = 0.0002$ are not unique and any combination along the curve from point A1 to A2 will satisfy the constraints. Furthermore, we must continually check that the weight is indeed minimized. For example, the optimization routine will often find false minima along the g_1 constraint line that satisfies all the constraints, but is not a global minimum of $W = 0.2$ N.

6.11 COMPUTER DATABASE DESIGN SELECTION PROCEDURE

A polymer material property database is provided with this text in Microsoft Excel format. At present, 48 materials are included in the database. The user can easily add materials to the database if desired. Table 6.8 gives the names and descriptions of the polymers in the files Polymer.xls. Some of the names are the common names, for example, LDPE. Others were created to be descriptive, for example, glass-reinforced epoxy as EPGR. The mechanical property names are shown in Table 6.9.

TABLE 6.8
Names and Descriptions of Polymer Materials in Database Files

Polymer ID	Description	Name
1	Polyethylene Low–Med-Density Linear	LDPE
2	Polypropylene Homopolymer unfilled	Pp
3	Polystyrene Homopolymer High–Med Flow	PS
4	Polyvinyl Chloride PVC Acetate MC, Rigid	PVC
5	Polyamide (Nylon) Type 6 Molding and Extrusion Compound	PA6
6	Polycarbonate Unfilled Mold, Extrusion Resign High Viscosity	PC
7	Epoxy, Bisphenol Molding Compound Glass Fiber-Reinforced	EPGR
8	Melamine Formaldehyde Cellulose Filled (Thermoset)	MF0
9	Phenolic Molding Compound, Wood Flour and Mineral Filled	PFMC
10	Polyurethane Liquid Casting Resin (Thermoset)	PUR
11	ABS, Flame Retardant Grade, Molding and Extrusion	ABS
12	ABS, Injection Molding Grade Medium Impact	ABSM
13	ABS, Injection Molding Grade High Impact	ABSH
14	ABS, Injection Molding Grade 20% Glass-Reinforced Homopolymer	ABSR
15	Polyoxymethylene (Acetal) Homopolymer	POM
16	Polyoxymethylene (Acetal) Impact Modified Homopolymer	POMI
17	Polyoxymethylene (Acetal) 20% Glass-Reinforced Homopolymer	POMR
18	Acrylic, Molding and Extrusion Grade	PMMA
19	PMMA, Impact Modified	PMMH
20	Epoxy, Glass Fiber Reinforced	EPGR
21	Epoxy, Casting and Resins and Compounds, Unfilled	EP
22	Modified Polytetrafluoroethylene PE-TFE, Unfilled	PTFE
23	Modified Polytetrafluoroethylene PE-TFE, 25% Glass Fiber Reinforced	PTFR
24	Melamine Formaldehyde (Amino) Glass Fiber Reinforced	MGFR
25	Phenolic, High-Strength Glass Fiber Reinforced	PFGR
26	Phenolic, Casting Resins, Unfilled (Thermoset)	PF
27	Polyamide (Nylon), Type 6, 30–35% Glass Fiber Reinforced	PA6R
28	Polyamide (Nylon), Type 6, High-Impact Modified Copolymer	PA6H
29	Polyamide (Nylon), Type 6, Impact Modified 30% Glass Fiber Reinforced	PA6I
30	Polyamide (Nylon), Type 66, Molding Compound	PA66-mold
31	Polyamide (Nylon), Type 66, 30–35% Glass Fiber Reinforced	PA66-Hi

TABLE 6.8 (continued)
Names and Descriptions of Polymer Materials in Database Files

Polymer ID	Description	Name
32	Polyamide (Nylon), Type 66, High Impact Rubber Modified Compound	PA66-GFR
33	Polyamide-imide, Unfilled Compound and Injection Molding Compound	PAI
34	Polyamide-imide, 30% Glass Fiber Reinforced	PAI-GFR
35	Polycarbonate, 30% Glass Fiber Reinforced	PCGR
36	Polycarbonate, Modified Blends	PCCO
37	Polystyrene(SAN) Copolymer	PSCO
38	Polystyrene(SAN) Copolymer, Glass Fiber Reinforced	PSCO-GFR
39	Polystyrene(SAN) Heat-Resistant Copolymer, Impact Modified	PSCO-HR
40	Polystyrene(SAN) Heat-Resistant Copolymer, 20% Glass Fiber Reinforced	PSCO-GFR-20
41	Polyurethane 10–20% Glass Fiber Reinforced, Molding Compounds	PUR-GFR
42	Silicone Molding Encapsulated Compound, Mineral, and/or Glass Filled	SI-F
43	Silicone/Polyimide Pseudo-Interpenetrating Networks Silicone/Nylon 12	SI
44	Polysulfone, 30% Glass Fiber Reinforced	PSO
45	Polyethersulfone, 20% Glass Fiber Reinforced	PESO-GFR 20
46	Polyethersulfone, unfilled	PESO
47	Polyethersulfone, 30% Glass Fiber Reinforced	PESO-GFR 30
48	PVC Molding Compound, 15% Glass Fiber Reinforced	PVCR

TABLE 6.9
Names and Descriptions of Physical and Mechanical Properties in Database Files

Property	Codes
Glass transition temperature (°C)	T_{g_lo}-T_{g_hi}
Melting temperature (°C)	T_{m_lo}-T_{m_hi}
Elongation to break (%)	ε_{b_lo}-ε_{b_hi}
Yield strength (MPa)	S_{y_lo}-S_{y_hi}
Ultimate tensile strength (MPa)	S_{ut_lo}-S_{ut_hi}
Ultimate compressive strength (MPa)	S_{uc_lo}-S_{uc_hi}
Flexure strength (Pa)	S_{fs_lo}-S_{fs_hi}
Tensile modulus (GPa)	E_{t_lo}-E_{t_hi}
Compression modulus (GPa)	E_{c_lo}-E_{c_hi}
Flexure modulus at 23°C (GPa)	$E_f(23)_{lo}$-$E_f(23)_{hi}$
Izod impact (J/cm)	S_{izod_lo}-S_{izod_hi}
Deflection temperature (at 1.8 MPa) (°C)	$T_d(1.8)_{lo}$-$T_d(1.8)_{hi}$
Specific gravity	sg_{lo}-sg_{hi}

6.11.1 Example Problem of Impact of a Beam

The cantilever beam problem considered in Chapter 5, where the performance criteria were toughness and stiffness is considered here as an example of using the polymer material database as a computer selection procedure. Recall that the minimum width of the beam corresponded to maximizing the product of $T \times E$, where T is the toughness or Izod impact strength (ft lbs/in.), and E was the flexure modulus (psi).

$$E_f(23)_{lo} \times s_{izod\_lo} > 3000 \tag{6.54}$$

It is decided to search the data base with:

$$E_f(23)_{lo} > 500, \quad s_{izod\_lo} > 2.0, \quad \text{and} \quad sg_{lo} < 1.6. \tag{6.55}$$

HOMEWORK PROBLEMS

6.1. Compare the three PF's of Table 6.1 and 6.2 for circular shafts under bending based on
1. A strength basis
2. A stiffness basis

(Hint: Need diameter ratio d_2/d_1.)

6.2. Prepare a material comparison table for the PF-GP, PF-MI, and PF-high impact (PF-HI) materials for a thermal gradient through a beam on a deflection basis (thermal bowing).

6.3. Compare a square cross section to a circular cross section ($h = b \neq d$) for Example 6.1.

For the following optimal design problems an additional constraint that can be placed on this solution is the inclusion of lateral buckling. However, discussion of lateral buckling is beyond the scope of the text. Dependent on the loading conditions this can become a factor. In order to avoid this and bound the solution, the ratio of area dimensions should be less than 10.

6.4. A cantilever beam ($L = 9$ in.) with a concentrated transverse load of 75 lb and a concentrated tensile axial load of 50 lb. Ensure that these loads can be carried indefinitely.

Compare hollow cross-sections of rectangular, circular and an inverted isosceles triangle. The minimum wall thickness should be greater than 5% of any dimension. Note that this allows for thick-walled sections!

The objective of this project is to minimize weight. Obviously, this is a two-part problem of material selection and optimization.

Write a short report (be formal) describing the selection process and the alternative material candidates. In order to facilitate checking of your solution utilize the following dimensions for sample calculations (factor of safety = 2):

Rectangular tube (base = 1 in., height = 2 in., thickness = 0.1 in.)
Circular tube (D_0 = 1.5 in., D_i = 1.4 in.)
Inverted triangular tube (base = 1 in., height = 1.5 in., thickness = 0.1 in.)

6.5. Select a polymeric material from the polymer database to meet the following constraints:

A cantilever beam ($L = 18$ in.) with a concentrated transverse end load of 50 lb, a concentrated torque of 25 in.-lb, and an axial load of 200 lb. Limit the vertical deflection to 0.375 in., the axial deflection to 0.05 in. and the rotational deflection to 5°. The heat deflection temperature at 264 psi should be greater than 475°F.

Investigate both solid and hollow ($0.05 < t/b < 0.1$ and $0.05 < t/h < 0.1$, t is constant) rectangular cross sections.

The objective of this project is to minimize weight. Obviously, this is a two-part problem of material selection and optimization.

Write a short report (be formal) describing the selection process and the alternative material candidates. Include a graphical solution for the solid cross section.

REFERENCES

1. Arora, J. S., *Introduction to Optimum Design,* McGraw-Hill, New York, 1989.
2. Vanderplaats, G. N., *Numerical Optimization Techniques for Engineer Design with Applications,* McGraw-Hill, New York, 1984.
3. *DOT, Design Optimization Tools*, CR 1990, VMA Engineering, Goleta, CA, CR 1990.
4. Flugge, W., *Stresses in Shells,* Springer-Verlag, Heidelberg, 1962.
5. *Modern Plastics Encyclopedia*, McGraw-Hill, Inc., New York, 1989.

7 Design Applications of Some Polymers

The design applications of polymers are, of course, limitless. Therefore, to keep this text within reasonable space limits, a few polymer applications are selected for illustrative purposes. A general description of these selected polymers, as well as some detailed design calculations, will be given.

7.1 PHENOLIC RESINS WITH FILLERS

Phenolic (PF) resins have a wide range of applications—from the use of a general purpose (GP) phenolic as a housing of an electric drill to the use of odorless grades for handles of cooling pots, to its use as an adhesive in a resin-bonded diamond grinding wheel. PF are generally classified into three groups: general purpose (GP), medium impact (MI), and high impact (HI) phenolic.

In the GP PF, the filler is a fine wood flour. Acids are used to remove the lignin and sap so that the filler is pure cellulose. The properties of the GP PF include: high surface gloss, good mold filling, reasonable mechanical strength, low cost, good heat resistance –250°F continuous and up to 400°F for short times, water and chemical resistance, and good electrical insulation at common voltages.

For the MI and HI compounds, fibers are added to increase the impact strength. Cotton fibers and paper pulp fibers increase the impact strength of the MI PF by five times. Rayon yard, nylon filament, or glass fibers are used to increase the impact resistance of the HI PF up to a factor of 20. However, the fiber-reinforced compounds suffer from poorer surface finish and they require higher molding pressures. The composition of a typical PF mixture is listed in Table 7.1.[1]

Mineral fillers are used where higher temperatures are encountered. Asbestos is used up to 450°F. Fibered asbestos increases the impact strength as well. Mica is used to reduce electrical losses up to 400°F in insulators in radios and TVs. Carbon, silica, and quartz fibers in space vehicles and missiles permit operation at temperatures up to 30,000°F for short times. These fibers are used in rocket exhaust nozzles and re-entry nose cones in laminate form. However, these very high-temperature composites are very expensive. Metallic powders are also used as fillers: lead powder as an x-ray barrier, iron powder in electric motor brushes, and molybdenum powder in low-friction applications. Because of their strength-to-weight ratio and damping properties, Nylon fibers are used in aircraft parts such as pulleys, cables, and spacers. Powdered Teflon PF are used in bearing applications such as the bushings for the guide rods on a hydraulic press. A summary of other uses of molded PF is given in Table 7.2.[1]

TABLE 7.1
Composition of Typical PFs

Component	Composition: Parts by Weight
Resin (binder)	60
Filler	50
Fiber	80
Hardening agent (curative)	15
Plasticizer	5
Flake nylon or buna-N-rubber	5
Dye	5
Lubricant	1

Source: Adapted from Milby, R. V., *Plastics Technology*, McGraw-Hill, New York, NY, 1973, 581pp.

The Society of Plastics Industry (SPI) has adopted an SPI grade number or notation for PF illustrated as follows for the following example:

$$\text{PF } 22\ 03\ 3 = 22\text{-}03\text{-}3,$$

where 22 indicates heat-distortion temperature = 220°F, 03 indicates Izod impact strength = 0.3 ft-lb/in., and 3 indicates minimum tensile strength = 3 ksi.

TABLE 7.2
Applications for Molded PFs

Type	Use
	Molded PFs
G.P.	Pot handles, knobs, camera parts, iron handles, closures, appliances, switches
Impact resistant	Gears, gunstocks, portable tools, pulleys, welding rod holders
Electrical grade	Condensors, spacers, insulators
Heat resistant	Ashtrays, rocket blast tubes, igniter parts, pump parts, rocket nozzles
Chemical resistant	Photo-developing equipment, labware, washing machine parts
Special use	Oil-less bearings, slide cams, x-ray parts, cores, bushings
	Laminated PFs
Paper base	Terminal strip, panels, washers
Fabric base	Gears, bearings, helmets, radio mounting board
Asbestos base	High temperature applications, such as valve parts, bearings
Decorative grade	Serving tray

Source: Adapted from Milby, R. V., *Plastics Technology*, McGraw-Hill, New York, NY, 1973, 581pp.

TABLE 7.3
Comparison of PC and PF Polymers

Polymer	Specific Gravity (g)	Impact Strength (ft-lb/in.)	Price (P, $/lb)
PF-GP	1.32–1.55	0.24–0.50	0.40
PF-MI	1.36–1.43	0.60–8.0	1.00
PF-HI	1.75–1.90	10–33	2.00
PC	1.20	12–16	1.20

7.2 POLYCARBONATE

The PC thermoplastic is available in sheet, rod, ball, and slab forms or shapes. In its natural state it is clear and transparent, but 80 colors are available. Since it is ductile, it can be nailed, sawed, punched, cold drawn, sheared, drilled, and riveted without cracking (if care is taken). As far as toughness is concerned, it is 50 times stronger in impact than glass. It has dimensional stability up to 215–250°F. PC can withstand water environments, but it will not hold up in chlorine, strong acids, alkalies, or high-octane gasoline. However, a 30% glass-filled PC has been used for the carburetor body of power chain saws. PC is compared to the PFs in Table 7.3. It is lower in weight, and is competitive with PF-HI in impact strength at a lower cost. (Costs are relative and fluctuate with available demand, and inflation.)

7.3 EXAMPLE DESIGN WITH PC: FAN IMPELLER BLADE

This example and some others in this chapter are taken from Miller,[2] but the example designs are modified and extended here.

Consider the use of PC for the blade of a household fan to run at 10,000 rpm for 10 h a day at an operating temperature of 140°F.[2] The design problem is to limit the creep of the blade to maintain a clearance between the impeller and the shroud. The fan impeller blade is shown in Figure 7.1. A free body diagram of a radial element enables us to determine the centrifuged force F at any radius r in the blade.

The body force R from rotation is

$$R = mr\omega^2. \tag{7.1}$$

Substitution of the mass m in terms of density and the cross-sectional area A gives

$$R = \frac{\rho A dr}{g} r\omega^2. \tag{7.2}$$

Equilibrium of the element $A\,dr$ requires

$$dF + R = 0, \quad \text{or} \quad \frac{dF}{dr} = -\frac{\rho A r\omega^2}{g}. \tag{7.3}$$

The solution is

$$F = -\frac{\rho A r^2 \omega^2}{2g} + C. \tag{7.4}$$

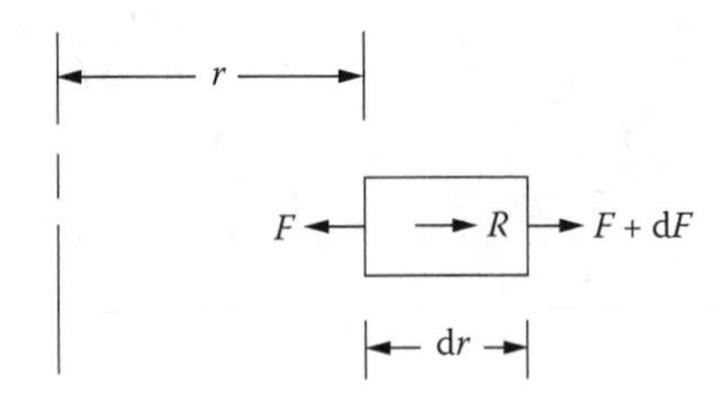

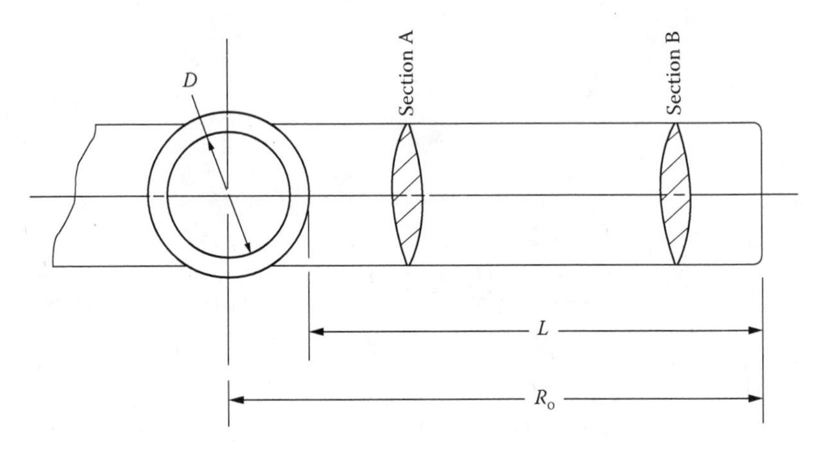

FIGURE 7.1 Fan impeller blade.

Use of the boundary condition that $F = 0$ at the outer edge of the blade, $F = 0$ when $r = r_0$, yields

$$F = \frac{\rho A \omega^2}{2g}(r_o^2 - r^2). \tag{7.5}$$

For $r_o = 5.0$ in., the maximum stress at an inside radius of $r_i = r_o - L = 5.0 - 4.0 = 1.0$ in. is

$$\sigma_{max} = \frac{F_{max}}{A} = \frac{\rho\omega^2}{2g}(25 - 1) = \frac{12\rho\omega^2}{g} \tag{7.6}$$

For the given conditions

$$\sigma_{max} = 0.0432\left(\frac{\text{lb}}{\text{in.}^3}\right)\frac{12}{386.4}\left(\frac{\text{sec}^2}{\text{in.}}\right)\left(\frac{2\pi 10{,}000}{60}\right)^2\left(\frac{\text{rad}}{\text{s}}\right)^2 \tag{7.7}$$

or

$$\sigma_{max} = 1470\,\text{psi}. \tag{7.8}$$

Check the strength

$$\sigma_{max} < S_u = 9 \text{ to } 10\,\text{ksi} \quad \text{and} \quad \sigma_{max} < S_y = 8 \text{ to } 10\,\text{ksi}. \tag{7.9}$$

7.3.1 Creep Strain

From isochronous stress–strain data (see Chapter 2) in Figure 7.2, the creep strain at 10 h at 140°F is

$$\varepsilon = 0.75\%. \tag{7.10}$$

Next check the crazing line in Figure 7.2. Crazing is the initiation of small cracks. This phenomenon develops in polymers under creep and also under exposure to solvents, or both. Note that the stress of 1470 psi at 10 h is left of the crazing line which must be assured.

What is the elongation of the blade? The stress is not uniform in the blade, so that $\varepsilon = \Delta L/L$ does not apply. Therefore, consider elongation of a differential length dr:

dr dδ

$$\varepsilon = \frac{d\delta}{dr} = \frac{\sigma}{E} = \frac{F}{AE}. \tag{7.11}$$

Integrating

$$\delta = \int_{r_1}^{r_o} d\delta = \int_{r_1}^{r_o} \frac{F}{AE}\, dr, \tag{7.12}$$

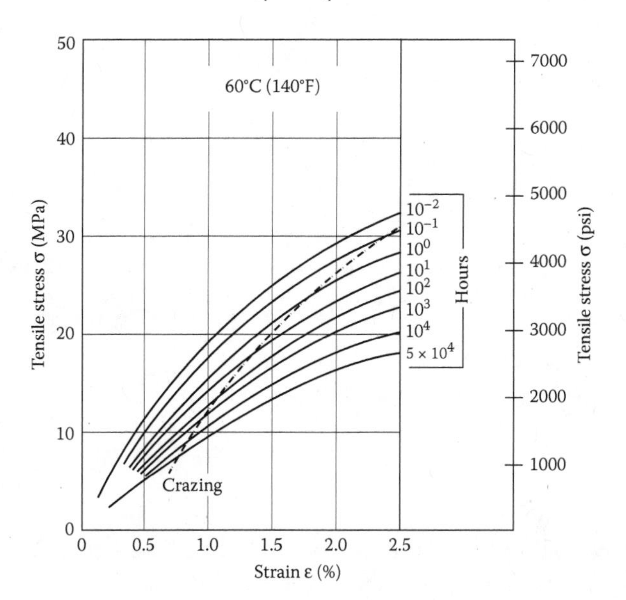

FIGURE 7.2 PC isochronous stress–strain curves at 140°F. (Adapted from Miller, E. ed., *Plastics Products Design Handbook, Part A, Materials and Components*, Marcel Dekker, New York, 1981, ISBN 0-8247-1339-7.)

$$\delta = \frac{\rho\omega^2}{2gE}\int_{r_1}^{r_o}(r_o^2 - r^2)\,dr, \tag{7.13}$$

$$\delta = \frac{\rho\omega^2}{2gE}\left(r_o^3 - \frac{r_o^3}{3} - r_o^2 r_i + \frac{r_i^3}{3}\right). \tag{7.14}$$

For the PC polymer, from Figure 7.2,

$$J(t) = \frac{1}{E} = \frac{\varepsilon}{\sigma_o} = \frac{0.0075}{1470} = 5.1^{(10^{-6})}, \tag{7.15}$$

thus,

$$\delta = 0.018\,\text{in}. \tag{7.16}$$

The total clearance required for the design then is δ the creep deflection plus the deflection of the hub plus the molding tolerance.

The result (Equation 7.16) is conservative (an upper bound), since the maximum stress (1470 psi) was used in Equation 7.15. An improved model, a finite-element model shown below, can be used to obtain a more accurate calculation:

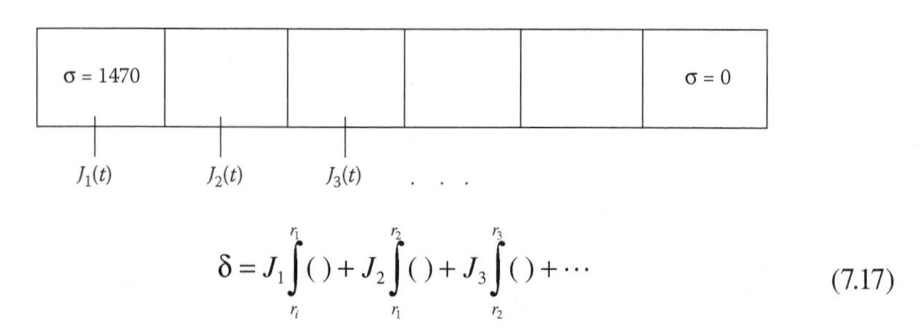

$$\delta = J_1\int_{r_i}^{r_1}() + J_2\int_{r_1}^{r_2}() + J_3\int_{r_2}^{r_3}() + \cdots \tag{7.17}$$

A material comparison table can also be constructed for this problem if alternate materials were to be considered. The rating ratios will include

$$\frac{F_2}{F_1} = \frac{\rho_2}{\rho_1}, \quad \frac{\delta_2}{\delta_1} = \left(\frac{\rho_2}{\rho_1}\right)\left(\frac{E_1}{E_2}\right), \quad \frac{W_2}{W_1} = \frac{\rho_2}{\rho_1}. \tag{7.18}$$

7.3.2 Impact Failure

There is another potential failure mode that must be considered in design of fan impeller blades, and that is impact failure if the blade were to strike an obstacle. Assume impact at the tip of the blade. Then the blade can be taken as a cantilever beam loaded at the end. Assume a uniform cross section, and assume the kinetic energy of rotation of the blade before impact is totally absorbed as strain energy in bending the beam.

The kinetic energy is

$$KE = \frac{1}{2}\int v^2 \mathrm{d}m, \tag{7.19}$$

where the velocity from rotation is $v = \Omega r$ and $\mathrm{d}m = \rho A\,\mathrm{d}r/g$. Thus,

$$KE = \frac{1}{2}\frac{\rho A\,\Omega^2}{g}\int_{r_i}^{r_o} r^2\,\mathrm{d}r = \frac{1}{2}\frac{\rho A\,\Omega^2}{g}\frac{(r_o^3 - r_i^3)}{3}. \tag{7.20}$$

For the dimensions and conditions for the prior example, this expression gives

$$KE = 635\frac{\text{in.lbs}}{\text{in.}^3} \tag{7.21}$$

for a blade volume of AL = 4.0 A. If we use the maximum Izod value of 18 ft-lb/in. for PC from Table 5.1 and if we assume a thickness of blade of 0.1 in., then the Izod toughness per volume would be

$$T = \frac{18(12)}{4(0.1)} = 540\ \text{in.lbs/in.}^3 \tag{7.22}$$

If the stress concentration is the same in the blade as in the Izod test the blade will fail in impact. If there were no stress concentration in the blade it would survive. If there is stress concentration then either a material with higher impact strength would have to be found, or the rotational speed could be decreased. Decreasing the speed by 1/2 will decrease the kinetic energy *KE* by 1/4.

7.4 EXAMPLE DESIGN WITH PC: SNAP/FIT DESIGN

Consider a snap/fit-type latch for a closure on a door or box or similar device.[2] Consider the configuration shown in Figure 7.3 and model it as a cantilever beam where the deflection δ must be at least equal to the height h if the latch is to work:

$$\delta = h. \tag{7.23}$$

A check of the properties of PC shows that

$$\varepsilon \leq 1.5\%, \tag{7.24}$$

to avoid permanent plastic strain. From Figure 7.3, choose the configuration with $a/b = 1$ so $C = 0.667$. Assume $L = 0.65$ in. and $t = 0.08$ in., then

$$\delta = h = \frac{C\varepsilon L^2}{t} = \frac{0.667(0.015)(0.65)^2}{0.08} = 0.053\ \text{in.} \tag{7.25}$$

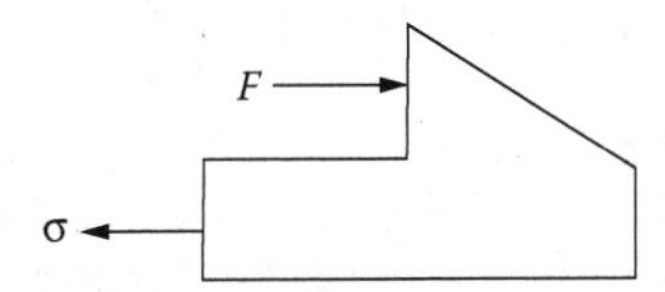

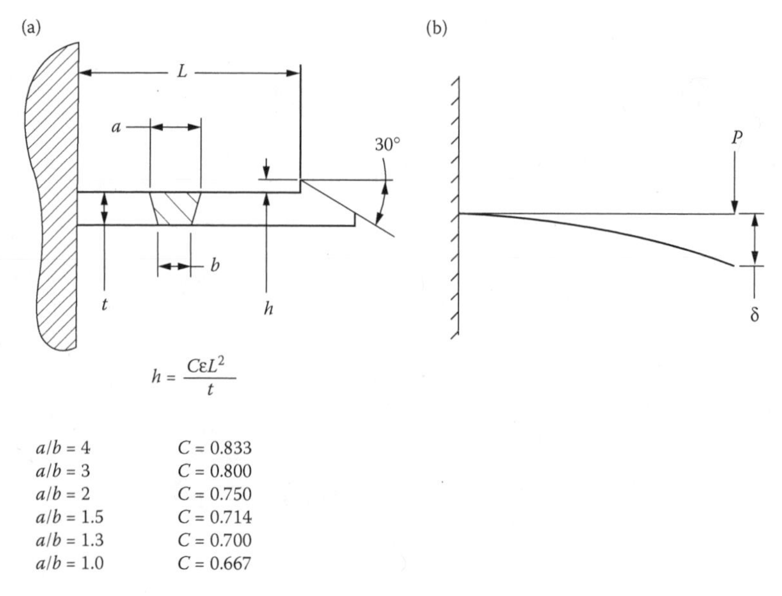

FIGURE 7.3 PC snap/fit design. (a) Gemetry and (b) cantilever model.

This is the analysis for assembly. Next consider a stress analysis for loading during use. Assume a design load of $F = 15$ lbs. Also assume dimensions of $a = b = 0.125$ in. Calculate the average axial stress σ:

$$\sigma = \frac{15}{0.125(0.08)} = 1500 \text{ psi.} \tag{7.26}$$

This stress corresponds to a design limitation of time at load to avoid crazing of about 5000 h at 140°F from Figure 7.2. Next, check the average shear stress

$$\tau = \frac{15}{0.125(0.053)\sqrt{3}} = 1307 \text{ psi.} \tag{7.27}$$

If a design shear stress from maximum shear theory is taken as one-half the design tensile stress, a tensile stress of 2 (1300) = 2600 psi should be used in Figure 7.2. This would limit the design use further to only 10 h at 140°F.

Consider a more accurate stress analysis including the bending stress from F rather than only the average stress σ. What about the effect of a stress concentration at the corner? Should bending creep data be used rather than tensile creep data? If the latch were subject to repeated openings over short times—what about a fatigue analysis?

7.5 EXAMPLE DESIGN OF PVC PIPE

PVC is the fastest-growing pipeline material. In the United States, its use grew from 1,770,000 lbs in 1980 to 3.7 billion lbs in 1989 to about 5.5 billion lbs in 2000.[3] On a cost-to-performance basis, PVC is better than the next contender PE by a factor of 2 or

3 to 1. The advantages of PVC include high tensile strength, high modulus E, HI strength, excellent chemical resistance, lightweight, less energy in manufacture, less energy in installation, little training needed in installation, no heavy equipment needed, high life expectancy, easily cut in the field, good flow characteristics, and low build-up of crud on smooth interior walls. PVC is not as brittle as clay, concrete, or cast iron pipe, and it does not corrode like iron pipe. It has many uses as rural water pipe, irrigation pipe, municipal water pipe, sewer pipe (up to 30 in.), and other industrial piping. However, it should be noted that caution is being used with regard to the PVC pipe for drinking water. Recent research indicates that the PVC pipe may contain human carcinogens.

The GP grades include BF Goodrich 103 EPF-76, Tenneco 225PG, Georgia Pacific 1091, Shintech SE950, Borden 100RE, Diamond 450, Hooker 9418, and Air Products 1230P. The ASTM D-1784 class specification for class 12454B rigid PVC is defined as follows:

Class: 12454 B
Identification:

1. PVC homopolymer

Property and Minimum Value:

2. Impact strength (Izod) (0.65 ft-lb/in.)
3. Tensile strength (7000 psi)
4. Modulus of elasticity in tension (400,000 psi)
5. Deflection temperature under load (158°F)

B Chemical Resistance

Other ASTM designation order numbers are listed in Table 7.4.
The ASTM Standard D-2241, Table 7.5, uses a standard dimension ratio (SDR) for piping, where

$$\text{SDR} = \frac{D_o}{t}, \quad (7.28)$$

where D_o is the outside diameter and t is the wall thickness. The International Standards Organization has their ISO design formula for working pressure defined as follows:

$$P = \frac{2S}{\text{SDR} - 1}, \quad (7.29)$$

where

$$\sigma = \frac{S}{\eta} = \text{design stress}, \quad (7.30)$$

S = strength and η = factor of safety. A design stress of $\sigma = 2000$ psi is recommended for 12454 B pipe. This corresponds to a factor of safety of 2.2 for preventing creep rupture at 100,000 h or more life as indicated in Table 7.6.

TABLE 7.4
Requirements of Rigid PVC Compounds, ASTM D-1784

Designation Order No.	Property and Unit	0	1	2	3	4	5	6	7	8
1	Base resin	Unspecified	PVC homopolymer	Chlorinated PVC	Ethylene chloride copolymer	Propylene vinyl chloride copolymer	Vinyl acetate vinyl chloride copolymer	Alkyl vinyl ether vinyl chloride copolymer		
2	Impact strength (Izod) min:									
	J/m of notch	Unspecified	34.7	34.7	80.1	266.9	533.8	800.7		
	ft-lb/in. of notch		0.65	0.65	1.5	5.0	10.0	15.0		
3	Tensile strength min:									
	MPa	Unspecified	34.5	34.5	41.4	48.3	55.2			
	psi		5000	5000	6000	7000	8000			
4	Modulus of elasticity in tension, min:	Unspecified	1930	1930	2206	2482	2758	3034		
	MPa		280,000	280,000	320,000	360,000	400,000	440,000		
	psi									
5	Deflection temperature under load, min:	Unspecified	55	55	60	70	80	90	100	110
	1.82 MPa (264 psi):		131	131	140	158	176	194	212	230
	°C									
	°F									

Note: The minimum property value will determine the cell number although the maximum expected value may fall within a higher cell.

TABLE 7.5
Physical Dimension of PVC Pressure-Rated Pipe of SDRs

Nominal Size (in.)	Diameter		Minimum Wall Thickness
	D_o average (in.)	D_i inside (in.)	t (in.)
	SDR 64 (63 psi water pressure rating)		
6	6.625	6.417	0.104
8	8.625	8.355	0.135
10	10.750	10.414	0.168
12	12.750	12.352	0.199
	SDR 41 (100 psi)		
3-1/2	4.000	3.804	0.098
4	4.500	4.280	0.110
5	5.563	5.291	0.136
6	6.625	6.301	0.162
8	8.625	8.205	0.210
10	10.750	10.226	0.262
12	12.750	12.128	0.311

From mechanics of materials the reader should recall that the hoop stress (circumferential stress) in a thin cylinder is

$$\sigma_\theta = \frac{PD}{2t}. \tag{7.31}$$

If D is taken as the mean diameter, then

$$D = D_o - t. \tag{7.32}$$

For $\sigma_\theta = S$, Equation 7.31 reduces to Equation 7.38.

TABLE 7.6
Creep Rupture Strength for 12454 B PVC Pipe Material

Time to Failure at Various Hoop Stresses

Time (h)	Hoop Stress (1000 psi)
0.1	8
1.0	7
10	6
100	5.5
1000	5
10,000	4.8 extrapolated
100,000	4.4 extrapolated

Example 7.1

Consider a 6 in. SDR 64 pipe. From Table 7.5: $D_o = 6.625$, $t = 0.104$, and

$$P = 2\left(\frac{2000}{\left(\frac{6.625}{0.104} - 1\right)}\right) = 63.8\,\text{psi}. \tag{7.33}$$

This is close to the rated pressure of 63 psi given in Table 7.5. The hoop stress, however, is not the only consideration in piping design. One should be aware of thermal expansions in piping:

$$\Delta L = \alpha L \Delta T \quad \text{if free,} \tag{7.34}$$

and

$$\sigma = E\alpha\Delta T \quad \text{if fixed.} \tag{7.35}$$

Water hammer can cause a pressure increase in pipe:

$$\Delta P = 0.0134V(\Delta V), \tag{7.36}$$

where V = water velocity (ft/s).

The designer must also take into account external stresses in piping. Bending stresses can develop from the weight of the piping material plus the water.

7.6 DESIGN WITH FLUOROCARBON RESINS

Considered next are self-lubricated plastic materials, namely fluorocarbon resins, fabrics, and coatings.[2] Dupont's Teflon, the thermoplastic PTFE, has a low coefficient of friction, μ, and a large temperature range from –450°F to +500°F. The PTFE types are the most common and are available in sheet rod or film. Perfluoro alkoxy alkane (PFA) resins have comparable properties to PTFE. Another fluorinated ethylene propylene is a copolymer, that is, easier to mold, but it has a higher coefficient of friction μ and a temperature limit of 390°F.

The pure resins are limited to light loads and low draft velocities because of cold-flow problems and low thermal conductivity values (k). Therefore, various fillers are added such as glass, graphite, bronze, and M_oS_2. These resins and the filled compounds have a pressure–velocity limit of 8000–10,000 psi fpm. Unfilled PTFE has a static $\mu = 0.05$–0.08, and a dynamic $\mu = 0.10$–0.13. The μ values are larger with fillers ranging from 0.08 up to 0.50. The coefficient of friction, μ decreased with pressure under dynamic conditions.

A wear factor K is used in self-lubricated bearing design. The radial wear t in a bearing is given by

$$t = KPVT, \tag{7.37}$$

where t = wear in inch units, K = wear factor, P = pressure (psi), V = velocity (fpm), and T = time (h). The average pressure is calculated from

$$P = \frac{W}{DL}, \tag{7.38}$$

where W = radial load, D = diameter of bearing, and L = length. The power loss P_L in a bearing is given by

$$P_L = \frac{\mu WND\pi}{720} \text{(ft-lb/s)}, \tag{7.39}$$

where N = angular velocity (rpm).

The running clearance in a bearing will change with heat buildup and thermal expansion. Because of high coefficient of thermal expansion values for polymers, they have not been used in electric motors because of small running clearance requirements; that is,

$$\alpha_{\text{polymer}} > \alpha_{\text{steel}}, \alpha_{\text{bronze}} \tag{7.40}$$

$5(10^{-5}) > 0.8(10^{-5}), 1.2(10^{-5})$°F.

The total running clearance is composed of a shaft allowance (tolerance) plus an installation allowance plus thermal expansion α allowance. The shaft allowance is about 0.005 in. on a 1.0 in.-diameter shaft and about 0.003 in. on a 10.0 in. shaft. The installation allowance is about 0.002 in. The thermal expansion allowance is calculated from the bearing temperature which has contributions from

$$T_{\text{bearing}} = T_{\text{ambient}} + \Delta T_{\text{friction}} - \Delta T_{\text{dissipation}}. \tag{7.41}$$

Because polymer bearings are generally restrained by metal, the "rule of thumb" from handbooks says the volume expansion due to 3α must be used in the calculation. But this is not exact because the geometry and the Poisson effect must be taken into account. The equation used on page 262 of the *Plastics Products Design Handbook*,[2]

$$\Delta R = t(3\alpha)\Delta T,$$

is incorrect. The factor is not 3, and it is found using elasticity theory for cylindrical geometry that ΔR should be proportional to R not t.

For a bearing only constrained at the outer diameter

$$\Delta R_i = -2R_i \frac{\alpha \Delta T}{(K^2+1) - \mu(K^2-1)}, \tag{7.42}$$

where R_i is the inside radius of the bearing, $K = D_o/D_i = R_o/R_i$ = wall ratio, ΔT is the difference in temperature between the bearing and the metal housing, and μ is Poisson's ratio.

For a bearing constrained in the axial direction as well, which is a plane strain condition in the theory of elasticity, Equation 7.39 can be modified with the substitution of μ' in place of μ, where

$$\mu' = \mu/(1-\mu). \tag{7.43}$$

Example 7.2

Calculate the thermal expansion of a polymer bearing for $\alpha = 5 \times 10^{-5}$, $R_o = 0.6$, R_i = 0.5 in., $\Delta T = 50$°F, and $\mu = 0.4$. For a bearing only constrained at the outer diameter,

$$\Delta R_{\text{i}} = -2(0.5)\frac{5(10^{-5})(50)}{(1.44+1)-0.4(1.44-1)} = -0.0011\ \text{in.}$$

For a bearing constrained in the axial direction as well,

$$\mu' = 0.4/(1-0.4) = 0.667 \quad \text{and} \quad \Delta R_{\text{i}} = -0.00116\ \text{in.}$$

The inside diameter change of the bearing will be twice this or $\Delta D = -0.0023$ in.

HOMEWORK PROBLEMS

7.1. Consider the design of the fan impeller blade shown Figure 7.1. The blade is made of PC material with the isochronous stress–strain data given in Figure 7.4. The fan is to be designed to run for 10 h/day at a temperature of 140°F.

Assume $r_{\text{o}} = 6.0$ in., $L = 5.0$ in., and $\rho = 0.0432\ \text{lb}_{\text{m}}/\text{in.}^3$

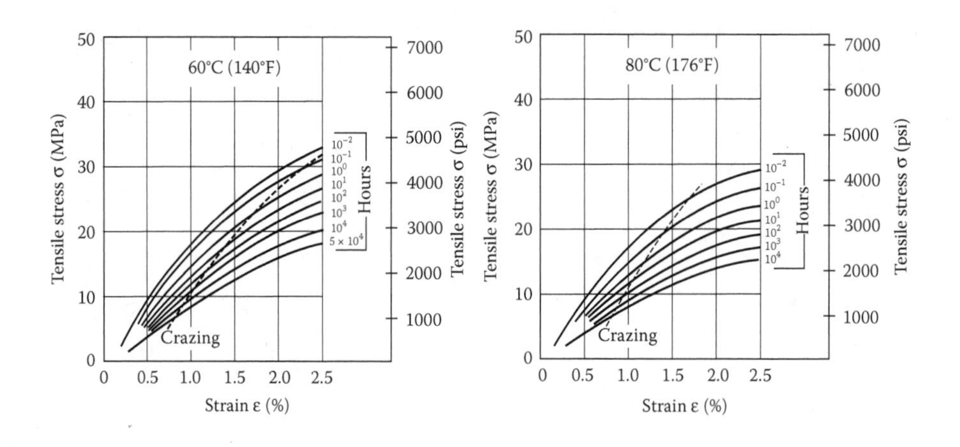

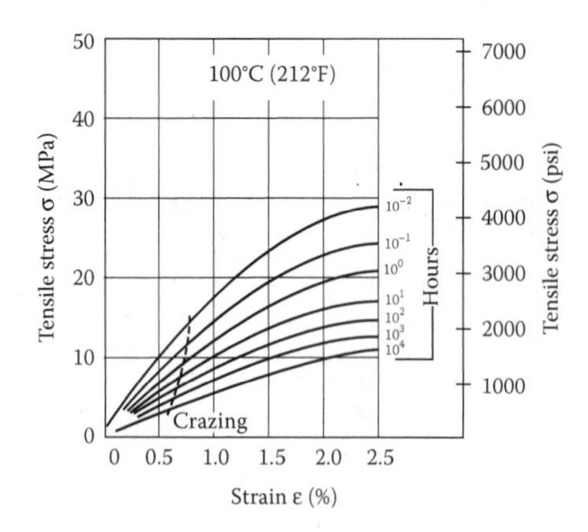

FIGURE 7.4 Isochronous stress–strain curves for PC. (Adapted from Miller, E. ed., *Plastics Products Design Handbook, Part A, Materials and Components*, Marcel Dekker, New York, 1981, ISBN 0-8247-1339-7.)

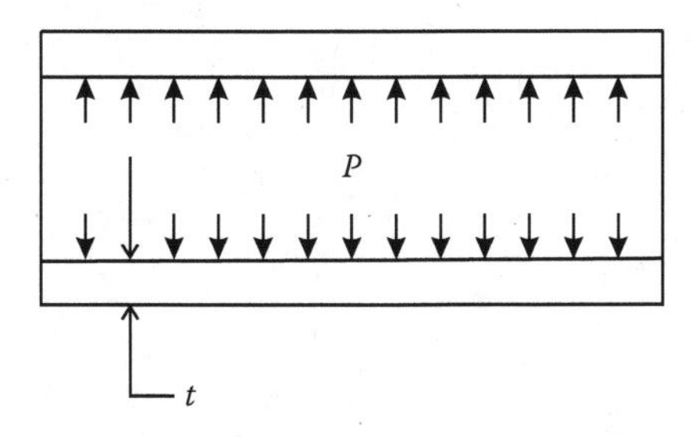

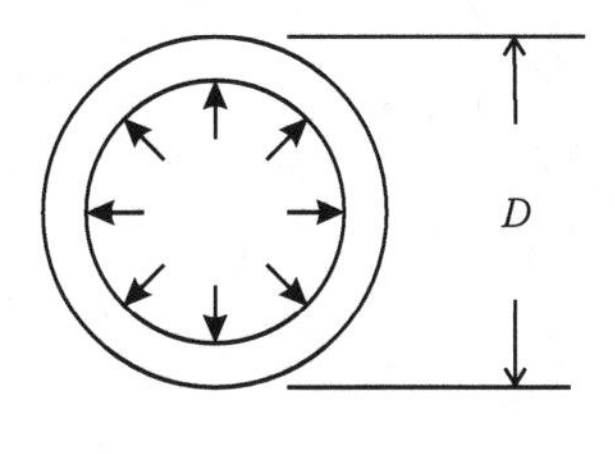

FIGURE 7.5 Cylindrical pipe.

1. For a factor of safety of 2 on the crazing stress, and for a strain limit of 1% ($\varepsilon \leq 1\%$), determine the maximum allowable speed.
2. Instead of limiting the rotational speed, can the cross-sectional area of the blade be increased to satisfy the factor of safety on the crazing stress (yes/no)? If (yes), by how much should the area be increased?

7.2. The PC material is used in the design of a 3.0 in. diameter cylindrical pipe (see Figure 7.5) with an internal pressure $P = 50$ psi at 200°F. Determine the thickness T required if the maximum stress must be limited to avoid crazing (stress cracking), and if the change in diameter due to creep is limited to 0.02 in.

7.3. Using the equations for stress in thick-walled cylinders from the theory of elasticity, or from a design textbook derive Equation 7.42. Hint: Use Hooke's law in the form

$$\varepsilon_\theta = \frac{1}{E}(\sigma_\theta - \mu\sigma_r) + \alpha\Delta T = u/r.$$

Let $\sigma_r = -p$ at $r = R_o$ and $u = 0$ at R_o, where u = radial displacement. Then solve for $u = \Delta R_i$ at $r = R_i$.

REFERENCES

1. Milby, R. V., *Plastics Technology*, McGraw-Hill, New York, NY, 1973, 581pp.
2. Miller, E. ed., *Plastics Products Design Handbook, Part A, Materials and Components*, Marcel Dekker, New York, 1981, ISBN 0-8247-1339-7.
3. Facts and Figures, www.ChemIndustry.com, pubs.acs.org, June 25, 2001.

8 Composite Material Mechanics

8.1 INTRODUCTION

Polymers and composite components have the advantage of being lighter than metal componets. This is important in aircraft and automotive applications where weight advantage corresponds to a fuel savings. Polymers also have good chemical resistance and are easily fabricated and joined. On the other hand, polymers lack thermal stability and have only moderate resistance to environmental degradation. They also possess relatively poor mechanical properties in comparison with metals. The practice of combining fillers and other reinforcing materials with plastics to enhance their performance characteristics and at the same time mitigate some of their less desirable characteristics is part of the broad technology of fabricated composite materials.

In this chapter, we will develop some of the basic concepts of the mechanics of composite materials needed to better understand the behavior and design of engineering components made from fiber-reinforced plastics. The first part of the chapter introduces the nomenclature and definition of terms needed to describe the manufacture and mechanical behavior of composite materials with emphasis on continuous glass fiber-reinforced thermoset polymeric materials. The second part of the chapter develops methods for the mechanical analysis of elastic and strength behavior of single–ply continuous fiber-reinforced composite materials, with an introduction to analysis of multiple plies, or composite laminates. Finally, experimental testing methods are described for determining the elastic and strength properties of single–ply fiber-reinforced composites.

8.2 COMPOSITE MATERIAL NOMENCLATURE AND DEFINITIONS

The terminology used to describe and define composite materials and their behavior is introduced in this section. Many of the terms are familiar from earlier study of mechanics of materials, while other terms specific to composites may be unfamiliar. A collection of these terms in one location is provided for easy reference.

Homogeneous body: The material properties of a homogeneous body are independent of position in the body. Thus, physical properties from a representative material sample may be obtained from any location in a homogeneous body, provided the orientation of the sample is fixed relative to a set of axes attached to the body. In this chapter, the term "material properties" includes the "mechanical properties" discussed in Chapters 2 and 3 as well as typical material properties such as density and so on.

Isotropic body: The measured material properties of an isotropic body are *independent of the orientation* of the representative material sample taken to characterize its behavior (i.e., independent of the axis of testing). A steel that is not cold worked is an example of a material which is both isotropic and homogeneous. Thus, a simple tensile test specimen may be chosen from any location and for any orientation in the piece of steel to obtain its tensile properties.

Orthotropic body: The material properties of an orthotropic material are different in three mutually perpendicular directions (material principal directions). The material properties for any other orientation in an orthotropic material can be obtained by appropriate transformation of the material poverties obtained for the principal material directions. Thus, the properties of an orthotropic material are a function of orientation. Planes perpendicular to the principal material directions are called planes of material symmetry.

Anisotropic body: The material properties at a point in an anisotropic body are different for each orientation of a material sample taken at this point. Thus, there are no planes of material symmetry in this type of material, and all material properties are a function of orientation.

Heterogeneous body: A heterogeneous material is made up of distinct and separate materials. Metal alloys, which are made up of different materials that are blended together, may be considered to form a single homogeneous material at the macroscopic level. Composites, on the other hand, consist of two or more physically distinct material regions at the macroscopic level. The amounts of the materials are combined in a controlled way to achieve a mixture having more useful behavioral characteristics than any of the constituent materials would achieve on their own. Composite materials are heterogeneous.

Matrix material: This text considers composite materials that are polymer-based compounds. The polymer base or resin is called the matrix, in which reinforcing material is added to enhance the performance characteristics of the composite material. In general, the matrix has low strength and stiffness in comparison with the reinforcement which is stiffer, stronger, and more brittle. The function of the matrix is to bind the reinforcement (usually fibers) in a desired distribution and orientation, and to transmit the external loading to them through the reinforcement/matrix interface. Both thermoplastic and thermoset resins are used as matrix material for polymeric composites.

Fiber-reinforcing material: A reinforcement is a strong material bonded into a matrix to improve its mechanical properties. Materials selected for fiber reinforcement should have high strength and stiffness and low density. Some elements having these characteristics are carbon, boron and silicon, and ceramic compounds such as SiO_2, SiC, Al_2O_3, BN, and Si_3N_4. These are all very brittle solids of low density which, when drawn into fine fibers, become strong, stiff materials. The most important fibers developed for modern composites are glass (SiO_2), carbon, boron, and highly drawn polymers such as polyparaphenylene terephthalamide (Kevlar 49).

Lamina (layer, ply): The arrangement of unidirectional or multidirectional fibers in a matrix to form a thin layer is called a layer, ply, or lamina of composite material. In general, several such layers are bonded together, with layer orientations chosen to form a single multilayered sheet having optimum material performance characteristics tailored for specific loading and environmental conditions. Three common types of lamina are unidirectional, woven, and random mat shown schematically in Figure 8.1.

Laminate: A stack of laminae bonded together to form a single multilayered sheet of composite material is called a laminate.

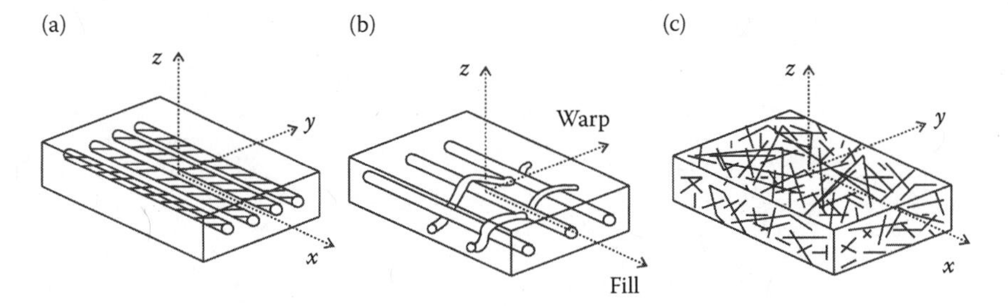

FIGURE 8.1 Typical reinforcement configuration. (a) Unidirectional, (b) woven, and (c) random.

Ply orientation (Unidirectional plies): A lamina having straight parallel fibers is called a unidirectional ply. Let the L, T-axes, lying parallel (longitudinal) and transverse to the ply fiber direction, respectively, serve to describe the orientation of the lamina with respect to an arbitrary reference axes, x, y, lying in the plane of the laminate (see Figure 8.2). The angle θ between the reference axis x and the ply fiber direction L is used to specify the ply orientation. The angle θ is positive for counterclockwise rotation. Thus, Figure 8.2 shows a $+\theta$ ply, a $-\theta$ ply, a 0° ply, and a 90° ply.

Stacking sequence: The stacking sequence is the order in which plies of varying orientation are stacked one on top of the other. In this text the stacking sequence is specified starting from the top layer of the laminate.

Cross ply laminate–angle ply laminate: Two commonly encountered stacking sequences are shown in Figure 8.3. A cross-ply laminate is made by stacking alternating 90° and 0° plies. An angle ply laminate is obtained by alternate stacking of $\pm\theta$ plies.

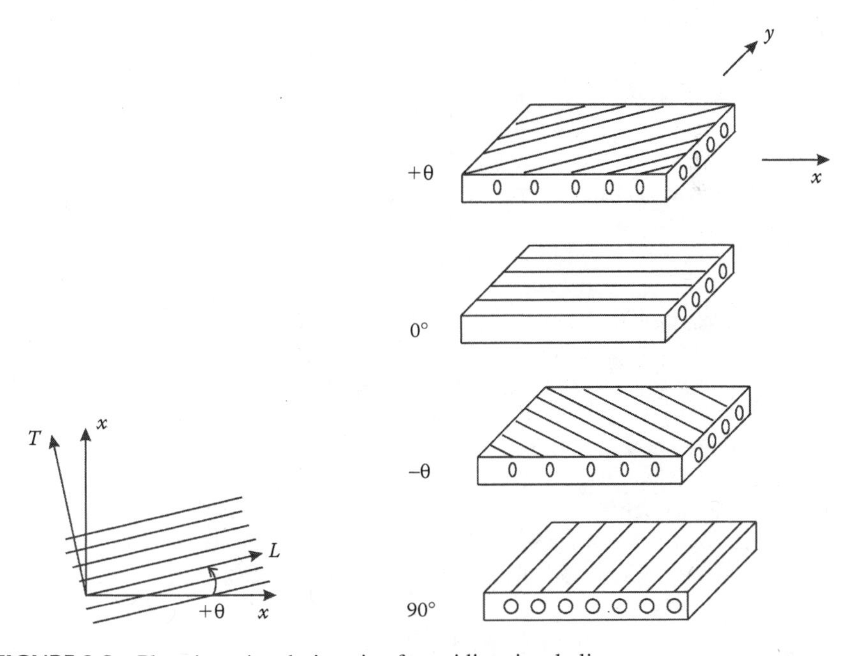

FIGURE 8.2 Ply orientation designation for unidirectional plies.

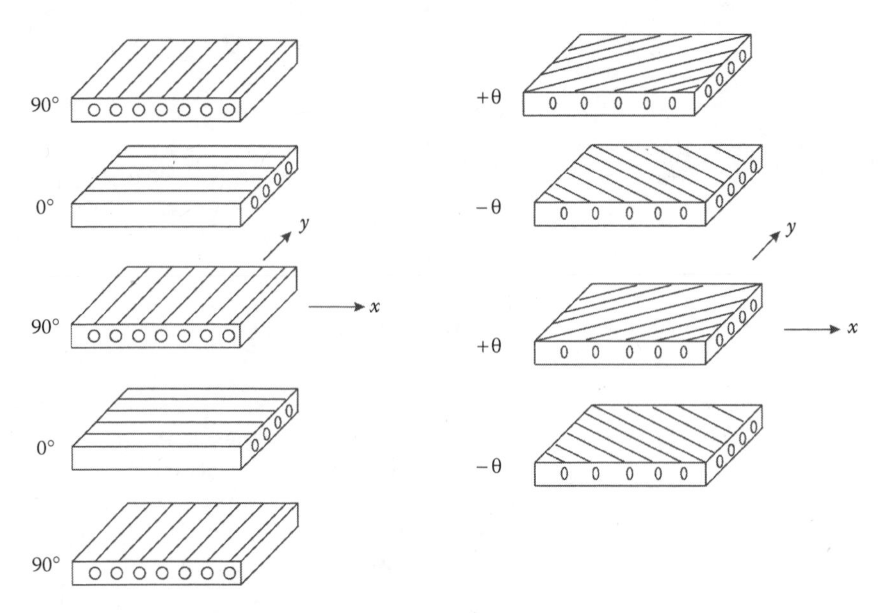

FIGURE 8.3 Cross-ply and angle-ply laminate configuration.

Laminate orientation code: The following code is designed to specify laminate composites as concisely as possible as related to: (a) the orientation of each ply relative to the laminate reference axis (x-axis), (b) the number of laminae at each orientation, and (c) the stacking sequence of the laminae in the laminate. Key features of the code are listed below.[1]

1. Each lamina is denoted by a number representing its orientation angle in degrees.
2. Individual adjacent laminae are separated in the code by a slash if their orientation angles are different. Adjacent laminae having the same angle are denoted by an appropriate numerical subscript.
3. The laminae are listed in sequence from one laminate face to the other, with brackets indicating the beginning and end of the code.
4. Laminates that are symmetric about their geometric midplane require a listing of only the plies located on one side of the plane of symmetry, with a subscript s placed after the closing bracket to denote symmetry. If the symmetric laminate has an odd number of layers then a bar is placed over the last number in the code that represents the layer that is shared by both halves of the laminate.
5. Repeating sequences of laminae within a laminate are called subsets and are enclosed in parentheses with a numerical subscript placed after the closing parenthesis to denote the number of repetitions.

The laminate code is demonstrated in Table 8.1 by showing selected laminates and indicating the code(s) designation for each example are shown. There is often more than

TABLE 8.1
Examples of Laminate Code

Laminate	Code	Laminate	Code	Laminate	Code
45°		45°		45°	
0°		0°		–45°	$[\pm 45/\mp 30/0]$
90°		45°	$[45/0/45/90_2/30]$	–30°	
45°		90°		30°	
0°	$[(45/0/90)_2]_s$	90°		0°	
90°	or	30°			
90°	$[(45/0/90)]_{2s}$				
0°				45°	
45°		90°		0°	
90°		0°		–60°	$[45/0/-60_2/30]$
0°		0°		–60°	
45°		45°	$[90/0_2/45]_s$	30°	
		45°			
		0°			
45°		0°		45°	
0°		90°		45°	
90°				–45°	$[45_2/-45_2/0]$
45°				–45°	
0°	$[(45/0/90)_4]$	0°	$[0/45/90]_s$	0°	
90°		45°			
45°	or	90°		45°	
0°	$[45/0/90]_4$	45°		–45°	
90°		0°		–45°	
45°				45°	$[\pm\mp\pm 45/0]$
0°				45°	
90°				–45°	
				0°	

one way to write the laminate code for a specified laminate. The more concise version of the code is preferred. Defined in Table 8.2 is the notation used for designating engineering elastic properties and strengths of fibers, matrix, and unidirectional lamina.

8.3 ANALYSIS OF COMPOSITE STRUCTURES

The mechanical analysis of fiber-reinforced composite materials, which follows, is approached at three different levels:

1. The micromechanics of a single layer (or ply) of composite material
2. The macromechanics of a single layer of the composite
3. The macromechanics of several layers making up a laminated composite panel structure

TABLE 8.2
Notation for Composite Lamina and Constituent Material Properties

Fiber Properties
E_f = Young's modulus
μ_f = Poisson's ratio
G_f = Shear modulus
V_f = Volume fraction of fibers = (volume of fibers/total volume of composite)
X_f = Tensile strength along the axis of fibers
Matrix Properties
E_m = Young's modulus
G_{LT} μ_m = Poisson's ratio
G_m = Shear modulus
V_m = Volume fraction of matrix = (volume of matrix/total volume of composite) = $(1 - V_f)$
X_m = Isotropic tensile strength
S_m = Shear strength
Macroscopic Properties of Unidirectional Lamina (subscripts L and T refer to the in-plane longitudinal and transverse directions of the lamina)
E_L = Young's modulus of ply along fiber direction
E_f = Young's modulus of ply transverse to fiber direction
μ_{LT} = Major Poisson's ratio
μ_{TL} = Minor Poisson's ratio
G_{LT} = Shear modulus of ply
V_f = Volume fraction of fibers = (volume of fibers/total volume of composite)
X_L = Tensile strength in direction of fibers
X'_L = Compressive strength in direction of fibers
X_T = Tensile strength transverse to direction of fibers
X'_T = Compressive strength transverse to direction of fibers
S = Shear strength in plane of lamina

In micromechanical analysis a representative volume element from the composite layer is replaced by an equivalent heterogeneous element in which the fibers are lumped together as shown in Figure 8.4. This simplification maintains the concept of heterogeneity, but focuses on average stresses in the fiber matrix. The micromechanical and physical properties of the matrix and fibers are then related to the average (or effective) macroscopic mechanical and physical properties of the composite layer. These are important relationships in understanding how the constituent materials used in the design of a composite material can be chosen to improve or alter, in a desired way, the macroscopic response of the composite layer.

In the macroscopic analysis of a single ply of material, a representative volume from the heterogeneous composite lamina is replaced by an equivalent homogeneous orthotropic element (Figure 8.4). The behavior of the equivalent element is the same

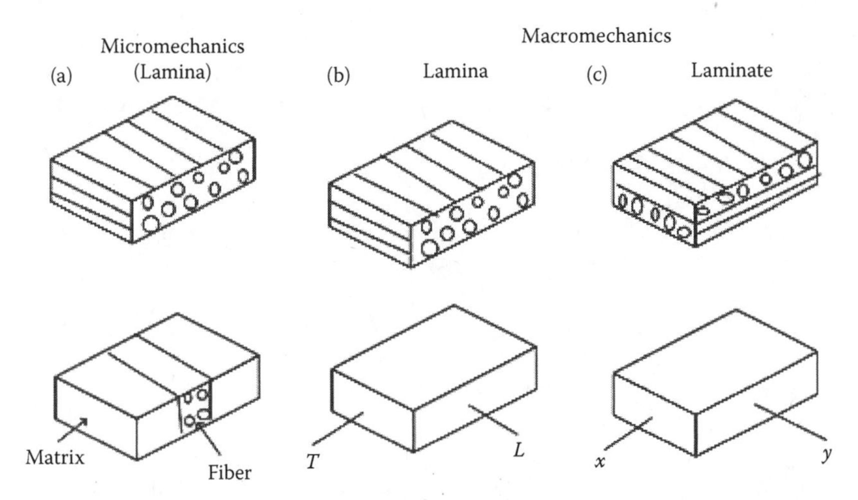

FIGURE 8.4 Representative volume elements for micromechanics and macromechanics analysis of composite materials. (a) Equivalent heterogeneous element, (b) equivalent homogeneous orthotropic element, and (c) equivalent homogeneous anisotropic element.

as the macroscopic response of the composite layer when subjected to the same thermomechanical loads. This analysis gives the stress–strain (or constitutive) laws of the plies in terms of the macroscopic properties of the lamina. These are important relationships in designing experimental tests for determining the macroscopic material properties of a single–ply, and also for relating ply stresses to ply failure when multiple plies are stacked to form a laminated structure.

Finally, in the macromechanics of a laminate, a representative volume made up of several layers of composite material, is replaced by an equivalent homogeneous anisotropic element as shown in Figure 8.4. Constitutive laws are then derived which relate the overall (macroscopic) deformation of the laminate to the applied stress resultants, or thermomechanical loads. These relationships provide a means of determining the strain distribution in a laminated composite structure when subjected to known loads and displacements. The strain distribution can then be used to calculate stresses in individual plies by using the stress–strain laws obtained from the macromechanics of a single layer.

In design with composite materials the engineer must be able to work back and forth through these three levels of analysis in order to select the best constituent materials for composite layers and to arrange the layer with the desired orientation and stacking sequence to achieve the optimum response of a laminated structure for a given loading environment.

8.3.1 Micromechanics of a Unidirectional Fiber-Reinforced Composite Layer (Lamina)

The objective of micromechanics analysis is to determine the elastic moduli (stiffness and compliance) of a composite material in terms of the elastic moduli of the constituent materials. An additional objective is to determine the strength of fiber-reinforced composites in terms of the fiber and the matrix strengths and their relative volumes.

The mechanics of materials approach is taken in developing micromechanical relationships between constituent and composite material properties. The basic assumptions inherent in this approach are the following:

1. The fibers are homogeneous, linearly elastic, isotropic, regularly spaced, and perfectly aligned
2. The matrix is homogeneous, linearly elastic, and isotropic
3. The lamina is considered macroscopically homogeneous and orthotropic, linearly elastic, and initially stress free
4. The bonding between fiber and matrix is assumed perfect with no voids existing between them

A representative heterogeneous volume is chosen from the composite lamina, which is the smallest volume element over which stresses and strains may be considered macroscopically uniform (Figure 8.4). To conceptually simplify the analysis, the fibers are shown lumped together occupying a portion of the representative element equivalent to the volume fraction of the fibers in the composite. This representation leads to equations which relate composite mechanical properties to constituent mechanical properties as a function of volume fraction.

The magnitude of the elastic moduli obtained for an anisotropic material will depend on the orientation of the coordinates used to describe the material elastic response. However, if the material elastic moduli are known for coordinates aligned with the principal material directions, then the elastic moduli for any other orientation can be determined through appropriate transformation equations. Thus, only four elastic constants are needed in order to fully characterize the in-plane macroscopic elastic response of an orthotropic lamina. The reference coordinates in the plane of the lamina are aligned with longitudinal axis (L) parallel to the fibers, and the transverse axis (T) perpendicular to the fibers. The engineering orthotropic elastic moduli of the lamina defined earlier are

E_L = longitudinal Young's modulus
E_T = transverse Young's modulus
G_{LT} = shear modulus
μ_{LT} = major Poisson's ratio

The micromechanics analysis which follows will use the mechanics of materials approach to obtain equations relating these composite elastic properties to the engineering elastic moduli of the fibers and matrix (E_F, G_f, μ_f, and E_m, G_m, μ_m). The key feature of the mechanics of materials approach is the simplifying assumption that planes perpendicular to the axis of the fibers remain plain during elastic deformation.

8.3.1.1 Determination of Apparent Longitudinal Young's Modulus

Consider first the case where the representative volume defined earlier is subjected to uniaxial loading in the longitudinal (fiber) direction only as shown in Figure 8.5. Then, the macroscopic (average) stresses acting on the element are

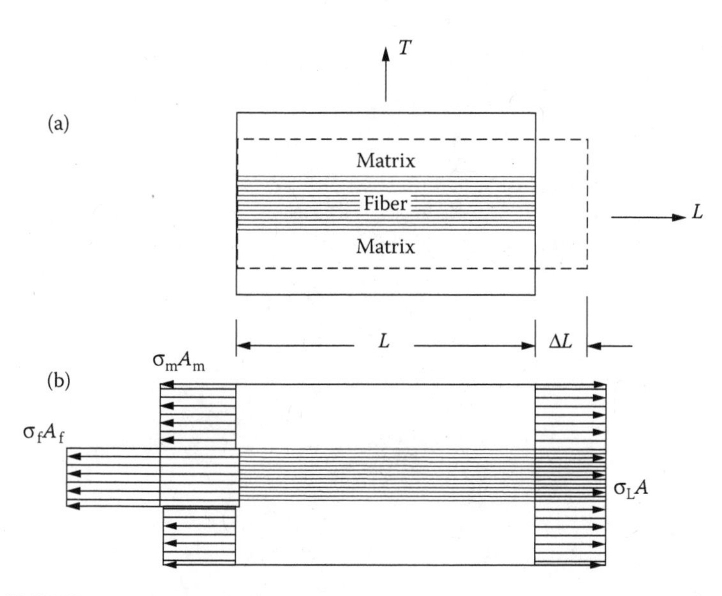

FIGURE 8.5 Mechanics of materials approach in viewing deformation of representative heterogeneous volume elements. (a) Deformation assumption and (b) force balance.

$$\sigma_L = \frac{P}{A}, \quad \sigma_T = 0, \quad \tau_{LT} = 0. \tag{8.1}$$

Now let the macroscopic strain of the element in the longitudinal direction be ε_L. Then, by definition the apparent (macroscopic) Young's modulus of the element, E_L, is given by the equation

$$E_L = \frac{\sigma_L}{\varepsilon_L}. \tag{8.2}$$

Now, the mechanics of materials deformation assumption states that the fibers and matrix elongate by the same amount in the longitudinal direction. Therefore, they experience the same longitudinal strain. That is,

$$\varepsilon_{fL} = \varepsilon_{mL} = \varepsilon_L = \frac{\Delta L}{L}. \tag{8.3}$$

Moreover, since the macroscopic transverse and shearing stresses, σ_T and τ_{LT}, are zero, the fibers and matrix must also be in a state of uniaxial longitudinal stress. Hence, Young's moduli for the fibers and matrix are related to their respective stresses and strains as follows:

$$E_f = \frac{\sigma_f}{\varepsilon_{fL}} = \frac{\sigma_f}{\varepsilon_L}, \quad E_m = \frac{\sigma_m}{\varepsilon_{mL}} = \frac{\sigma_m}{\varepsilon_L}. \tag{8.4}$$

Next, consider static equilibrium of the representative volume in Figure 8.5b. Balancing forces in the longitudinal direction gives

$$\sigma_L A = \sigma_f A_f + \sigma_m A_m. \tag{8.5}$$

Dividing through by the area A gives

$$\sigma_L = \sigma_f \frac{A_f}{A} + \sigma_m \frac{A_m}{A}. \quad (8.6)$$

Now, suppose the length of the representative element is L. Then the total volume of the element is $V = AL$, and $A_f L$ is the volume of that point of the element made up of fibers. Therefore, the fiber volume ratio is defined as

$$V_f = \frac{A_f L}{AL} = \frac{A_f}{A}. \quad (8.7)$$

Similarly, the matrix volume ratio is

$$V_m = \frac{A_m L}{AL} = \frac{A_m}{A}, \quad (8.8)$$

where $V_m = 1 - V_f$.

Finally, substituting Equations 8.7 and 8.8 into Equation 8.6 gives the result

$$\sigma_L = \sigma_f V_f + \sigma_m V_m = \sigma_f V_f + \sigma_m (1 - V_f). \quad (8.9)$$

Dividing throughout by ε_L and substituting Equations 8.2 and 8.4 into Equation 8.9 gives the desired equation for the macroscopic modulus E_L as a function of fiber volume fraction and the elastic moduli of the constituent materials making up the composite:

$$E_L = E_f V_f + E_m V_m = E_f V_f + E_m (1 - V_f). \quad (8.10)$$

Equation 8.10 is referred to as the rule of mixtures for the apparent Young's modulus of the composite in the direction of the fibers. The rule of mixtures predicts a simple linear variation of E_L ranging in value from E_m to E_f as the fiber volume fraction goes from 0 to 1 (Figure 8.6).

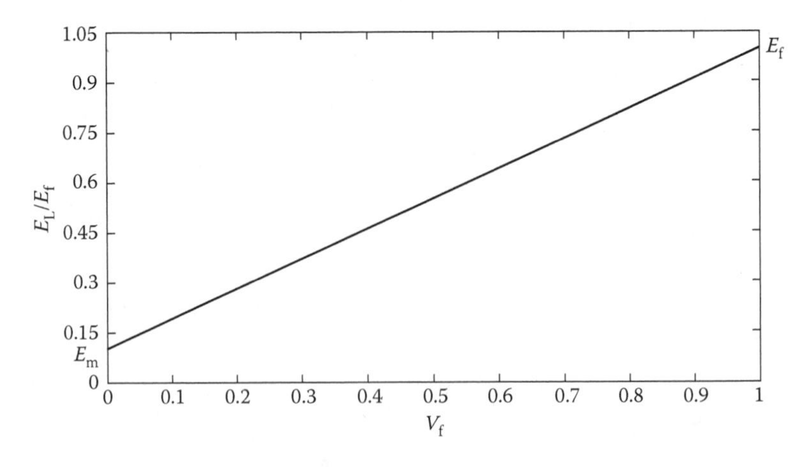

FIGURE 8.6 Variation of E_L with fiber volume fraction.

8.3.1.2 Determination of Major Poisson's Ratio of Unidirectional Lamina

The major Poisson's Ratio of an orthotropic lamina is defined as the negative of the ratio of transverse strain to longitudinal strain when the representative element is subjected to uniaxial loading in the longitudinal direction such that

$$\mu_{LT} = -\frac{\varepsilon_T}{\varepsilon_L}, \tag{8.11}$$

given σ_L not zero and $\sigma_T = 0$, and $\tau_{LT} = 0$.

Since, under conditions of Equation 8.11 the fibers and matrix are also subjected to uniaxial loading in the longitudinal direction, the respective fiber and matrix Poisson ratios are defined as

$$\mu_f = -\frac{\varepsilon_{fT}}{\varepsilon_{fL}} = -\frac{\varepsilon_{fT}}{\varepsilon_L}, \quad \mu_m = -\frac{\varepsilon_{mT}}{\varepsilon_{mL}} = -\frac{\varepsilon_{mT}}{\varepsilon_L}. \tag{8.12}$$

Now, the transverse strain of the representative element (Figure 8.7) is given by the expression

$$\varepsilon_T = \frac{\Delta W}{W}, \tag{8.13}$$

where

$$\Delta W = \Delta W_f + \Delta W_m = \varepsilon_{fT} W_f + \varepsilon_{mT} W_m \tag{8.14}$$

Substitution of Equation 8.14 into Equation 8.13 and noting that

$$\frac{W_f}{W} = V_f \quad \text{and} \quad \frac{W_m}{W} = V_m$$

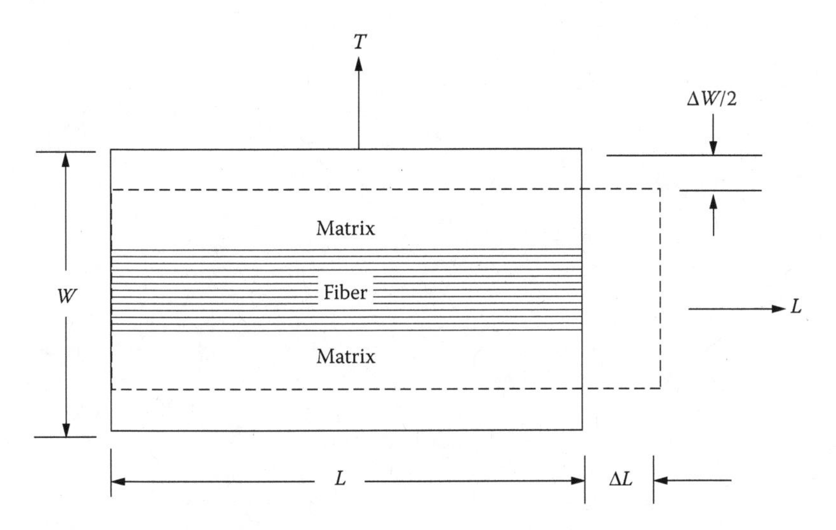

FIGURE 8.7 Transverse strain due to uniaxial longitudinal stress.

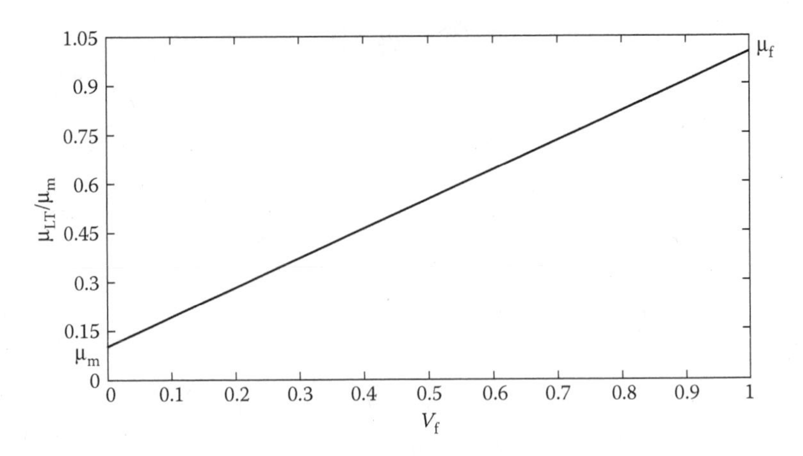

FIGURE 8.8 Variation of μ_{LT} with fiber volume fraction.

yields the result

$$\varepsilon_T = \varepsilon_{fT} V_f + \varepsilon_{mT} V_m. \tag{8.15}$$

Putting the definitions of Equations 8.11 and 8.12 into Equation 8.15 gives the desired expression relating the macroscopic major Poisson's ratio to the fiber and matrix Poisson's ratios as a function of the fiber volume fraction:

$$\mu_{LT} = \mu_f V_f + \mu_m V_m = \mu_f V_f + \mu_m (1 - V_f). \tag{8.16}$$

Hence, the major Poisson's ratio also obeys the rule of mixtures, predicting a linear variation in μ_{LT} ranging in value from μ_m to μ_f as the fiber volume fraction goes from 0 to 1 (Figure 8.8).

8.3.1.3 Apparent Transverse Young's Modulus

In order to find the transverse Young's Modulus, the representative element is subjected to a transverse uniaxial stress σ_T with $\sigma_L = 0$, and $\tau_{LT} = 0$, as shown in Figure 8.9. Since σ_T is the only macroscopic stress acting on the element, then the apparent transverse Young's modulus is

$$E_T = \frac{\sigma_T}{\varepsilon_T}. \tag{8.17}$$

The total transverse elongation of the element W due to σ_T is the combined transverse stretching of fibers and matrix:

$$\Delta W = \Delta W_f + \Delta W_m,$$

or

$$W \varepsilon_{TW} = \varepsilon_{fT} W_f + \varepsilon_{mT} W_m,$$

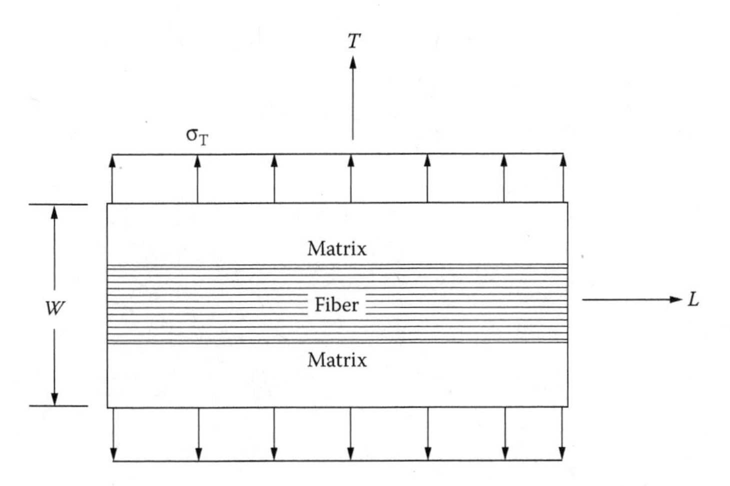

FIGURE 8.9 Representative element subjected to uniaxial stress in the transverse direction.

or

$$\varepsilon_T = \varepsilon_{fT} V_f + \varepsilon_{mT} V_m. \tag{8.18}$$

In order to relate the fiber and matrix strains to the constituent material properties, note that transverse normal stresses across the fiber–matrix interface must be continuous, that is,

$$\sigma_T = \sigma_{fT} = \sigma_{mT}. \tag{8.19}$$

Assuming that fibers and matrix are also subjected to transverse stress only,* then Hooke's Law for these materials gives

$$\varepsilon_{fT} = \frac{\sigma_T}{E_f}, \quad \varepsilon_{mT} = \frac{\sigma_T}{E_m}. \tag{8.20}$$

Substituting the results of Equation 8.20 and Equation 8.17 into Equation 8.18 gives the desired result

$$\frac{1}{E_T} = \frac{V_f}{E_f} + \frac{V_m}{E_m},$$

or

$$\frac{E_T}{E_m} = \frac{1}{V_f(E_m/E_f) + V_m}. \tag{8.21}$$

Equation 8.21 is referred to the inverse rule of mixtures, which predicts a nonlinear variation in apparent transverse modulus E_T that ranges from E_m to E_f as the

* This simplifying assumption violates the mechanics-of-materials postulate made earlier that fibers and matrix stretch the same amount in the fiber direction (i.e., that planes remain plane). A more exact analysis would include the effect of longitudinal stresses in the fibers and matrix (in Equation 8.20) required to satisfy this deformation constraint.

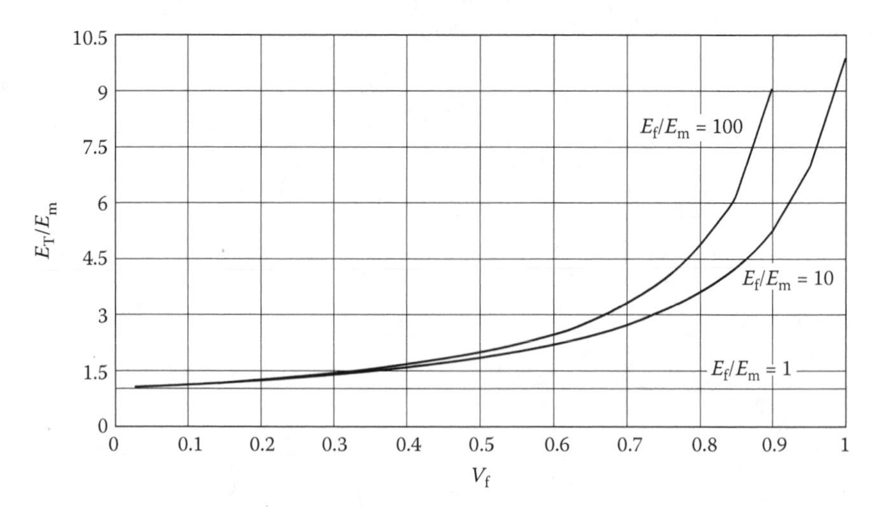

FIGURE 8.10 Variation of E_T with fiber volume fraction.

fiber volume fraction goes from 0 to 1. This variation is plotted in Figure 8.10 for cases where the fiber-to-matrix stiffness $E_f/E_m = 10$ and 100.

Adding 40% by volume of reinforcing fibers to the matrix for the cases plotted would result in only about 50% increase in transverse stiffness of the composite, but would increase longitudinal stiffness by ~400% (Figure 8.6). Hence, the addition of fiber reinforcement to a polymer matrix can have a very pronounced effect on its stiffness and directional properties, particularly in the fiber direction.

8.3.1.4 Apparent Shear Modulus

Consider finally the case where the representative element is subjected to a state of pure shear τ_{LT} with $\sigma_T = 0$ and with $\sigma_L = 0$, as shown in Figure 8.11. Equilibrium considerations require that the fiber and matrix portion of the area also be subjected to pure shear

$$\tau_f = \tau_m = \tau_{LT}. \tag{8.22}$$

Hooke's Law for the constituent materials reduces to

$$\gamma_f = \frac{\tau_f}{G_f} = \frac{\tau_{LT}}{G_f}, \quad \gamma_m = \frac{\tau_m}{G_m} = \frac{\tau_{LT}}{G_m}. \tag{8.23}$$

Similarly, the orthotropic macroscopic response of the representative element is

$$\gamma_{LT} = \frac{\tau_{LT}}{G_{LT}}. \tag{8.24}$$

The total shearing deformation is the sum of the fiber and matrix shearing deformation shown in Figure 8.11, that is,

$$\Delta = \Delta_f + \Delta_m. \tag{8.25}$$

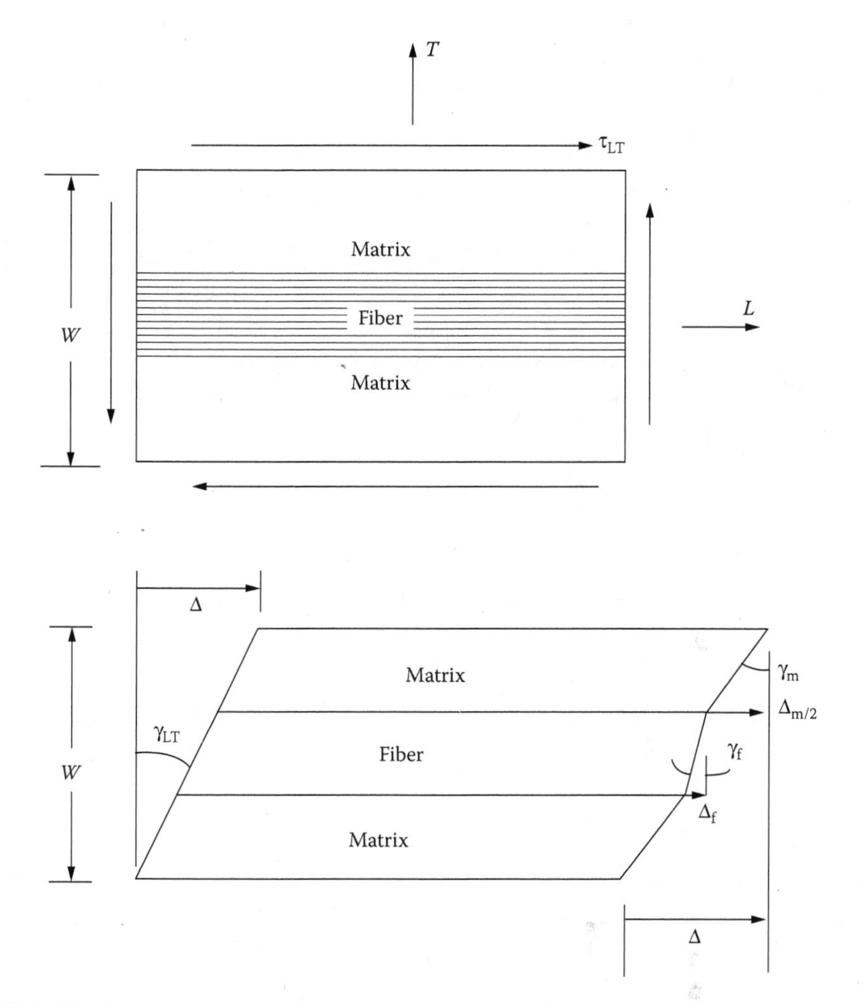

FIGURE 8.11 Representative element subjected to pure shear, τ_{LT}.

For small shearing strains, Equation 8.25 can be written as

$$\gamma_{LT}W = \gamma_f W_f + \gamma_m W_m.$$

Dividing by W, this equation becomes

$$\gamma_{LT} = \gamma_f V_f + \gamma_m V_m. \tag{8.26}$$

Finally, substituting Equations 8.23 and 8.24 into Equation 8.26 gives the desired equation relating the macroscopic shear modulus G_{LT} to the constituent shear moduli as a function of fiber volume fraction:

$$\frac{1}{G_{LT}} = \frac{V_f}{G_f} + \frac{V_m}{G_m},$$

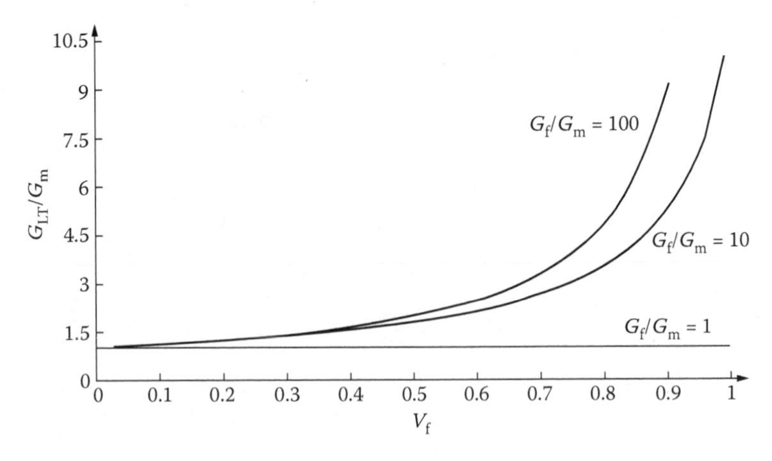

FIGURE 8.12 Variation of G_{LT} with fiber volume fraction.

or

$$\frac{G_{LT}}{G_m} = \frac{1}{V_f(G_m/G_f)+V_m}. \tag{8.27}$$

The shear modulus, like the transverse Young's modulus, obeys the inverse rule of mixtures. Equation 8.27 is plotted in Figure 8.12.

8.3.1.5 Summary of Results from Micromechanics Analysis of Lamina Elastic Moduli

The four main sets of equations relating the macroscopic engineering elastic properties of the lamina to the constituent elastic properties as a function of fiber volume fraction are grouped together below:

Longitudinal modulus (rule of mixtures): $E_L = E_f V_f + E_m V_m$

Major Poisson's ratio (rule of mixtures): $\mu_{LT} = \mu_f V_f + \mu_m V_m$

Transverse modulus (inverse rule): $\frac{1}{E_T} = \frac{V_f}{E_f} + \frac{V_m}{E_m}$

Shear modulus (inverse rule): $\frac{1}{G_{LT}} = \frac{V_f}{G_f} + \frac{V_m}{G_m}$

For most practical fiber-reinforced polymeric composites, the fibers are much stiffer than the matrix:

$$E_f \gg E_m \quad \text{and} \quad G_f \gg G_m.$$

Also, practical values of the fiber volume ratio range between

$$0.1 < V_f < 0.6.$$

Thus, it is apparent that the composite longitudinal Young's modulus and major Poisson's ratio are strongly influenced by the fiber elastic response whereas the composite transverse Young's modulus and shear modulus behavior is dominated by the matrix elastic response, except at large fiber volume fractions.

8.3.1.6 Prediction of Tensile Strength in Fiber Direction

The prediction of macromechanical strength properties of a unidirectional fiber-reinforced composite lamina using micromechanics models has met with less success than the elastic moduli predictions of the earlier sections. The structural designer will most likely rely primarily on results from mechanical tests that measure the macromechanical strength properties of the composite lamina directly. Nevertheless, it is instructive to look at the micromechanics model for tensile strength in the fiber direction of a lamina to gain a better understanding of how the composite functions.

Consider the typical case of a unidirectional fiber-reinforced composite lamina in which the fibers are brittle (do not deform plastically), and the matrix can sustain much larger longitudinal strains before rupture than the fibers. A schematic showing the deformation and strength behavior of the lamina is shown in Figure 8.13. As a longitudinal tensile load is slowly applied to the ply, the composite lamina deforms in the following stages.

Stage 1: Both fibers and matrix deform together elastically. During this stage of deformation the macroscopic ply stress is determined by the rule of mixtures, derived earlier in Equation 8.9:

$$\sigma_L = \sigma_f V_f + \sigma_m (1 - V_f).$$

Stage 2: When the longitudinal strain in the composite reaches the ultimate strain of the fibers, ε_{fu}, the fibers will break.

Stage 3: If the fiber volume ratio is small ($V_f < V_{min}$), the remaining intact matrix material will be sufficient to carry the applied load (strength of the ply is matrix controlled), and the load can be increased further until the stress in the matrix reaches its ultimate strength X_{mu}. In this case, the longitudinal strength of the ply, X_L, will be given by

$$X_L = X_{mu} (1 - V_f). \tag{8.28}$$

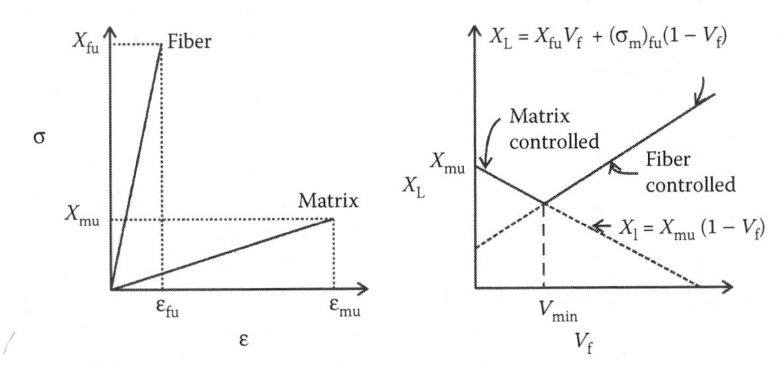

FIGURE 8.13 Schematic showing deformation and strength behavior based on micromechanics of brittle fiber lamina.

Stage 4: If the fiber volume ratio is large ($V_f > V_{min}$), when the fibers break, the remaining matrix material cannot sustain the applied load and the ply will fail (strength of the ply is fiber controlled). In this case the longitudinal strength of the ply can be derived from Equation 8.9, which yields

$$X_L = X_{fu}V_f + (\sigma_m)_{fu}(1 - V_f), \tag{8.29}$$

where $(\sigma_{mu})_{fu}$ is the stress in the matrix when subjected to a strain equal to the ultimate fiber strain (Figure 8.13a). Since both fiber and matrix are linear elastic up to failure, then

$$X_{fu} = E_f\varepsilon_{fu}, \quad (\sigma_m)_{fu} = E_m\varepsilon_{fu}. \tag{8.30}$$

Putting these results back into Equation 8.29 gives the ply longitudinal strength in terms of fiber strength and elastic moduli of the fiber and matrix:

$$X_L = X_{fu}\left[V_f + \frac{E_m}{E_f}(1 - V_f)\right]. \tag{8.31}$$

The fiber volume ratio V_{min} occurs at the intersection of the matrix-controlled and fiber-controlled strength curves shown in Figure 8.13b. Hence, setting Equations 8.28 and 8.31 as equal, and solving for $V_f = V_{min}$, gives the result

$$V_{min} = \frac{X_{mu} - (E_m/E_f)X_{fu}}{X_{mu} + (1 - (E_m/E_f))X_{fu}}. \tag{8.32}$$

In Figure 8.13b, the addition of fiber reinforcement does not increase the longitudinal strength of the composite lamina until the volume ratio of the fibers reaches a critical value V_{cr}, which is larger than V_{min} determined in Equation 8.32. An expression for this critical value is obtained by setting $X_L = X_{mu}$ in Equation 8.31 and solving for $V_f = V_{cr}$

$$V_{cr} = \frac{X_{mu} - (E_m/E_f)X_{fu}}{(1 - (E_m/E_f))X_{fu}}. \tag{8.33}$$

The fibers begin to act effectively as a reinforcing element to strengthen the composite only when the fiber volume ratio exceeds V_{cr}, which is important to understand when designing composite materials. For most polymeric composites V_{cr} and V_{cr} are very small, of the order of 1%.[1]

8.3.2 Macromechanics of a Unidirectional Fiber-Reinforced Composite Layer of Lamina

In macromechanical analysis of a fiber-reinforced lamina we replace the heterogeneous composite material by an equivalent homogeneous anisotropic lamina. A representative volume element is taken from this equivalent layer for purposes of analyses (Figure 8.4c). The stresses acting on the element are "smeared out" to show that the average, or macroscopic state of stress acting on the element. The objective of macromechanical analysis is to establish stress–strain relations (Hooke's Law) for the material in terms of compliance and stiffness using averaged (apparent) mechanical properties of the lamina. The most useful form of the stress–strain relations involves engineering constants. The engineering constants are particularly helpful in describing the composite behavior because they are defined in terms of simple tests that can be performed in the laboratory. For example, with an isotropic material, values for both Young's modulus, E and Poisson's ratio, μ can be determined from a single tensile test. These values can be used to calculate the shear modulus G for the isotropic material from the relation

$$G = \frac{E}{2(1+\mu)}.$$

These three engineering constants are sufficient to define the stress–strain relationships of an isotropic material. Hence, one simple test provides all material properties needed to completely define the mechanical response of a linear elastic isotropic material. A larger number of tests will be needed to obtain the engineering constants required to define the macroscopic elastic response of an anisotropic lamina.

8.3.2.1 Stress–Strain Relationships for Isotropic Materials

A review of Hooke's Law for isotropic materials before we consider anisotropic materials is instructive. For an isotropic linear elastic material at constant temperature, Hooke's Law in Cartesian coordinates (1, 2, 3) has the familiar form:

$$\varepsilon_1 = \frac{\sigma_1 - \mu(\sigma_2 + \sigma_3)}{E}, \quad \varepsilon_2 = \frac{\sigma_2 - \mu(\sigma_3 + \sigma_1)}{E}, \quad \varepsilon_3 = \frac{\sigma_3 - \mu(\sigma_1 + \sigma_2)}{E},$$

$$\gamma_{12} = \frac{\tau_{12}}{G}, \quad \gamma_{23} = \frac{\tau_{23}}{G}, \quad \gamma_{31} = \frac{\tau_{31}}{G}. \tag{8.34}$$

When studying composite materials, it is convenient to express the strain–stress equations using matrix algebra. This approach will be used throughout the remainder of the chapter. Equations 8.34 can thus be written as

$$\begin{Bmatrix} \varepsilon_1 \\ \varepsilon_2 \\ \varepsilon_3 \\ \gamma_{12} \\ \gamma_{23} \\ \gamma_{31} \end{Bmatrix} = \begin{bmatrix} \frac{1}{E} & -\frac{\mu}{E} & -\frac{\mu}{E} & 0 & 0 & 0 \\ -\frac{\mu}{E} & \frac{1}{E} & -\frac{\mu}{E} & 0 & 0 & 0 \\ -\frac{\mu}{E} & -\frac{\mu}{E} & \frac{1}{E} & 0 & 0 & 0 \\ 0 & 0 & 0 & \frac{1}{G} & 0 & 0 \\ 0 & 0 & 0 & 0 & \frac{1}{G} & 0 \\ 0 & 0 & 0 & 0 & 0 & \frac{1}{G} \end{bmatrix} \begin{Bmatrix} \sigma_1 \\ \sigma_2 \\ \sigma_3 \\ \tau_{12} \\ \tau_{23} \\ \tau_{31} \end{Bmatrix} = [S] \begin{Bmatrix} \sigma_1 \\ \sigma_2 \\ \sigma_3 \\ \tau_{12} \\ \tau_{23} \\ \tau_{31} \end{Bmatrix}. \tag{8.35}$$

Equations 8.35 are further simplified if we consider the case of plane stress, where the stresses are applied in the (1, 2) plane. Equation 8.35 then becomes

$$\begin{Bmatrix} \varepsilon_1 \\ \varepsilon_2 \\ \gamma_{12} \end{Bmatrix} = \begin{bmatrix} \frac{1}{E} & -\frac{\mu}{E} & 0 \\ -\frac{\mu}{E} & \frac{1}{E} & 0 \\ 0 & 0 & \frac{1}{G} \end{bmatrix} \begin{Bmatrix} \sigma_1 \\ \sigma_2 \\ \tau_{12} \end{Bmatrix} = [S] \begin{Bmatrix} \sigma_1 \\ \sigma_2 \\ \tau_{12} \end{Bmatrix}. \tag{8.36}$$

$[S]$ is called the compliance matrix, relating the in-plane strain vector $\{\varepsilon\}$ to the applied in-plane stress $\{\sigma\}$. Inversion of Equations 8.36 gives

$$\begin{Bmatrix} \sigma_1 \\ \sigma_2 \\ \tau_{12} \end{Bmatrix} = \begin{bmatrix} \frac{E}{1-\mu^2} & \frac{\mu E}{1-\mu^2} & 0 \\ \frac{\mu E}{1-\mu^2} & \frac{E}{1-\mu^2} & 0 \\ 0 & 0 & G \end{bmatrix} \begin{Bmatrix} \varepsilon_1 \\ \varepsilon_2 \\ \gamma_{12} \end{Bmatrix} = [Q] \begin{Bmatrix} \varepsilon_1 \\ \varepsilon_2 \\ \gamma_{12} \end{Bmatrix}, \tag{8.37}$$

where $[Q]$ is called the stiffness matrix. It should be obvious that the stiffness and compliance matrices are related, one being the inverse of the other that is,

$$[Q] = [S]^{-1} \quad \text{and} \quad [S] = [Q]^{-1}. \tag{8.38}$$

These matrices are symmetric, that is, terms on opposite sides of the downward sloping diagonal of the matrix are identical. Equation 8.36 through Equation 8.38 confirm experience that for isotropic materials there is no coupling between normal stress and shear strain, nor between shearing stress and normal strain. However, we will find that decoupling between normal stress and shear response is not representative of anisotropic materials.

8.3.2.2 Anisotropic Materials: Contracted Notation

In an anisotropic material, coupling can exist between *all* stresses and *all* strains. Hence, the generalized Hooke's law for a linearly elastic anisotropic material, expressed using matrix algebra, has no zeros appearing in the compliance matrix:

$$\begin{Bmatrix} \varepsilon_1 \\ \varepsilon_2 \\ \varepsilon_3 \\ \varepsilon_4 \\ \varepsilon_5 \\ \varepsilon_6 \end{Bmatrix} = \begin{bmatrix} S_{11} & S_{21} & S_{31} & S_{41} & S_{51} & S_{61} \\ S_{12} & S_{22} & S_{32} & S_{42} & S_{52} & S_{62} \\ S_{13} & S_{23} & S_{33} & S_{43} & S_{53} & S_{63} \\ S_{14} & S_{24} & S_{34} & S_{44} & S_{54} & S_{64} \\ S_{15} & S_{25} & S_{35} & S_{45} & S_{55} & S_{65} \\ S_{16} & S_{26} & S_{36} & S_{46} & S_{56} & S_{66} \end{bmatrix} \begin{Bmatrix} \sigma_1 \\ \sigma_2 \\ \sigma_3 \\ \sigma_4 \\ \sigma_5 \\ \sigma_6 \end{Bmatrix}. \tag{8.39}$$

This expression can be written in Cartesian tensor notation and matrix notation, respectively, as

TABLE 8.3
Contracted Notation

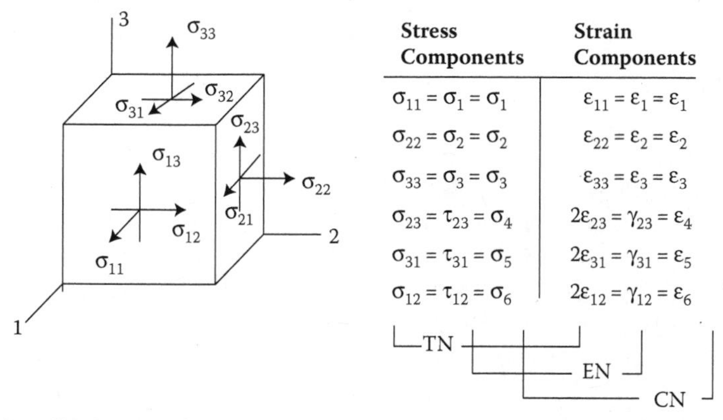

Stress Components	Strain Components
$\sigma_{11} = \sigma_1 = \sigma_1$	$\varepsilon_{11} = \varepsilon_1 = \varepsilon_1$
$\sigma_{22} = \sigma_2 = \sigma_2$	$\varepsilon_{22} = \varepsilon_2 = \varepsilon_2$
$\sigma_{33} = \sigma_3 = \sigma_3$	$\varepsilon_{33} = \varepsilon_3 = \varepsilon_3$
$\sigma_{23} = \tau_{23} = \sigma_4$	$2\varepsilon_{23} = \gamma_{23} = \varepsilon_4$
$\sigma_{31} = \tau_{31} = \sigma_5$	$2\varepsilon_{31} = \gamma_{31} = \varepsilon_5$
$\sigma_{12} = \tau_{12} = \sigma_6$	$2\varepsilon_{12} = \gamma_{12} = \varepsilon_6$

Note: CN, Contracted notation; EN, Engineering notation; TN, Tensor notation.

$$\varepsilon_{ij} = S_{ij}\sigma_j \quad \text{and} \quad \{\varepsilon\} = [s]\{\sigma\}.$$

Contracted notation (CN) has been introduced in the equations, where it is convenient for computer solution to use single-digit subscripts to designate stress and strain terms. The relationships between (1) tensor or elasticity notation, (2) engineering notation (EN), and (3) CN are defined in Table 8.3.

For an anisotropic lamina under plane stress, Equation 8.39 reduces to

$$\begin{Bmatrix} \varepsilon_1 \\ \varepsilon_2 \\ \gamma_{12} \end{Bmatrix} = \begin{bmatrix} S_{11} & S_{12} & S_{16} \\ S_{21} & S_{22} & S_{26} \\ S_{16} & S_{26} & S_{66} \end{bmatrix} \begin{Bmatrix} \sigma_1 \\ \sigma_2 \\ \tau_{12} \end{Bmatrix}. \tag{8.40}$$

Note that EN is used for the strain and stress vectors, while CN is retained in the compliance matrix in Equation 8.40. Also, advantage has been taken of symmetry of the compliance matrix to reduce the number of independent compliance terms, that is, we assumed

$$S_{16} = S_{61}, \quad S_{26} = S_{62}, \quad \text{and} \quad S_{12} = S_{21}.$$

The unusual mixture of notation in Equation 8.40 is typical of what has evolved as standard notation in engineering tests and literature on composite materials. With this notation S_{16} and S_{26} represent coupling between shear strain and normal stress in the 1, 2 plane. The term S_{66} couples shearing strain to shearing stress in the 1, 2 plane.

8.3.2.3 Orthotropic Lamina: Hooke's Law in Principal Material Coordinates

In the macromechanical analysis of a thin unidirectionally reinforced composite lamina, a representative volume element of the composite is replaced by an equivalent homogeneous orthotropic element as shown in Figure 8.14. The orientation of

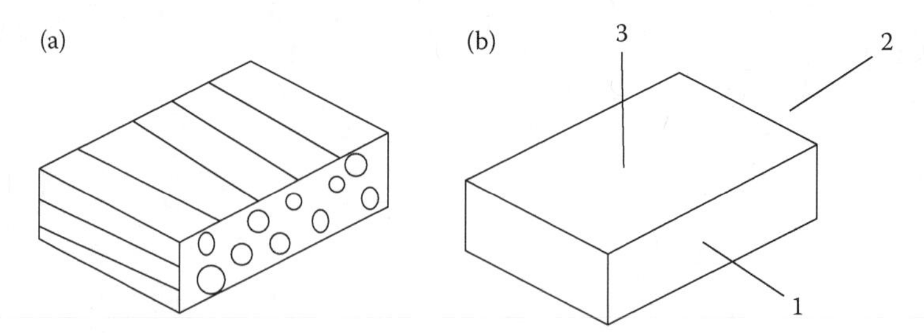

FIGURE 8.14 Representative volume element from a unidirectional lamina. (a) Micromechanics, representative heterogeneous element and (b) macromechanics, equivalent homogeneous orthotropic element.

the element is chosen so that the Cartesian coordinates in the plane of the lamina 1, 2 are directed along and perpendicular to the fibers, respectively. These are referred to as the principal material axes of the lamina. When the stress and strain of an orthotropic lamina are given in principal material coordinates, there is no coupling between normal stresses and shearing strains. Moreover, in the analysis of thin composite structures each ply of the composite can be assumed in a state of plane stress. Thus, Hooke's Law for the orthotropic lamina is obtained by setting S_{16} and S_{26} equal to 0 in Equation 8.40 to give

$$\begin{Bmatrix} \varepsilon_1 \\ \varepsilon_2 \\ \gamma_{12} \end{Bmatrix} = \begin{bmatrix} S_{11} & S_{12} & 0 \\ S_{21} & S_{22} & 0 \\ 0 & 0 & S_{66} \end{bmatrix} \begin{Bmatrix} \sigma_1 \\ \sigma_2 \\ \tau_{12} \end{Bmatrix}. \tag{8.41}$$

Alternatively, designating the coordinates directed along and perpendicular to the fibers as L and T, Equation 8.41 becomes

$$\begin{Bmatrix} \varepsilon_L \\ \varepsilon_T \\ \gamma_{LT} \end{Bmatrix} = \begin{bmatrix} S_{11} & S_{12} & 0 \\ S_{21} & S_{22} & 0 \\ 0 & 0 & S_{66} \end{bmatrix} \begin{Bmatrix} \sigma_L \\ \sigma_T \\ \tau_{LT} \end{Bmatrix}. \tag{8.42}$$

Equation 8.42 can be inverted by premultiplying both sides of the equation by $[S]^{-1}$, and noting that $[S]^{-1}[S] = 1$ and $[S]^{-1} = [Q]$. Thus,

$$\begin{Bmatrix} \sigma_L \\ \sigma_T \\ \tau_{LT} \end{Bmatrix} = \begin{bmatrix} Q_{11} & Q_{12} & 0 \\ Q_{21} & Q_{22} & 0 \\ 0 & 0 & Q_{66} \end{bmatrix} \begin{Bmatrix} \varepsilon_L \\ \varepsilon_T \\ \gamma_{LT} \end{Bmatrix}. \tag{8.43}$$

It is apparent from Equations 8.42 and 8.43 that four material elastic properties (compliance or stiffness) are needed to characterize the in-plane behavior of a linear elastic orthotropic lamina. It is convenient to define these material properties in terms of measured engineering constants (Young's moduli, E_L and E_T, shear modulus G_{LT}, and Poisson's ratios μ_{LT} and μ_{TL}). The longitudinal Young's modulus is determined with a tensile specimen cut from the lamina with loading axis parallel to the fibers a $\theta = 0°$ tensile specimen) (see Figure 8.2 for the definition of θ). The slope of the resulting stress–strain curve $\sigma_L(\varepsilon_L)$ defines the longitudinal modulus E_L:

$$E_L \equiv \frac{\sigma_L}{\varepsilon_L}. \tag{8.44}$$

The major Poisson's ratio μ_{LT} is obtained from the same tensile specimen by measuring transverse and longitudinal strains and taking their ratio, such that

$$\mu_{LT} \equiv -\frac{\varepsilon_T}{\varepsilon_L}. \tag{8.45}$$

If for the longitudinal tensile test, the stresses $\sigma_L \neq 0$ with $\sigma_T = 0$ and $\tau_{LT} = 0$, are substituted into Equation 8.42 and solved for strains, the results are

$$\varepsilon_L = S_{11}\sigma_L, \quad \varepsilon_T = S_{21}\sigma_L, \quad \gamma_T = 0. \tag{8.46}$$

Combining Equation 8.46 with Equation 8.44 and Equation 8.45 gives

$$S_{11} = \frac{1}{E_L} \quad \text{and} \quad S_{21} = -\frac{\mu_{TL}}{E_L}. \tag{8.47}$$

In a similar manner, the transverse Young's modulus E_T and minor Poisson's ratio μ_{TL} can be determined from a simple tensile test where the loading axis of the test specimen is taken perpendicular to the fiber direction of the lamina (a 90° tensile specimen). Proceeding with an analysis similar to that above will give

$$S_{22} = \frac{1}{E_T} \quad \text{and} \quad S_{12} = -\frac{\mu_{LT}}{E_T}. \tag{8.48}$$

The major and minor Poisson's ratios are not independent material properties, because $S_{12} = S_{21}$, then

$$\frac{\mu_{LT}}{E_L} = \frac{\mu_{TL}}{E_T}. \tag{8.49}$$

In practice, the major Poisson's ratio is *measured* using a longitudinal tensile specimen (Equation 8.45), and the minor Poisson's ratio is *computed* using Equation 8.49.

Finally, a pure shear stress is applied to a 0° test specimen, such that

$$\tau_{LT} = \tau, \quad \sigma_L = 0, \quad \text{and} \quad \sigma_T = 0. \tag{8.50}$$

A plot of shear stress versus shear strain will have a slope equal to the shear modulus. Hence, putting the results of Equation 8.50 into Equation 8.42 gives an expression relating compliance to the measured engineering shear modulus,

$$S_{66} = \frac{1}{G_{LT}}. \tag{8.51}$$

Summarizing the results obtained above, the compliance matrix for an orthotropic lamina can be expressed in terms of engineering elastic constants as

$$[S]=\begin{bmatrix} \frac{1}{E_L} & -\frac{\mu_{TL}}{E_T} & 0 \\ -\frac{\mu_{LT}}{E_L} & \frac{1}{E_T} & 0 \\ 0 & 0 & \frac{1}{G_{LT}} \end{bmatrix}. \tag{8.52}$$

The experimental procedures for determining engineering elastic constants for unidirectional composite lamina will be discussed in Section 8.3. The stiffness matrix can be expressed in terms of engineering elastic constants by taking the inverse of Equation 8.52:

$$[Q]=[S]^{-1}=\begin{bmatrix} \frac{E_L}{1-\mu_{LT}\mu_{TL}} & \frac{\mu_{LT}E_T}{1-\mu_{LT}\mu_{TL}} & 0 \\ \frac{\mu_{TL}E_L}{1-\mu_{LT}\mu_{TL}} & \frac{E_T}{1-\mu_{LT}\mu_{TL}} & 0 \\ 0 & 0 & G_{LT} \end{bmatrix}. \tag{8.53}$$

Hence, the four measured engineering constants E_L, E_T, G_{LT}, and μ_{LT} are sufficient to describe the linear elastic macromechanical behavior of a thin orthotropic composite lamina subjected to plane stress.

8.3.2.4 Stress (Strain) Relationships for Off-Axis Orientation

The orthotropic stress and strain relationships of Equations 8.42 and 8.43 were defined in principal material directions, for which there is no coupling between extension and shear behavior. However, the coordinates natural to the solution of the problem generally will not coincide with the principal directions of orthotropy. For example, consider a simply supported beam manufactured from an angle-ply laminate. The principal material coordinates of each ply of the laminate make angles $\pm\theta$ relative to the axis of the beam. In the beam problem stresses and strains are usually defined in the beam coordinate system (x, y), which is off-axis relative to the lamina principal axes (L, T).

Recall that the Hooke's law for an isotropic material remains the same regardless of the orientation of the stress element being considered. For example, if the state of plane stress is known at a point in the (x, y) plane (i.e., σ_x, σ_y, τ_{xy} is known at a point), the state of strain at the point (ε_x, ε_y, γ_{xy}) can be determined using Equation 8.34. Similarly, if we know the stress $\sigma_{x'}$, $\sigma_{y'}$, $\tau_{xy'}$ relative to a new set of coordinates $(x', y'$) rotated relative to the (x, y) axes, the strains $\varepsilon_{x'}$, $\varepsilon_{y'}$, $\gamma_{xy'}$ can be determined using the same equations. This "sameness" ("iso" comes from the Greek word meaning the same) does not hold for anisotropic materials. Suppose the state of plane stress at a point in an orthotropic lamina has been determined for the (x, y) coordinates, where the lamina principal material coordinates L and T are rotated relative to the reference axes x and y (see Figure 8.15). Now decoupling between shearing strain and normal stress

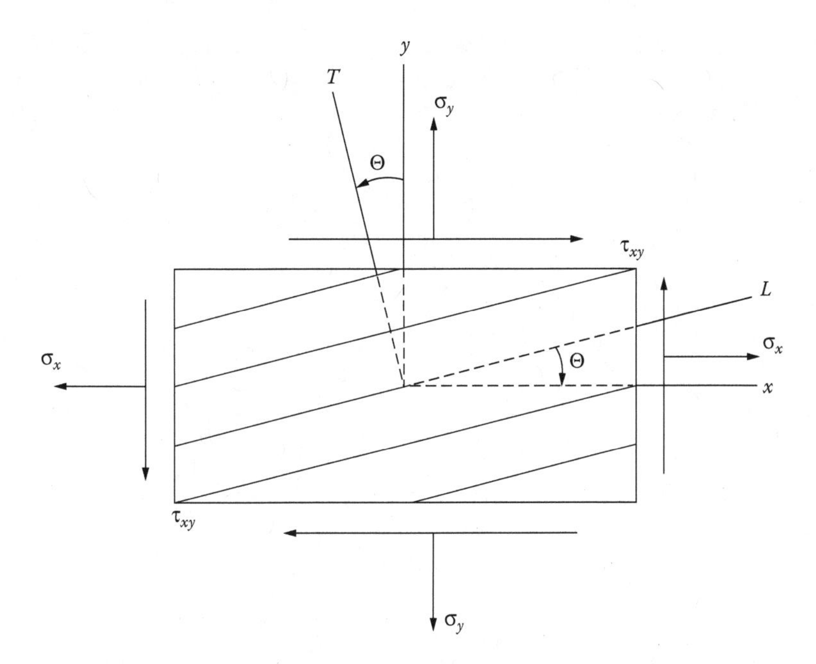

FIGURE 8.15 Positive rotation of principal material axes (L, T) from arbitrary reference axes (x, y).

no longer exists, and the stress (strain) relations for the arbitrary reference axes take on the following forms:

$$\begin{Bmatrix} \varepsilon_x \\ \varepsilon_y \\ \gamma_{xy} \end{Bmatrix} = \begin{bmatrix} \bar{S}_{11} & \bar{S}_{12} & \bar{S}_{16} \\ \bar{S}_{21} & \bar{S}_{22} & \bar{S}_{26} \\ \bar{S}_{16} & \bar{S}_{26} & \bar{S}_{66} \end{bmatrix} \begin{Bmatrix} \sigma_x \\ \sigma_y \\ \tau_{xy} \end{Bmatrix}, \tag{8.54}$$

or alternatively,

$$\begin{Bmatrix} \sigma_x \\ \sigma_y \\ \tau_{xy} \end{Bmatrix} = \begin{bmatrix} \bar{Q}_{11} & \bar{Q}_{12} & \bar{Q}_{16} \\ \bar{Q}_{21} & \bar{Q}_{22} & \bar{Q}_{26} \\ \bar{Q}_{16} & \bar{Q}_{26} & \bar{Q}_{66} \end{bmatrix} \begin{Bmatrix} \varepsilon_x \\ \varepsilon_y \\ \gamma_{xy} \end{Bmatrix}. \tag{8.55}$$

The matrices $[S]$ and $[Q]$ appearing in Equations 8.42 and 8.43 are called the reduced compliance and stiffness matrices, while $[\bar{S}]$ and $[\bar{Q}]$ in Equations 8.54 and 8.55 are known as the transformed reduced compliance and stiffness matrices. In general, the lamina mechanical properties, from which compliance and stiffness can be calculated, are determined experimentally in principal material directions and provided to the designer in a material specification sheet by the manufacturer. Thus, a method is needed for transforming stress–strain relations from off-axis to the principal material coordinate system.

To begin the analysis, recall the stress transformation equations from elementary mechanics of materials. In a first course on mechanics of materials the transformation equations are related to Mohr's circle of stress, which leads to graphical procedures for stress transformation. In the present analysis the transformation equations are written using matrix algebra, consistent with the approach taken in texts dealing with composite materials.

Consider the case where the principal material axes, L and T, are rotated counterclockwise through an angle θ relative to the body reference axes (x, y) as shown in Figure 8.15. The stress transformation equations relating known stresses in the x, y coordinate system to stresses in the L, T coordinate system are

$$\begin{Bmatrix} \sigma_L \\ \sigma_T \\ \tau_{LT} \end{Bmatrix} = \begin{bmatrix} \cos^2\theta & \sin^2\theta & 2\sin\theta\cos\theta \\ \sin^2\theta & \cos^2\theta & -2\sin\theta\cos\theta \\ -\sin\theta\cos\theta & \sin\theta\cos\theta & \cos^2\theta - \sin^2\theta \end{bmatrix} \begin{Bmatrix} \sigma_x \\ \sigma_y \\ \tau_{xy} \end{Bmatrix}. \tag{8.56}$$

Note in particular that the transformation equations have nothing to do with material properties or stress–strain relations, but are merely a rotation of stresses.

Similarly, the strain transformation equations are

$$\begin{Bmatrix} \varepsilon_L \\ \varepsilon_T \\ \frac{1}{2}\gamma_{LT} \end{Bmatrix} = \begin{bmatrix} \cos^2\theta & \sin^2\theta & 2\sin\theta\cos\theta \\ \sin^2\theta & \cos^2\theta & -2\sin\theta\cos\theta \\ -\sin\theta\cos\theta & \sin\theta\cos\theta & \cos^2\theta - \sin^2\theta \end{bmatrix} \begin{Bmatrix} \varepsilon_x \\ \varepsilon_y \\ \frac{1}{2}\gamma_{xy} \end{Bmatrix}. \tag{8.57}$$

Note that strains transform with the same transformation equation as stresses provided that the tensor definition of shear strain is used (see Table 8.3).

The transformation matrix appearing in Equations 8.56 and 8.57 can be written in a simpler form as follows:

$$[T] = \begin{bmatrix} m^2 & n^2 & 2nm \\ n^2 & m^2 & -2nm \\ -nm & nm & m^2 - n^2 \end{bmatrix}, \tag{8.58}$$

where $m = \cos\theta$ and $n = \sin\theta$. The inverse of $[T]$ is

$$[T]^{-1} = \begin{bmatrix} m^2 & n^2 & -2nm \\ n^2 & m^2 & 2nm \\ nm & -nm & m^2 - n^2 \end{bmatrix}. \tag{8.59}$$

The transformation of stress is commonly written as

$$\begin{Bmatrix} \sigma_L \\ \sigma_T \\ \tau_{LT} \end{Bmatrix} = [T] \begin{Bmatrix} \sigma_x \\ \sigma_y \\ \tau_{xy} \end{Bmatrix}, \quad \text{or} \quad \begin{Bmatrix} \sigma_x \\ \sigma_y \\ \tau_{xy} \end{Bmatrix} = [T]^{-1} \begin{Bmatrix} \sigma_L \\ \sigma_T \\ \tau_{LT} \end{Bmatrix}, \tag{8.60}$$

and transformation of strain as

$$\begin{Bmatrix} \varepsilon_L \\ \varepsilon_T \\ \frac{1}{2}\gamma_{LT} \end{Bmatrix} = [T] \begin{Bmatrix} \varepsilon_x \\ \varepsilon_y \\ \frac{1}{2}\gamma_{xy} \end{Bmatrix}, \quad \text{or} \quad \begin{Bmatrix} \varepsilon_x \\ \varepsilon_y \\ \frac{1}{2}\gamma_{xy} \end{Bmatrix} = [T]^{-1} \begin{Bmatrix} \varepsilon_L \\ \varepsilon_T \\ \frac{1}{2}\gamma_{LT} \end{Bmatrix}. \tag{8.61}$$

Finally, we introduce Reuter's matrix, which is defined as

$$[R] = \begin{bmatrix} 1 & 0 & 0 \\ 0 & 1 & 0 \\ 0 & 0 & 2 \end{bmatrix}, \quad \text{and} \quad [R]^{-1} = \begin{bmatrix} 1 & 0 & 0 \\ 0 & 1 & 0 \\ 0 & 0 & \frac{1}{2} \end{bmatrix}. \tag{8.62}$$

Reuter's matrix allows us to relate tensor and engineering strain vectors as follows:

$$\begin{Bmatrix} \varepsilon_L \\ \varepsilon_T \\ \gamma_{LT} \end{Bmatrix} = [R] \begin{Bmatrix} \varepsilon_L \\ \varepsilon_T \\ \frac{1}{2}\gamma_{LT} \end{Bmatrix}, \quad \text{or} \quad \begin{Bmatrix} \varepsilon_L \\ \varepsilon_T \\ \frac{1}{2}\gamma_{LT} \end{Bmatrix} = [R]^{-1} \begin{Bmatrix} \varepsilon_L \\ \varepsilon_T \\ \frac{1}{2}\gamma_{LT} \end{Bmatrix}. \tag{8.63}$$

The purpose of introducing the concepts developed in Equation 8.56 through Equation 8.63 has been to provide the tools needed to present a straightforward derivation of the transformed reduced compliance and stiffness matrices [S] and [Q] based on matrix algebra. The derivation makes use of the following sequence of operations to obtain stress–strain relations in the reference (x, y) coordinate system:

$$\begin{Bmatrix} \varepsilon_x \\ \varepsilon_y \\ \gamma_{xy} \end{Bmatrix} = [R] \begin{Bmatrix} \varepsilon_x \\ \varepsilon_y \\ \frac{1}{2}\gamma_{xy} \end{Bmatrix} = [R][T]^{-1} \begin{Bmatrix} \varepsilon_L \\ \varepsilon_T \\ \frac{1}{2}\gamma_{LT} \end{Bmatrix} = [R][T]^{-1}[R]^{-1} \begin{Bmatrix} \varepsilon_L \\ \varepsilon_T \\ \gamma_{LT} \end{Bmatrix}. \tag{8.64a}$$

$$= [R][T]^{-1}[R]^{-1}[S] \begin{Bmatrix} \sigma_L \\ \sigma_T \\ \tau_{LT} \end{Bmatrix} = [R][T]^{-1}[R]^{-1}[S][T] \begin{Bmatrix} \sigma_x \\ \sigma_y \\ \tau_{xy} \end{Bmatrix} = [\bar{S}] \begin{Bmatrix} \sigma_x \\ \sigma_y \\ \tau_{xy} \end{Bmatrix}. \tag{8.64b}$$

$[\bar{S}]$ is the transformed reduced compliance matrix needed for off-axis representation of stresses and strains in an orthotropic lamina. In a similar fashion, an equation for $[\bar{Q}]$, the transformed reduced stiffness matrix, can be derived:

$$\begin{Bmatrix} \sigma_x \\ \sigma_y \\ \tau_{xy} \end{Bmatrix} = [T]^{-1}[Q][R][T][R]^{-1} \begin{Bmatrix} \varepsilon_x \\ \varepsilon_y \\ \gamma_{xy} \end{Bmatrix} = [\bar{Q}] \begin{Bmatrix} \varepsilon_x \\ \varepsilon_y \\ \gamma_{xy} \end{Bmatrix} \tag{8.65}$$

If the matrix terms defining [S] and [Q] in Equations (8.64) and (8.65) are combined, then the following individual transformed compliance and stiffness components are obtained that are needed in the off-axis stress–strain relations (Equations 8.54 and 8.55):

$$\begin{aligned}
\bar{S}_{11} &= S_{11}\cos^4\theta + S_{22}\sin^4\theta + (2S_{12}+S_{66})\cos^2\theta\sin^2\theta \\
\bar{S}_{12} &= (S_{11}+S_{22}-S_{66})\cos^2\theta\sin^2\theta + S_{12}(\cos^4\theta+\sin^4\theta) \\
\bar{S}_{22} &= S_{11}\sin^4\theta + S_{22}\cos^4\theta + (2S_{12}+S_{66})\cos^2\theta\sin^2\theta \\
\bar{S}_{66} &= 4(S_{11}+S_{22}-2S_{12})\cos^2\theta\sin^2\theta + S_{66}(\cos^2\theta-\sin^2\theta)^2 \\
\bar{S}_{16} &= (2S_{11}-S_{66}-2S_{12})\cos^3\theta\sin\theta - (2S_{22}-2S_{12}-S_{66})\cos\theta\sin^3\theta \\
\bar{S}_{26} &= (2S_{11}-S_{66}-2S_{12})\cos\theta\sin^3\theta - (2S_{22}-2S_{12}-S_{66})\cos^3\theta\sin\theta
\end{aligned} \tag{8.66}$$

and

$$\begin{aligned}
\bar{Q}_{11} &= Q_{11}\cos^4\theta + Q_{22}\sin^4\theta + (2Q_{12}+4Q_{66})\cos^2\theta\sin^2\theta \\
\bar{Q}_{12} &= (Q_{11}+Q_{22}-4Q_{66})\cos^2\theta\sin^2\theta + Q_{12}(\cos^4\theta+\sin^4\theta) \\
\bar{Q}_{22} &= Q_{11}\sin^4\theta + Q_{22}\cos^4\theta + (2Q_{12}+4Q_{66})\cos^2\theta\sin^2\theta \\
\bar{Q}_{66} &= (Q_{11}+Q_{22}-2Q_{12}-2Q_{66})\cos^2\theta\sin^2\theta + Q_{66}(\cos^4\theta+\sin^4\theta) \\
\bar{Q}_{16} &= (Q_{11}-2Q_{66}-Q_{12})\cos^3\theta\sin\theta - (Q_{22}-Q_{12}-2Q_{66})\cos\theta\sin^3\theta \\
\bar{Q}_{26} &= (Q_{11}-2Q_{66}-Q_{12})\cos\theta\sin^3\theta - (Q_{22}-Q_{12}-2Q_{66})\cos^3\theta\sin\theta,
\end{aligned} \tag{8.67}$$

where

$$S_{11}=\frac{1}{E_L},\quad S_{12}=-\frac{\mu_{LT}}{E_L}=-\frac{\mu_{TL}}{E_T},\quad S_{22}=\frac{1}{E_T},\quad S_{66}=\frac{1}{G_{LT}} \tag{8.68}$$

$$Q_{11}=\frac{E_L}{(1-\mu_{LT}\mu_{TL})},\quad Q_{12}=\frac{\mu_{LT}E_T}{(1-\mu_{LT}\mu_{TL})}=\frac{\mu_{TL}E_L}{(1-\mu_{LT}\mu_{TL})},$$

$$Q_{22}=\frac{E_T}{(1-\mu_{LT}\mu_{TL})},\quad \text{and}\quad Q_{66}=G_{LT}.$$

The transformed reduced compliance and stiffness matrices [S] and [Q] relate off-axis stress and strain in an orthotropic lamina. Since these matrices are fully populated, the material responds to off-axis stresses as though it was fully anisotropic. Some consequences of the anisotropic nature of a unidirectional lamina are discussed in the following.

The way in which a fully anisotropic material deforms when subjected to loading is much different than the deformation characteristics of isotropic materials. Consider, for example, a tensile specimen cut from a unidirectional (orthotropic) lamina. If the loading axis is parallel to one of the material axes (i.e., parallel or perpendicular to the fiber direction) the specimen will deform without shear distortion as shown in Figure 8.16a. However, for off-axis loading the fibers tend to line up with the direction of loading resulting in a combination of normal and shear deformation as shown in Figure 8.16b. Hence, there is coupling between extension and shear as indicated by the stress–strain relations of Equations 8.54 and 8.55.

Example 8.1

Consider an example of off-axis loading as in Figure 8.15, where

$$\sigma_x = -3.5\,\text{MPa}, \quad \sigma_y = 7.0\,\text{MPa}, \quad \tau_{xy} = -1.4\,\text{MPa}, \quad \text{and} \quad \theta = 60°. \tag{8.68a}$$

The mechanical properties are given as

$$E_L = 14\,\text{GPa}, \quad E_T = 3.5\,\text{GPa}, \quad G_{LT} = 4.2\,\text{GPa}, \quad \mu_{LT} = 0.4, \quad \mu_{TL} = 0.1. \tag{8.68b}$$

What is required are the values for the strains ε_x, ε_y, and γ_{xy}.
The procedure to solve this problem is to

1. Transform the stresses to the L, T coordinates using Equation 8.56, that is,

$$\{\sigma\}_{LT} = [T]\{\sigma\}_{xy}. \tag{8.68c}$$

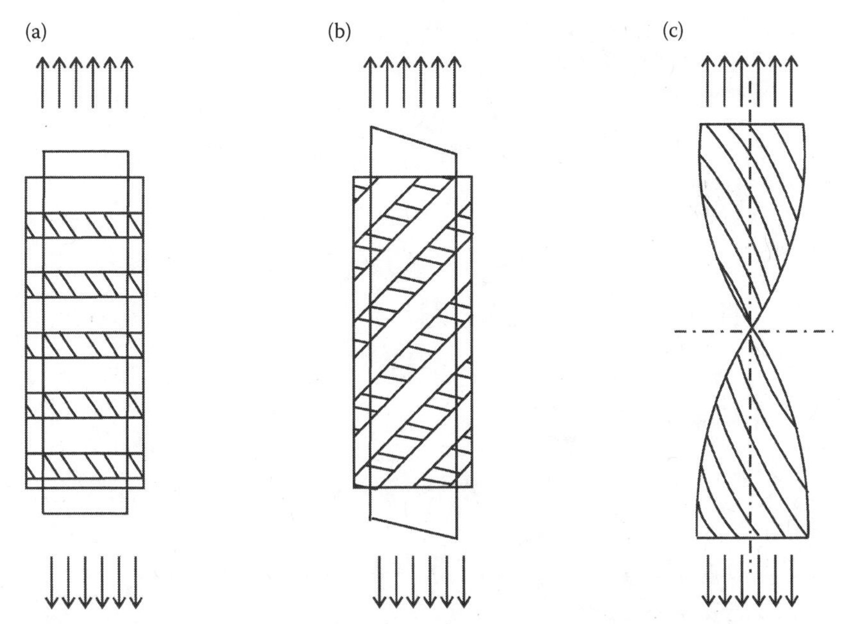

FIGURE 8.16 Deformation of orthotropic lamina subjected to simple tension: (a) load is parallel to principal material direction, (b) off-axis loading, (c) nonsymmetric angle-ply laminated made from identical unidirectional laminae.

2. Apply the constitutive Equation 8.42 to obtain strains in the L, T directions:

$$\{\varepsilon\}_{LT} = [S]\{\sigma\}_{LT}. \tag{8.68d}$$

3. Transform the strains back to the x, y directions using Equation 8.61, that is,

$$\{\varepsilon\}_{xy} = [T]^{-1}\{\varepsilon\}_{LT}. \tag{8.68e}$$

Following this procedure yields $m = \cos(60) = 0.5$, $n = \sin(60) = 0.866$, and

$$[T] = \begin{bmatrix} m^2 & n^2 & 2nm \\ n^2 & m^2 & -2nm \\ -nm & nm & m^2 - n^2 \end{bmatrix} = \begin{bmatrix} 0.250 & 0.750 & 0.866 \\ 0.750 & 0.250 & -0.866 \\ -0.433 & 0.433 & -0.500 \end{bmatrix}. \tag{8.68f}$$

Substitution into Equation 8.68c gives

$$\sigma_L = 3.16\,\text{MPa}, \quad \sigma_T = 0.34\,\text{MPa}, \quad \tau_{LT} = 5.24\,\text{MPa}. \tag{8.68g}$$

Substitution of the mechanical properties Equation 8.68b into Equation 8.52 gives

$$[S] = \begin{bmatrix} 0.0714 & -0.0286 & 0 \\ -0.0286 & 0.2857 & 0 \\ 0 & 0 & 0.2381 \end{bmatrix} \left(\frac{1}{\text{GPa}} \right). \tag{8.68h}$$

Substitution of Equation 8.68h into 8.68d gives

$$\varepsilon_L = 216(10^{-6}), \quad \varepsilon_T = 6.9(10^{-6}), \quad \text{and} \quad \gamma_{LT} = 1248(10^{-6}). \tag{8.68i}$$

The inverse of $[T]$ from Equation 8.59 is

$$[T]^{-1} = \begin{bmatrix} m^2 & n^2 & -2nm \\ n^2 & m^2 & 2nm \\ nm & -nm & m^2 - n^2 \end{bmatrix} = \begin{bmatrix} 0.250 & 0.750 & -0.866 \\ 0.750 & 0.250 & 0.866 \\ 0.433 & -0.433 & -0.500 \end{bmatrix}. \tag{8.68j}$$

Substitution of (8.68j) into (8.68e) gives

$$\varepsilon_x = -481(10^{-6}), \quad \varepsilon_y = 704(10^{-6}), \quad \text{and} \quad \gamma_{xy} = -443(10^{-4}). \tag{8.68k}$$

8.4 EXPERIMENTAL DETERMINATION OF ENGINEERING ELASTIC CONSTANTS

A tensile specimen cut from an angle-ply laminate, made by bonding two unidirectional laminae at angles $\pm\theta$ to the load direction, has deformation characteristics as shown amplified in Figure 8.16c. Fibers in adjacent plies of the laminate tend to line up with the load, causing out-of-plane twisting of the laminate. The possibility of out-of-plane deformation when subjected to in-plane loading must be taken into account in design of laminated composite structures. The anisotropic behavior of composites must also be taken into account in the design of experiments to determine lamina elastic moduli.

The stress (strain) laws (Equations 8.42, 8.43, 8.54, and 8.55), and the related equations expressing stiffness and compliance components in terms of engineering constants (Equation 8.66 through Equation 8.68 provide the basis for designing experiments to determining the elastic and strength properties of an orthotropic lamina). Material characterization tests require the design of experiments in which a test coupon taken from the lamina is subjected to a known (or calculated) state of uniform stress, either uniaxial or biaxial. Strain gage rosettes that are bonded to the test specimen measure the strains at a point in the region of uniform stress. The known values of stress and strain components are then substituted into the stress–strain equations, and values for the material constants are determined.

The procedure described above is straightforward in principle. However, in practice, great care must be taken in the test fixture design to assure that applied loads cause a uniform state of stress in the test specimen. Two types of tests that have been developed for this purpose include: (1) the simple tensile test for uniaxial states of stress, (2) the thin-walled tube subjected to combined torsion and internal pressure, for biaxial states of stress. Some theoretical aspects of the simple tensile test are developed in the sample problem which follows. For a more detailed discussion on experimental procedures for characterizing the material properties of composite materials, see Calsson and Pipes[2] and Whitney et al.[3]

Example 8.2

Consider a tensile specimen cut from a unidirectional lamina with the material direction at an angle to the load direction as shown in Figure 8.17. The grips are designed so that a pure tensile load P is transmitted to the test specimen (i.e., grips are free to rotate to accommodate shear distortion). Two strain gages (A) and (B) are bonded to the unloaded specimen with orientations shown in Figure 8.17. Derive an equation for the "apparent" Young's modulus E_x of the specimen in terms of engineering constants of the lamina (E_L, E_T, μ_{LT}, and G_{LT}).

SOLUTION

Let x be the loading axis of the test specimen, and y be the coordinate axis perpendicular to the load axis as shown in Figure 8.17. The central portion of the specimen, away from the loading grips, is assumed subjected to a state of uniform uniaxial tensile stress:

$$\{\sigma\} = \begin{Bmatrix} \sigma_x \\ 0 \\ 0 \end{Bmatrix}. \tag{8.69}$$

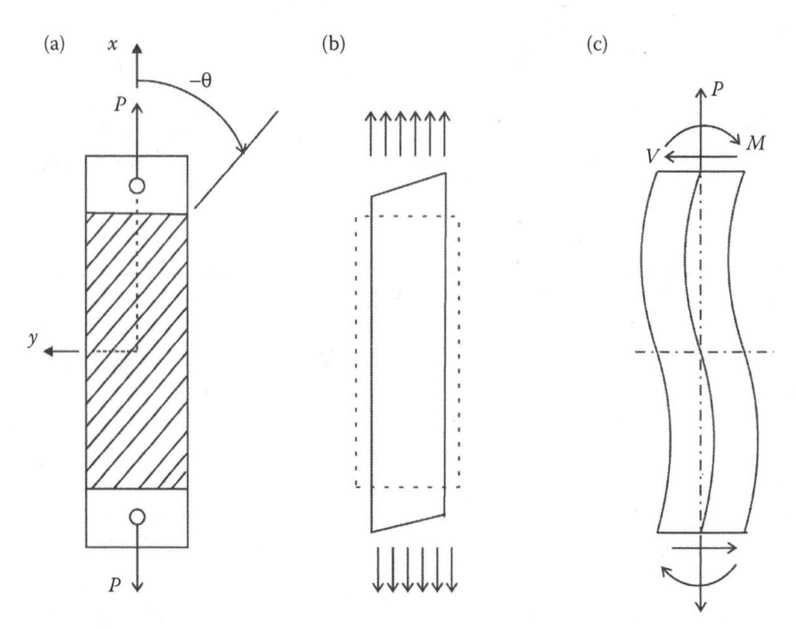

FIGURE 8.17 Influence of end constraints in the testing of anisotropic lamina. (a) Off-axis loading, (b) uniform state of stress, and (c) effect of clamped ends.

where $\sigma_x = P/A$, P is the applied load, and A is the specimen cross-sectional area. The strain is related to the stress using Equation 8.54:

$$\begin{Bmatrix} \varepsilon_A \\ \varepsilon_B \\ \gamma_{AB} \end{Bmatrix} = \begin{Bmatrix} \varepsilon_x \\ \varepsilon_y \\ \gamma_{xy} \end{Bmatrix} = \begin{bmatrix} \bar{S}_{11} & \bar{S}_{12} & \bar{S}_{16} \\ \bar{S}_{21} & \bar{S}_{22} & \bar{S}_{26} \\ \bar{S}_{16} & \bar{S}_{26} & \bar{S}_{66} \end{bmatrix} \begin{Bmatrix} \frac{P}{A} \\ 0 \\ 0 \end{Bmatrix}. \tag{8.70}$$

Solving explicitly for the strains ε_A and ε_B, and making use of Equations 8.66 and 8.68, gives the following results in terms of engineering constants:

$$\varepsilon_A = \left(\frac{\cos^4\theta}{E_L} + \frac{\sin^4\theta}{E_T} + \left(\frac{1}{G_{LT}} - \frac{2\mu_{LT}}{E_L} \right) \left(\cos^2\theta \sin^2\theta \right) \right) \frac{P}{A} \tag{8.71}$$

$$\varepsilon_B = \left(\left(\frac{1}{E_L} + \frac{1}{E_T} - \frac{1}{G_{LT}} \right) \cos^2\theta \sin^2\theta - \left(\frac{\mu_{LT}}{E_L} \right) \left(\cos^4\theta + \sin^4\theta \right) \right) \frac{P}{A}.$$

The apparent Young's modulus E_x is, by definition, the ratio $\sigma_x(\varepsilon_x)$ (slope of the stress–strain plot). Hence, for an off-axis test specimen subjected to uniaxial stress in the x-direction, we obtain the following equation for the modulus:

$$\frac{1}{E_x} = \frac{\varepsilon_A A}{P} = \left(\frac{\cos^4\theta}{E_L} + \frac{\sin^4\theta}{E_T} + \left(\frac{1}{G_{LT}} - 2\frac{\mu_{LT}}{E_L} \right) \cos^2\theta \sin^2\theta \right). \tag{8.72}$$

Special Case 1: $\theta = 0°$

If the loading axis is chosen to coincide with the fiber direction (0° specimen), Equation 8.71 is reduced to

$$\varepsilon_A = \frac{1}{E_L}\frac{P}{A} \quad \text{or} \quad E_L = \frac{P}{\varepsilon_A A},$$
$$\varepsilon_B = \frac{-\mu_{LT}}{E_L}\frac{P}{A} \quad \text{or} \quad \mu_{LT} = \frac{-\varepsilon_B}{\varepsilon_A}. \tag{8.73}$$

Hence, the longitudinal modulus E_L and the major Poisson's ratio μ_{LT} of a unidirectional lamina can be determined from a single experiment using an appropriately instrumented 0° tensile specimen. Two strain gages should be mounted back-to-back on opposite sides of the specimen to compensate for bending. Then if the pair of gages for either direction A or B are mounted in opposite arms of a Wheatstone bridge, the output will be proportional to the average strain ε_A or ε_B to be used in the prior equations.

Special Case 2: $\theta = 90°$

In this case a 90° specimen is subjected to uniaxial tension. Equation 8.71a simplifies to give

$$\varepsilon_A = \frac{1}{E_T}\frac{P}{A} \quad \text{or} \quad E_T = \frac{P}{\varepsilon_A A}. \tag{8.74}$$

Hence, the transverse modulus E_T can be determined from a single tensile test on a 90° specimen using strain gages mounted parallel to the axis of loading.

The values obtained for E_L, μ_{LT}, and E_T in these two tests can then be used to calculate a value for the minor Poisson's ratio μ_{TL} using Equation 8.49.

Special Case 3: $\theta = 45°$

In this case, a 45° tensile specimen is selected, and a pair of strain gages is mounted in the direction of the applied load. Solving Equation 8.72 for the shear modulus G_{LT} gives the result

$$\frac{1}{G_{LT}} = \frac{4}{E_x} - \frac{1}{E_L} - \frac{1}{E_T} + \frac{2\mu_{LT}}{E_L}. \tag{8.75}$$

In Equation 8.75, E_x is the slope of the stress (strain) plot ($\sigma_x(\varepsilon_x)$) obtained using experimental data from the 45° specimen. Hence, provided E_L, E_L, and μ_{LT} have already been determined from 0° and 90° tensile tests, the 45° test specimen provides a means to experimentally determine the shear modulus G_{LT} of the lamina.

The load in the first two cases given in Example 8.2 is applied along principal material directions. Hence, there is no coupling between extensional and shear deformation. As a result, it is a relatively simple task to design test fixtures to apply pure axial loading resulting in a uniform state of uniaxial stress in the section of the tensile specimen between the loading grips. However, coupling occurs between shear and extensional deformation for the 45° tensile specimen due to off-axis loading. The grips through which the tensile load is applied resist shear deformation and, hence, introduce in-plane couples and shear forces at the ends of the specimen as shown in Figure 8.17c. As a consequence of the end couples and shear, a uniform state of stress does not occur, nor is it possible to determine precisely the state of stress in the test specimen. Nevertheless, it has been demonstrated that if the length-to-width ratio of the test specimen is 10 or greater, and if the strain gage is placed at the middle of the test specimen, then the 45° specimen and Equation 8.75 still provide an accurate method of determining the lamina shear modulus G_{LT}.[4]

HOMEWORK PROBLEMS

8.1. A composite consists of aligned fibers in a rubber matrix. How much greater is the longitudinal Young's modulus than the transverse Young's modulus at a volume fraction of fibers of 0.6? The following properties are given:

$$E_f/E_m = 10^4, \quad E_f = 10^7 \text{ dyn/cm}^2.$$

8.2. What is the expected longitudinal tensile strength of an aligned fiber composite made by filament winding? The composite contains 65 volume percent fibers with a tensile strength of 250,000 psi. The matrix has a tensile strength of 12,000 psi.

8.3. A nylon fiber has uniaxial orientation in which the polymer chains are parallel to the fiber axis. Is $E_L > E_T$? (L = longitudinal, T = transverse.) Is G_{TT} greater than G_{LT}? Why?

8.4. Prove that μ_{LT} and μ_{TL} are not independent for an orthotropic fiber-reinforced polymer, that is, prove that $\mu_{LT}/E_L = \mu_{TL}/E_T$. This is known as the reciprocity relation. Under what conditions is it valid?

8.5. A fiber-reinforced polymer is loaded in uniaxial tension in a direction (x) at 45° to the fiber direction (L) as shown in Figure 8.17. Find all the strains in the x, y, z directions in terms of σ_x, E_L, E_T, μ_{LT}, μ_{TL}, and G_{LT}.

8.6. Prove that $\bar{Q}_{16} = 0$ for an isotropic material.

8.7. Given a 30° ply and the mechanical properties in Example 8.1 with an applied strain $\varepsilon_x = 200(10^{-6})$, calculate σ_x, σ_y, τ_{xy} assuming $\mu_{LT} = 0.25$.

8.8. Take the situation of two laminae butted and adhered to each other with the orientation as shown. If a stress is applied to this configuration, what is the resulting deformation perpendicular to the applied loading? More importantly, why? Assume that even though the laminae are adhered, any of the three deformations is possible a short distance away from the bond. The correct result can be proven mathematically. Perhaps surprisingly, more often than not, your intuition on this problem will fail you.

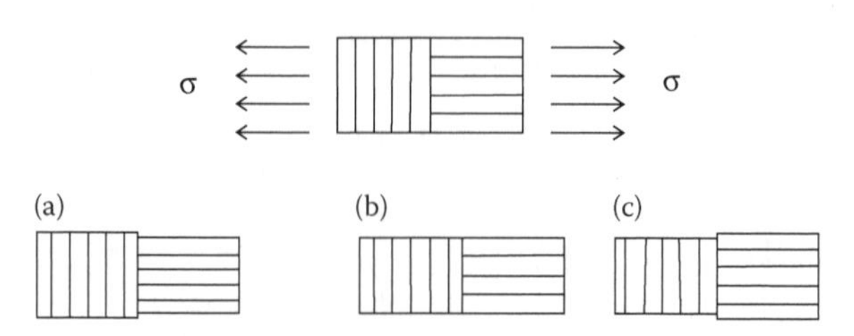

BIBLIOGRAPHY

Gibson, R. F., *Principles of Composite Material Mechanics*, McGraw-Hill, New York, NY, 1994.

Halpin, J. C., *Primer of Composite Materials*, Technomic Publishing Company, Lancaster, PA, 1984.

Hull, D., *An Introduction to Composite Materials*, Cambridge University Press, New York, NY, 1981.

Jones, J. M., *Mechanics of Composite Materials*, CRC Press, Boca Raton, FL, 1989.

Mallik, P. K., *Fiber-Reinforced Composites: Materials, Manufacturing, and Design*, Marcel Dekker, New York, NY, 1988.

Piggott, M. R., *Loading-Bearing Fiber Composites*, Pergamon Press, Elmsford, NY, 1980.

Tsai, S. W., *Composites Design*, Think Composites, Dayton, OH, 1986.

Tsai, S. W. and H.T. Hahn, *Introduction to Composite Materials*, Technomic Publishing Company, Lancaster, PA, 1980.

Vinson, J. R. and R.L. Sierakowski, *The Behavior of Structures Composed of Composite Materials,* Martinus Nijhoof Publishers, Dordrecht, Holland, 1986.

REFERENCES

1. Agrawal, B. D. and L. J. Broutman, *Analysis and Performance of Fiber Composites*, J. Wiley & Sons, New York, NY, 1980.
2. Calsson, L. S. and R. B. Pipes, *Experimental Characterization of Advanced Composite Materials*, Prentice-Hall, New York, NY, 1987.
3. Whitney, J. M., I. M. Daniel, and R. B. Pipes, *Experimental Mechanics of Fiber Reinforced Composite Materials*, SEM Monograph No. 4, Society for Experimental Mechanics, 1984.
4. Pindera, M. J. and C.T. Herakovich, Shear characterization of unidirectional composites with off-axis tension test, *Experimental Mechanics*, 103–112, March 1986.

9 Composite Laminate Failure

9.1 STRENGTH PROPERTIES AND FAILURE THEORIES

The strength characteristics of a composite orthotropic lamina are much more difficult to define than those of an isotropic material. In both cases, the engineer is interested in identifying material strength parameters which can be determined experimentally using simple tests, where the strength parameters are used to define allowable stresses for purposes of design.

9.1.1 A Review of Failure Theories for Isotropic Materials

Isotropic materials have axes of principal stress that coincide with the axes of principal strain. Moreover, principal stresses and strains are the extreme values of these quantities irrespective of orientation or direction. It is of practical importance, therefore, to seek failure criteria (physical measures), expressible in terms of principal stresses or strains, for which the material will begin to break down or fail whenever the specified measure exceeds a critical value at any point in the material. Some criteria (measures) that have proven successful for predicting failure due to static loading of isotropic materials include: (1) maximum normal stress, (2) maximum shearing stress, (3) maximum normal strain, (4) maximum shearing strain, and (5) distortional strain energy (equivalent to octahedral shear stress). Each of these quantities is expressible in terms of principal stresses. A failure theory based on one of these criteria states that the material subjected to *any* state of combined stress will fail when the value of the specified criterion (measure) at any point in the material exceeds the critical value determined in the simple tensile test. The critical value of any of these criteria for a particular material of interest is obtained by subjecting the material to a simple tensile test and calculating its value at failure.

Consider for example, the maximum shear stress (or Tresca) theory of failure with respect to yielding, which can be stated as follows: "A material subjected to any combination of loads will yield whenever the maximum shear stress at any point in the material exceeds the value of the maximum shear stress in a simple tensile test at yield."

The maximum shear stress at any point can be expressed in terms of principal stresses by the equation

$$\tau_{max} = \frac{\sigma_1 - \sigma_3}{2}, \tag{9.1}$$

where σ_1 and σ_3 are the maximum and minimum principal stress components, respectively. For the case of uniaxial tension at yield, Equation 9.1 reduces to

$$\tau_{max} = \frac{\sigma_1}{2} = \frac{\sigma_x}{2} = \frac{S_y}{2}, \tag{9.2}$$

where S_y is the yield strength in simple tension.

Equations 9.1, 9.2 provide the means of putting the Tresca theory of failure in equation form as follows:

$$\frac{(\sigma_1 - \sigma_3)}{2} \leq \frac{S_y}{2}. \tag{9.3}$$

Alternatively, introducing a factor of safety η, we obtain the design equation

$$(\sigma_1 - \sigma_3) = S_y \eta. \tag{9.4}$$

In a similar fashion, the distortional energy density (or von Mises) theory of failure can be stated as follows: "A material subjected to any combination of loads will yield whenever the distortional energy density at any point in the material exceeds the value of the distortional energy density in a simple tensile test at yield." The distortional energy density U_d of a linear elastic isotropic material can be expressed in terms of principal stresses by the equation

$$U_d = ((\sigma_1 - \sigma_2)^2 + (\sigma_2 - \sigma_3)^2 + (\sigma_3 - \sigma_1)^2)\frac{(1+\mu)}{E}, \tag{9.5}$$

where σ_1, σ_2, and σ_3 are the principal stresses. For the case of uniaxial tension, Equation 9.5 reduces to

$$U_d = ((\sigma_1 - 0)^2 + (0 - 0)^2 + (0 - \sigma_1))\left(\frac{1+\mu}{E}\right) = 2\sigma_1^2\left(\frac{1+\mu}{E}\right). \tag{9.6}$$

Equations 9.5 and 9.6 provide the means of putting the von Mises failure theory in equation form as follows:

$$(\sigma_1 - \sigma_2)^2 + (\sigma_2 - \sigma_3)^2 + (\sigma_3 - \sigma_1)^2 = 2S_y^2 \text{ at yield.} \tag{9.7}$$

Alternatively, introducing a factor of safety η, we obtain the design equation

$$(\sigma_1 - \sigma_2)^2 + (\sigma_2 - \sigma_3)^2 + (\sigma_3 - \sigma_1)^2 = 2\left(\frac{S_y}{\eta}\right)^2. \tag{9.8}$$

Similar failure theory and design equations can be derived for the other failure criteria mentioned previously.

In each case, only one strength value, namely the *yield strength*, is needed to predict failure. If failure is sought with reference to fracture rather than ductile deformation, then the appropriate strength value chosen will be *ultimate strength*, S_u. The material properties needed in design for static strength of most isotropic engineering materials are yield strength and ultimate strength in tension and compression.

The suitability of a particular failure theory can only be determined by experiment for each material and for each type of failure being considered. It is important to remember that for each of the failure theories discussed above, a single material strength property is all that is needed to define the failure criterion. Moreover, the value of the strength property is obtained from a single tensile test run on a sample of the material.

9.1.2 Strength and Failure Theories for an Orthotropic Lamina

Principal stress and strain directions are defined as coordinate orientations for which the shearing components of stress and strain vanish. In general, the principal stress directions at a point in an orthogonal lamina will not line up with the lamina material direction. Hence, the principal (normal) stresses will appear as off-axis loading and will cause shearing strains due to coupling (Equations 8.45 and 8.55). As a consequence, the principal strain directions will not coincide with directions of principal stress. Hence, there is no advantage gained in seeking failure criteria for orthotropic lamina expressible in terms of principal stresses. Before proceeding further with the task of defining macrostructural strength properties of an orthotropic lamina, it is helpful to look at the microstructural failure mechanisms associated with failure of a unidirectional composite layer.

Figure 9.1 shows schematically five possible microstructural failure mechanisms for longitudinal compressive loading of a 0° test specimen. Other failure mechanisms have been identified which are associated with longitudinal tensile loading, and transverse tensile and compressive loading of unidirectional lamina. These mechanisms include: (1) fiber breakage, (2) fiber pullout, (3) matrix shear failure, (4) matrix tensile failure, (5) fiber splitting, (6) fiber/matrix de-bonding, and (7) compressive crush of fibers. The specific microstructural failure mechanism which triggers failure of a lamina under general loading will depend on the strength properties of the constituent materials, and the fiber/matrix interface bonding strength. Moreover, each of these properties can be affected by environmental and loading conditions, by the processing and manufacturing methods used in making the composite lamina, and also by the stacking sequence used in placing the lamina within a laminate. Because of the above complexities

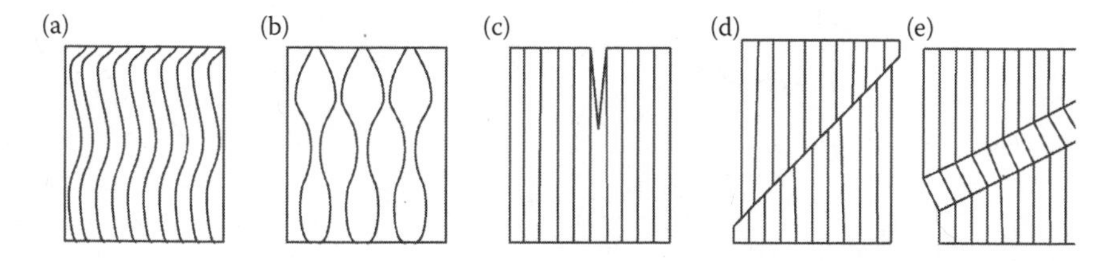

FIGURE 9.1 A schematic representation of five possible modes of failure associated with uniaxial test specimen: (a) fiber microbuckling in shear mode, (b) fiber microbuckling in extensional mode, (c) fiber/matrix splitting caused by transverse tensile failure, (d) fiber shear failure, and (e) fiber/matrix kink band formation.

development of a general failure criterion on the basis of microstructural behavior and strength data is difficult.

In Section 8.3, the lamina macrostructural elastic moduli have been defined for the principal material directions. The elastic properties are determined experimentally from uniaxial stress states. If these experiments are carried through to failure of the test specimen, then one also obtains the following macromechanical failure strength properties of the lamina defined in principal material coordinates:

X_L = longitudinal tensile strength
X'_L = longitudinal compressive strength
X_T = transverse tensile strength
X'_T = transverse compressive strength
S = shear strength

The five lamina strength properties defined above represent five distinct macrostructural failure modes associated with principal material directions of the lamina. Moreover, most engineering composites behave in a relatively brittle manner so that the strength properties correspond to fracture of the lamina. If one wanted to design a composite lamina to improve its strength properties, it would be necessary to examine the microstructural failure mechanisms which triggered the various failure modes. This information would help to understand how to alter constituent materials to postpone specific triggering mechanisms and hence, raise the value of the strength properties. On the other hand, if we are primarily interested in using available composite materials in design, then we use macrostructural failure theories based on the lamina strength properties.

One of the simplest failure theories for orthotropic lamina is the maximum stress theory of failure. This theory states,

Failure of the lamina will occur if any stress component in principal material directions exceeds the corresponding lamina strength value. That is to avoid failure,

$$\begin{array}{ll} \text{(tension)} & \text{(compression)} \\ \sigma_L \le X_L & \sigma_L \le X'_L \\ \sigma_T \le X_T & \sigma_T \le X'_T \\ \tau_{LT} \le S & \end{array} \qquad (9.9)$$

Example 9.1

Determine the allowable stress σ_x for uniaxial off-axis loading of the unidirectional fiber-reinforced composite lamina, shown in Figure 8.17. Assume that the lamina strength properties X_L, X'_L, X_T, X'_T, and S are known. Use the maximum stress theory of failure.

SOLUTION

In order to apply the maximum stress theory of failure, the stress components must be determined in principal material directions. Using the transformation Equation 8.56, the following relationships are obtained:

$$\begin{Bmatrix} \sigma_L \\ \sigma_T \\ \tau_{LT} \end{Bmatrix} = \begin{bmatrix} \cos^2\theta & \sin^2\theta & \sin\theta\cos\theta \\ \sin^2\theta & \cos^2\theta & -2\sin\theta\cos\theta \\ -\sin\theta\cos\theta & \sin\theta\cos\theta & \cos^2\theta - \sin^2\theta \end{bmatrix} \begin{Bmatrix} \sigma_x \\ 0 \\ 0 \end{Bmatrix} \tag{9.10}$$

Solving for the uniaxial stress σ_x in terms of principal material stress components and applying the maximum stress failure theory constraints shown in Equation 9.9 gives the following five conditions that must be met for allowable stress σ_x:

For tension:

$$\begin{aligned} \sigma_x &= \frac{\sigma_L}{\cos^2\theta} < \frac{X_L}{\cos^2\theta} \\ \sigma_x &= \frac{\sigma_T}{\sin^2\theta} < \frac{X_T}{\sin^2\theta} \\ \sigma_x &= \frac{\tau_{LT}}{\cos\theta\sin\theta} < \frac{S}{\cos\theta\sin\theta} \end{aligned} \tag{9.11a–c}$$

and for compression:

$$\begin{aligned} |\sigma_x| &< \frac{X'_L}{\cos^2\theta} \\ |\sigma_x| &< \frac{X'_T}{\sin^2\theta} \\ |\sigma_x| &< \frac{S}{\cos\theta\sin\theta} \end{aligned} \tag{9.11d–f}$$

The maximum allowable uniaxial stress, σ_x will be the smallest absolute value obtained using Equations 9.11a and 9.11b for a specified loading angle θ and lamina strength properties X_L, X'_L, X_T, X'_T, and S.

The maximum stress theory is illustrated in Figure 9.2a in which Equation 9.11 are plotted as a function of θ for an E-glass/epoxy composite with properties: $X_L = X'_L = 150$ ksi, $X_T = 4$ ksi, $X'_T = 20$ ksi, and $S = 6$ ksi. Experimental data are superimposed on the theoretical strength plots, with tension data denoted by solid circles and compression data by solid squares.[1] The theoretical cusps predicted by theory are not borne out by the experimental results, nor is there exact agreement in variation of stress with θ between theory and experiment.

One of the shortcomings of the maximum stress theory of failure is that there are no terms which account for interaction between stress components for the case of biaxial (or off-axis) loading. Another is that five independent equations must be satisfied. The Tsai–Hill theory of failure for anisotropic materials overcomes both of the above mentioned shortcomings. This theory can be expressed in terms of principal material stress components as follows:

$$\left(\frac{\sigma_L}{X_L}\right)^2 - \frac{(\sigma_L\sigma_T)}{X_L^2} + \left(\frac{\sigma_T}{X_T}\right)^2 + \left(\frac{\tau_{TL}}{S}\right)^2 = 1 \tag{9.12}$$

which we recognize from Equation 9.7 as extension of the von Mises theory from isotropic to orthotropic materials.

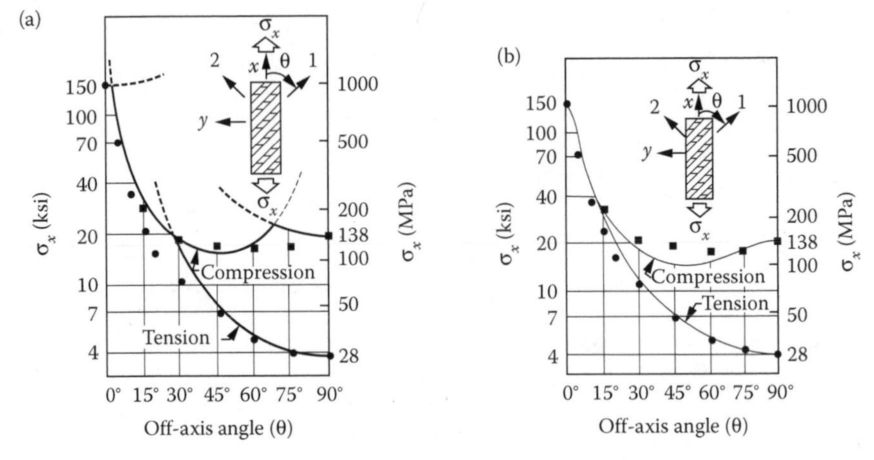

FIGURE 9.2 Theoretical curves for uniaxial strength of unidirectional E glass/epoxy composite lamina versus off-axis loading angle θ. Experimental tension (solid circles) and compression (solid squares) data are superposed on theoretical curves. (a) Maximum stress theory of failure and (b) Tsai-Hill theory of failure. (Adapted from Jones, R. M., *Mechanics of Composite Materials*, Taylor & Francis, London, 1998.)

For uniaxial off-axis loading, the results of Equation 9.10 can be substituted into Equation 9.12 to obtain

$$\left(\frac{1}{\sigma_x}\right)^2 = \frac{\cos^4\theta}{X_L^2} + \frac{\sin^4\theta}{X_T^2} + \left(\frac{1}{S^2} - \frac{1}{X_L^2}\right)\cos^2\theta\sin^2\theta \tag{9.13}$$

Equation 9.13 gives a single criterion for predicting failure for uniaxial loading. (Actually, two curves are represented by Equation 9.13.) For compression loading the tensile strengths X_L and X_T are replaced by the corresponding compressive strengths X'_L and X'_T). The results of Equation 9.13 are plotted in Figure 9.2b along with experimental data for E-glass/epoxy given earlier. The Tsai–Hill Theory of failure gives excellent agreement with experiment for this material, particularly for tensile failure. Note also that the variation in strength is smooth using this theory instead of having cusps.

Another form of a combined stress theory of failure is given by Tsai and Wu,[2] and by Goldenblat and Kopnov.[3] This failure criterion is expressed by

$$F = f_1\sigma_L + f_2\sigma_T + f_{11}\sigma_L^2 + f_{22}\sigma_L^2 + f_{66}\tau_{LT}^2 + 2f_{12}\sigma_L\sigma_T \tag{9.14}$$

where $F > 1$ indicates failure and $F \leq 1$ is safe; and where

$$f_1 = \frac{1}{X_L} - \frac{1}{X_L'}, \quad f_{11} = \frac{1}{X_L' X_L}, \quad \text{for } X_L' = |X_L'| > 0,$$
$$f_2 = \frac{1}{X_T} - \frac{1}{X_T'}, \quad f_{22} = \frac{1}{X_T X_T'}, \quad \text{for } X_T' = |X_T'| > 0, \tag{9.15}$$
$$f_{66} = \frac{1}{S^2}, \quad f_{12} = \frac{1}{2}(f_{11}f_{22})^{1/2}.$$

In contrast to the Tsai–Hill theory, Equation 9.12 where different values for the strength terms have to be used for tension and compression, the Tsai–Wu theory accommodates tension and compression in the one equation.

The example macroscopic failure criteria for composite lamina presented above should provide the reader with an understanding of the logic used in developing failure theories for composites. The design engineer who is actively engaged in analysis of composites should consider use of one of the many finite-element computer programs that include composite materials. Some of the failure theories above are included in these programs.

A survey of much of the past work that has been done in this important area of study is given in Tsai.[4] Other related topics of interest in the design of composites include the material response to impact on loading, fatigue under cyclic loading, fracture and crack propagation in composites, and the effects of environmental conditions on the material response.

The failure of laminate composites is a topic that has been researched for over 50 years. Currently, there are many competing theories on laminate failure. Typically, the failure theories are used to estimate the nominal layup of a laminate followed by experimental verification. The most noteworthy attempt to reconcile the various composite failure theories was The World-Wide Failure Exercise.[5] This exercise had its genesis in a meeting held in 1991.[6] After the meeting the authors conceived of the selection of test samples and biaxial loading cases for a pressurized tubular test specimen (see Table 9.1).

There were many researchers and software companies invited to participate in the exercise and a number did participate. For a full list of the participants and their theories, one must see the text. Suffice it to say that the theories ranged from the micromechanical to the macromechanical. In the end, the editors and organizers of selected five theories that fit various cases better than the majority of the other theories (see Table 9.2). In the final recommendation, they recommended Puck and Cuntze's theories for overall prediction, while Tsai's conservatism on most cases, meant that it could be used in conjunction with one of the other theories. All five of the theories were able to predict within ±50% of the failure strengths for 75% of the cases studied. Due to the complexity of the failure modes of two-dimensional loading, the lack

TABLE 9.1
Worldwide Failure Exercise Laminates

Laminate	Materials	Rationale
0 unidirectional lamina	E-glass/MY750, E-glass/LY556, T300/914C	Predictions of laminates are based on lamina results
90/30/−30/−30/30/90	Eglass/LY556	Number of failure modes were expected
90/45/−45/0/0/−45/45/90	AS4/3501–6 carbon/epoxy	Quasi-isotropic, typical of aerospace laminates, equal strengths expected in 0° and 90° directions
55/−55/−55/55		Widespread use in pipe applications, stress results in the compression–compression quadrant
0/90/90/0		Investigate matrix cracking
45/−45/−45/45		Analagous to 0/90/90/0 laminate under biaxial tension and shear

Source: Reprinted from *Failure Criteria in Fibre Reinforced Polymer Composites: The World-Wide Failure Exercise*, Hinton, M. J., A. S. Kaddour, and P. D. Soden, eds., Elsevier, London, © 2004, with permission from Elsevier.

TABLE 9.2
Meritorious Composite Failure Theories from the World-Wide Failure Exercise

Theory	Description
Bogetti[5,9,10]	Maximum strain theory in three dimensional. Includes nonlinear shear stress (strain) effect.
Cuntze[5,11,12]	Incorporates five failure mechanisms, two fiber and three inter-fiber, related to Puck's theory, interaction between failure mechanisms present.
Puck[5,13,14]	Utilizes independent failure criteria for fiber and interfiber fracture. Accounts for loss of stiffness after initial failure.
Tsai[5,15,16]	Essentially Tsai–Wu quadratic failure theory.
Zinoviev[5,17,18]	Maximum stress theory. Includes unloading of cracked lamina as well as geometric nonlinearity due to changes in the ply angle.

Source: Reprinted from *Failure Criteria in Fibre Reinforced Polymer Composites: The World-Wide Failure Exercise*, Hinton, M. J., A. S. Kaddour, and P. D. Soden, eds., Elsevier, London, © 2004, with permission from Elsevier.

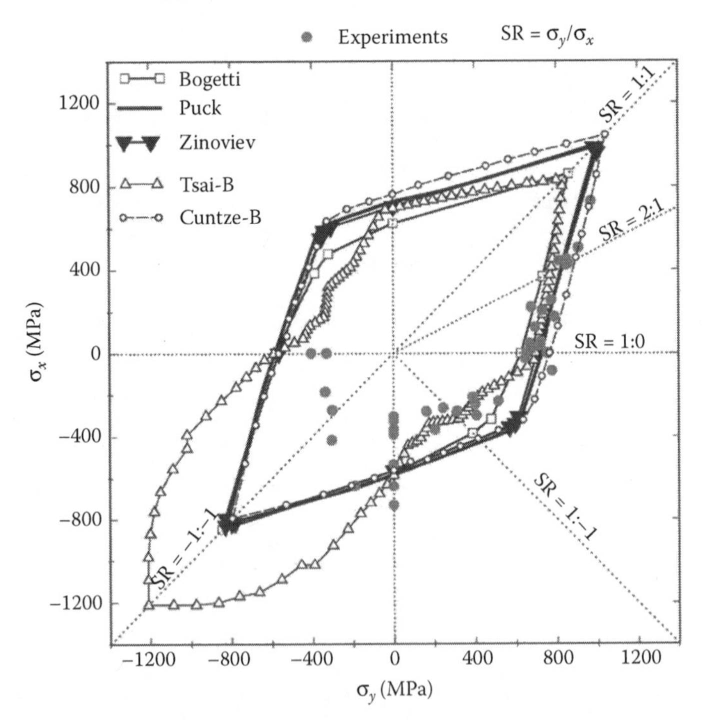

FIGURE 9.3 Comparison between the predicted and measured final failure stresses for (0°/±45°/90°) AS4/3501–6 carbon-fiber/epoxy laminates subject to biaxial loads (Test Case No 6-range of biaxial stress ratios). (Reprinted from *Failure Criteria in Fibre Reinforced Polymer Composites: The World-Wide Failure Exercise*, Hinton, M. J., A. S. Kaddour, and P. D. Soden, eds., Elsevier, London, © 2004, with permission from Elsevier.)

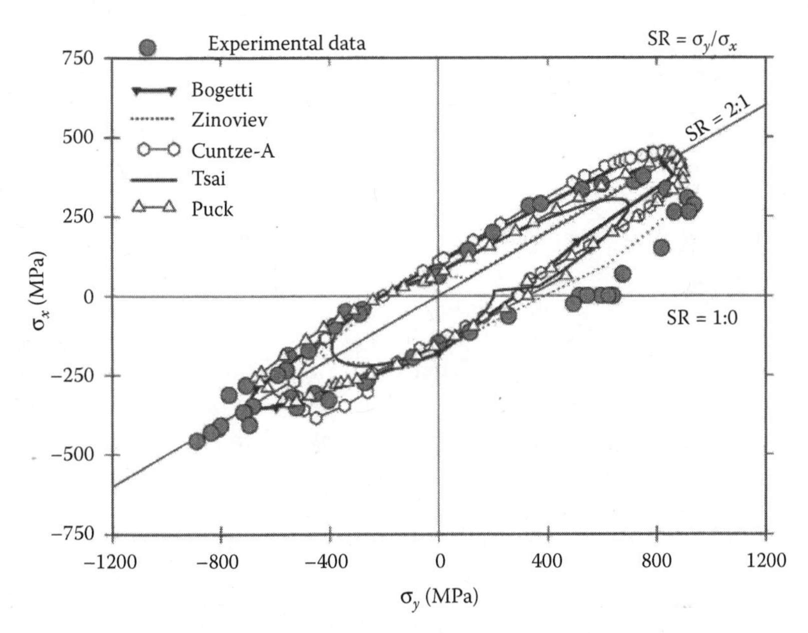

FIGURE 9.4 Comparison between the predicted and measured final failure stress for (±55°) E-glass/MY750 laminates subjected to biaxial loads (Test Case No 9-range of biaxial stress ratios). (Reprinted from *Failure Criteria in Fibre Reinforced Polymer Composites: The World-Wide Failure Exercise*, Hinton, M. J., A. S. Kaddour, and P. D. Soden, eds., Elsevier, London, © 2004, with permission from Elsevier.)

of universal prediction is not surprising. However, the Exercise provides the results of the various loading cases, that the reader can use to select the theory that best predicts the loading case desired to be analyzed. In addition, for static design the implication is that in order to be conservative the factor of safety should be greater than two for most situations. Figure 9.3 shows the worst fit of all the theories to the generated experimental data (note the failure envelopes in the compression–compression quadrant are dramatically nonconservative). Figure 9.4 shows the best fit of the theories for a given laminate and loading situation.

9.1.3 Failure by Fiber Pullout

The longitudinal tensile strength X_L of a lamina is usually determined by the fracture strength of the fibers if long fibers are used. For short, discontinuous fibers, failure by fiber pullout is more critical. How short is "short?" For short fibers, the tensile stress is largest at the midlength of the fiber and decreases toward zero at the ends. This tensile stress must be balanced by the shear stress τ at the bonding surface. The bond will fail when τ reaches the shear strength S_m of the matrix. Thus, from Figure 9.5, a critical length L_c of the fiber can be calculated from equilibrium of axial forces on the fiber[5]:

$$L_c = \frac{\sigma_f d_f}{2 S_m} \tag{9.16}$$

For $L > L_c$, the fiber is estimated to fail when $\sigma_f = X_{fu}$, where X_{fu} is the ultimate tensile strength of the fiber.

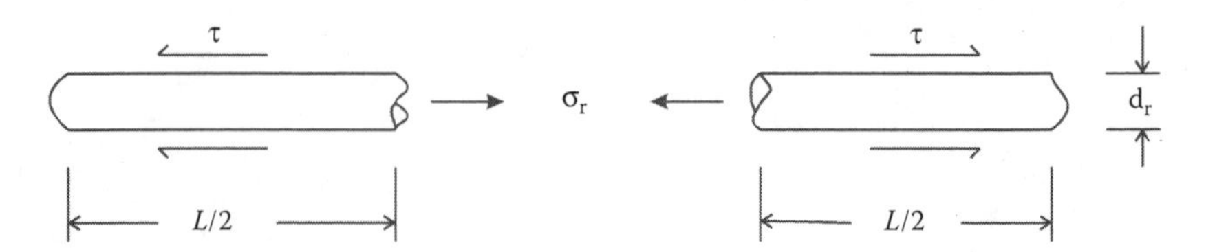

FIGURE 9.5 Stresses on short fibers.

For $L < L_c$, the matrix bond will fail in shear when $\tau = S_m$ of the matrix, and the fiber will pull out. This will also occur when there is less than perfect bonding, and the fiber debonds and pulls out. Contrary to what we might initially presume, less-than-perfect bonding is desired for better impact strength. Use of short fibers, designed to slip and pull out, gives a better polymer for energy absorption in impact situations. See Chapter 5 for impact properties.

9.2 STIFFNESS OF LAMINATED COMPOSITES

9.2.1 Sandwich Beam

Before we consider laminated plates with several layers, consider a laminated composite beam of the sandwich type shown in Figure 9.6a. Let the heights (thicknesses) of the layers be denoted by h_1 and h_2 as shown, where h_1 is the thickness of the faces and h_2 is the thickness of the core. Assume pure bending and assume that the bending strain is linear, that is, that planes remain plane as shown in Figure 9.6b. For linear elastic behavior

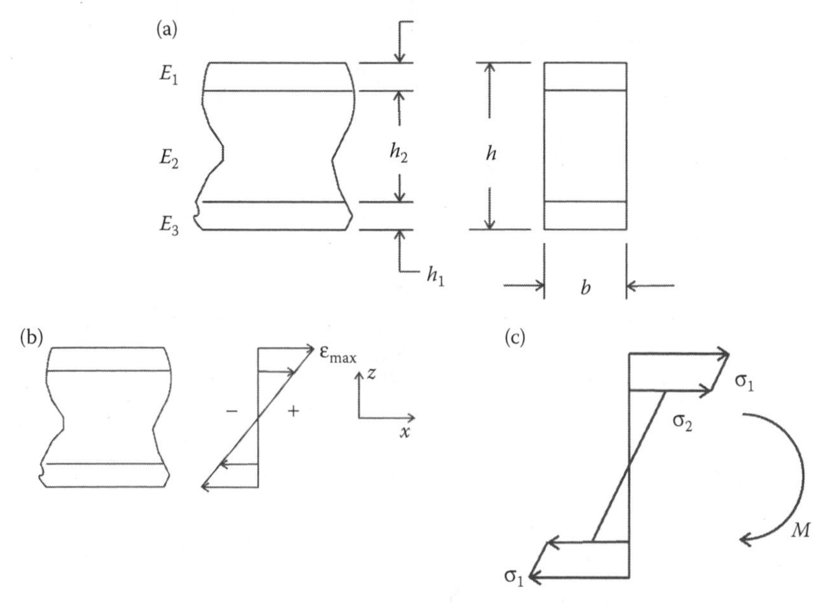

FIGURE 9.6 Bending of a sandwich beam. (a) Geometry, (b) strain distribution, and (c) stress distribution.

in each layer ($\sigma = E\varepsilon$) the longitudinal bending stress is discontinuous—it takes a jump at the interface as shown in Figure 9.6c.

The bending moment M on the cross section is found by integrating the differential force ($\sigma\, dA$) times the moment arm (z) over the cross-sectional area as follows:

$$M = \int \sigma z\, dA = 2 \int_0^{h/2} \sigma\, b z\, dz$$
$$M = 2b \int_0^{h_2/2} E_2 \varepsilon z\, dz + 2b \int_{h_2/2}^{h/2} E_1 \varepsilon z\, dz \tag{9.17}$$

Because the strain is assumed linear, we can write it as

$$\varepsilon = \varepsilon_{max} \frac{2z}{h} \tag{9.18}$$

Substitution of Equation 9.18 into Equation 9.17 and evaluation of the integrals give:

$$M = \frac{2\varepsilon_{max}}{h} E_1 I_{eff} \tag{9.19}$$

where

$$I_{eff} = \alpha I \tag{9.20}$$

$$\alpha = \frac{E_2}{E_1}\left(\frac{h_2}{h}\right)^3 + 1 - \left(\frac{h_2}{h}\right)^3 \tag{9.21}$$

$$I = \frac{bh^3}{12} \tag{9.22}$$

The bending stiffness of a homogeneous beam of material 1 is $E_1 I$. Thus, the effective bending stiffness of the sandwich beam has been reduced by the factor α, by replacing the core by material 2 with a lower modulus E_2 but with a lower density ρ_2. However, the stiffness-to-weight ratio has been increased as shown by Example 9.2.

Example 9.2

Assume $E_2 = 0.1E_1$ and $h_2/h = 0.9$. Equation 9.21 gives

$$\alpha = 0.1(0.9)^3 + 1 - (0.9)^3 = 0.344$$

and

$$I_{eff} = 0.344I$$

Next, assume $\rho_2 = 0.1\rho_1$. The weight of each material W_1 and W_2 are, respectively

$$W_1 = 2\rho_1 h_1 L \quad \text{and} \quad W_2 = \rho_2 h_2 L \tag{9.23a,b}$$

The total weight W_T of the sandwich beam is

$$W_T = W_1 + W_2 = \beta W \tag{9.24}$$

where

$$W = \rho_1 hbL$$

is the weight of a homogeneous beam of material 1, and

$$\beta = \frac{2h_1}{h} + \frac{\rho_2}{\rho_1}\frac{h_2}{h} \tag{9.25}$$

The stiffness-to-weight ratio of the composite beam is

$$\frac{E_1 I_{\text{eff.}}}{W_T} = \frac{E_1 \alpha I}{\beta W} = \frac{\alpha}{\beta}\frac{E_1 I}{W} \tag{9.26}$$

where $(E_1 I/W)$ is the stiffness-to-weight ratio for a homogeneous beam. For the example data,

$$\frac{E_1 I_{eff.}}{W_T} = \frac{0.344}{0.19}\frac{E_1 I}{W} = 1.81\frac{E_1 I}{W}$$

Thus, the example sandwich beam is 81% more effective on a stiffness-to-weight basis.

9.2.2 Orthotropic Plate

Before we consider a laminated plate, consider a homogeneous orthotropic plate made up of one layer, as shown in Figure 9.7. Unlike the beam the bending stress can now vary in the two directions L and T, or in general in the x and y directions, just like

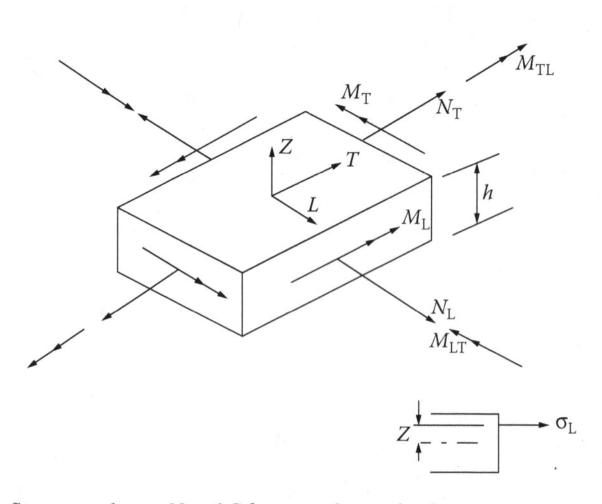

FIGURE 9.7 Stress resultants N and S for an orthotropic plate.

the average stresses. Assume that both stretching and bending occurs in the plate. In plate theory[7,8] force stress resultants are defined as follows:

$$N_L = \int \sigma_L \, dz = \int_{-h/2}^{h/2} \sigma_L \, dz \tag{9.27a}$$

$$N_T = \int \sigma_T \, dz = \int_{-h/2}^{h/2} \sigma_T \, dz \tag{9.27b}$$

$$N_{LT} = \int \sigma_{LT} \, dz = \int_{-h/2}^{h/2} \sigma_{LT} \, dz \tag{9.27c}$$

N_L = the average (membrane) normal stress force-resultant in the L direction per unit length in the T direction, N_T = the average (membrane) normal stress force resultant in the T direction per unit length in the L direction, and N_{LT} = the average (membrane) shear stress force resultant on the L normal plane in the T direction per unit length in the T direction.

Similarly, bending moment stress resultants are defined as[7,8]

$$M_L = \int \sigma_L \, z \, dz = \int_{-h/2}^{h/2} \sigma_L \, z \, dz \tag{9.28a}$$

$$M_T = \int \sigma_T \, z \, dz = \int_{-h/2}^{h/2} \sigma_T \, z \, dz \tag{9.28b}$$

$$M_{LT} = \int \sigma_{LT} \, z \, dz = \int_{-h/2}^{h/2} \sigma_{LT} \, z \, dz \tag{9.28c}$$

where

M_L = the bending moment resultant on the L normal edge in the T direction per unit length in the T direction,
M_T = the bending moment resultant on the T normal edge in the $-L$ direction per unit length in the T direction, and
M_{LT} = the twisting moment resultant on the L normal edge in the $-L$ direction per unit length in the T direction.

With the previous definitions, the stresses in the plate are given by

$$\sigma_L = \frac{N_L}{h} + \frac{12 M_L z}{h^3} \tag{9.29a}$$

$$\sigma_T = \frac{N_T}{h} + \frac{12 M_T z}{h^3} \tag{9.29b}$$

$$\sigma_{LT} = \frac{N_{LT}}{h} + \frac{12 M_{LT} z}{h^3} \tag{9.29c}$$

The above integral operators in Equation 9.27 and Equation 9.28 can be used as well to define stress resultants N_x, N_y, N_{xy}, M_x, M_y, and M_{xy} for any directions x and y.

Like in a beam the strains in the plate are assumed to vary linearly through the thickness. In the beam this corresponds to "planes remaining plane." In the plate this corresponds to "normals remaining normals." This is known as the Kirchoff hypothesis that says: "normals to the middle surface (reference surface) before deformation remain straight and normal to the middle surface after deformation." This is illustrated in Figure 9.8.

The longitudinal and transverse strain distributions through the thickness of the plate are given by

$$\varepsilon_L = (\varepsilon_L)_m + (\varepsilon_L)_b \quad \text{and} \quad \varepsilon_T = (\varepsilon_T)_m + (\varepsilon_T)_b \tag{9.30a,b}$$

where $(\varepsilon_L)_m$ and $(\varepsilon_T)_m$ are the average or membrane strains in the plate, and $(\varepsilon_L)_b$ and $(\varepsilon_T)_b$ are the bending strains in the plate, which are given by

$$(\varepsilon_L)_b = z\,\kappa_L, \quad (\varepsilon_T)_b = z\,\kappa_T, \tag{9.31a,b}$$

and

$$\kappa_L = \frac{1}{R_L}, \qquad \kappa_T = \frac{1}{R_T} \tag{9.32a,b}$$

where κ_L is the change in curvature in the L direction due to bending, and R_L is the radius of curvature under bending, and where κ_T is the change in curvature the T direction due to bending, and R_T is the radius of curvature under bending, and thus the normal strains are given by

$$\varepsilon_L = (\varepsilon_L)_m + z\,\kappa_L, \qquad \varepsilon_T = (\varepsilon_T)_m + z\,\kappa_T. \tag{9.33a,b}$$

With the Kirchoff hypothesis assumed above (i.e., normals remaining normal), the shear strain is assumed 0, $\gamma_{LT} \approx 0$ for an individual layer. This means the twist of a layer κ_{LT} is 0 also. However, the twist of a multilayered laminate can be non-0

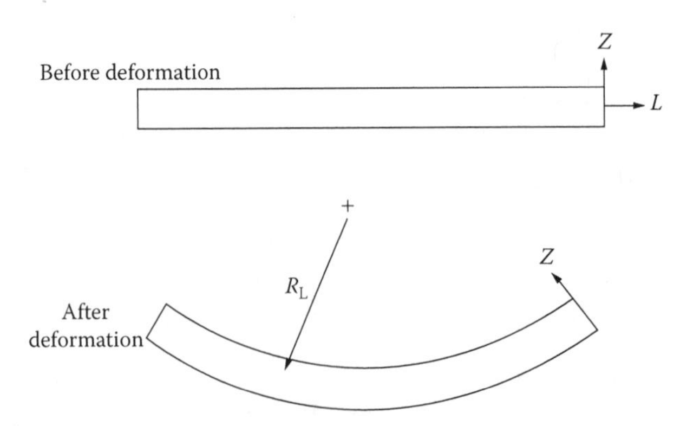

FIGURE 9.8 Deformation of a plate illustrating Kirchoff hypothesis of planes remaining plane.

because of different normal strains in the layers. Using the constitutive law (Equation 8.37), the extensional stiffness matrix of the plate $[A]$ is defined as

$$\{N\} = [A]\{\varepsilon\}_m \tag{9.34a}$$

where

$$[A] = h[Q] \tag{9.34b}$$

Similarly, a bending stiffness matrix $[D]$, relating the bending moments to the curvatures, is found when relating the bending stresses to the bending strains and integrating over the thickness using Equations 9.25:

$$\{M\} = [D]\{\kappa\} \tag{9.35a}$$

where

$$[D] = \frac{h^3}{12}[Q] \tag{9.35b}$$

9.2.3 Laminated Plates

Next consider a laminated plate made up of several layers as shown in Figure 9.9. The same assumption on linear strain through the thickness is made so that Equations

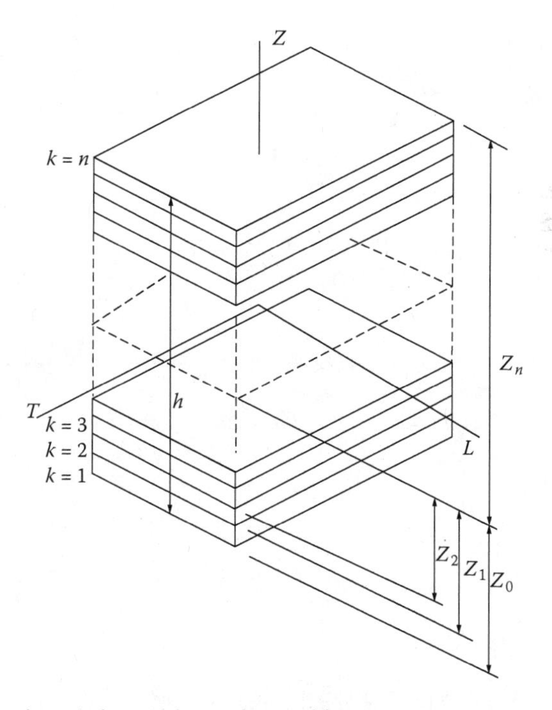

FIGURE 9.9 Laminated plate made up of several layers.

9.33 are still assumed valid for the laminated plate. However, the stress resultants for arbitrary lamina notation are now defined as

$$N_X = \sum_{k=1}^{n} \left\{ \int_{Z_{k-1}}^{Z_k} (\sigma_x)_k \, dz \right\} \tag{9.36}$$

$$M_X = \sum_{k=1}^{n} \left\{ \int_{Z_{k-1}}^{Z_k} (\sigma_x)_k \, z \, dz \right\} \tag{9.37}$$

Similar resultants for N_y, N_{xy}, M_y, and M_{xy} are obtained by summing integrals of σ_y, τ_{xy}, $z\sigma_y$, and $z\tau_{xy}$, respectively. Then it can be shown that

$$\{N\} = [A]\{\varepsilon\}_m + [B]\{\kappa\} \tag{9.38}$$

and

$$\{M\} = [B]\{\varepsilon\}_m + [D]\{\kappa\} \tag{9.39}$$

where $\{\varepsilon\}$ and $\{\kappa\}$ are now the average strain vector and resultant curvature vector for the composite plate, and where

$$A_{ij} = \sum_{k=1}^{n} \left(\bar{Q}_{ij}\right)(z_k - z_{k-1}), \tag{9.40a}$$

$$B_{ij} = \frac{1}{2} \sum_{k=1}^{n} \left(\bar{Q}_{ij}\right)(z_k^2 - z_{k-1}^2), \tag{9.40b}$$

$$D_{ij} = \frac{1}{3} \sum_{k=1}^{n} \left(\bar{Q}_{ij}\right)(z_k^3 - z_{k-1}^3) \tag{9.40c}$$

For a symmetric laminate, symmetric about the $z = 0$ plane, the $B_{ij} = 0$, and there is no coupling between extension and bending.

Since, the applied external loads to the laminate are known it is more useful to know the average extensional strain $\{\varepsilon\}_m$ and curvature $\{\kappa\}$, in order to calculate individual ply stresses:

$$\{\varepsilon\}_m = \left([B] - [D][B]^{-1}[A]\right)^{-1} \left(\{M\} - [D][B]^{-1}\{N\}\right) \tag{9.41a}$$

and

$$\{\kappa\} = \left([D] - [B][A]^{-1}[B]\right)^{-1} \left(\{M\} - [B][A]^{-1}\{N\}\right) \tag{9.41b}$$

However, it is easier to simply invert the stiffness matrix

$$\left\{\frac{N}{M}\right\}=\begin{bmatrix} A & B \\ B & D \end{bmatrix}\left\{\frac{\varepsilon_{\text{m}}}{\kappa}\right\} \tag{9.42}$$

to calculate the strain and curvature

$$\left\{\frac{\varepsilon_{\text{m}}}{\kappa}\right\}=\begin{bmatrix} A & B \\ B & D \end{bmatrix}^{-1}\left\{\frac{N}{M}\right\} \tag{9.43}$$

At this stage, the strains and curvatures are calculated either by Equations 9.41a and 9.41b or by Equation 9.43 that are known inputs. Once the global deformations are known the individual ply or lamina deformations can be calculated. Remember that since plane sections remain plane, the laminate analysis assumes that strain is both continuous and linearly distributed on a cross section of the laminate. The result of this is that stress is typically discontinuous due to the different reduced stiffness matrices of adjoining plies. For example, since the strain in the x-direction at the interface of plies is assumed to be the same, it follows that stress will be different. The total strain as a function of z is

$$\{\varepsilon(z)\}=\{\varepsilon_{\text{m}}\}+z\{\kappa\} \tag{9.44}$$

When the z location of the bottom and top of each ply are substituting into Equation 9.36 the resulting strains allow the calculation of stress via Equation 8.65. These stresses are then used to check for failure at both the top and bottom of each ply.

The laminate stiffness can be calculated according to the equations developed previously and used as input material properties to Finite Element computer programs that handle composite plates. There are also computer programs that are available to compute the stiffness for composite plates. A laminate analysis computer program is provided in Appendix D for use in the design projects. It is also provided on the accompanying CD as a MATLAB®. Note that the program can also be run with the GNU (a Unix like operating system) Octave, the analogous General Public License program. The algorithm for the laminate analysis is shown in Figure 9.10.

Example 9.3

Determine the bending stiffness matrix $[D]$ for an angle-ply laminate of plywood made up of 4 layers as shown in Figure 8.3 with $\theta = +45°, -45°, +45°$, and $-45°$. The laminate code is $[\pm 45]_2$. Each layer of wood is 0.125 in. thick. Thus, $z_0 = -0.250$, $z_1 = -0.250$, $z_3 = 0.125$, and $z_4 = 0.250$.

The longitudinal and transverse properties of the wood layers are given as

$$E_{\text{L}} = 1.0(10^6) \text{ psi}, \quad E_{\text{T}} = 0.380(10^6) \text{ psi}, \quad \text{G}_{\text{LT}} = 0.16(10^6), \quad \text{and} \quad \mu_{\text{LT}} = 0.40$$

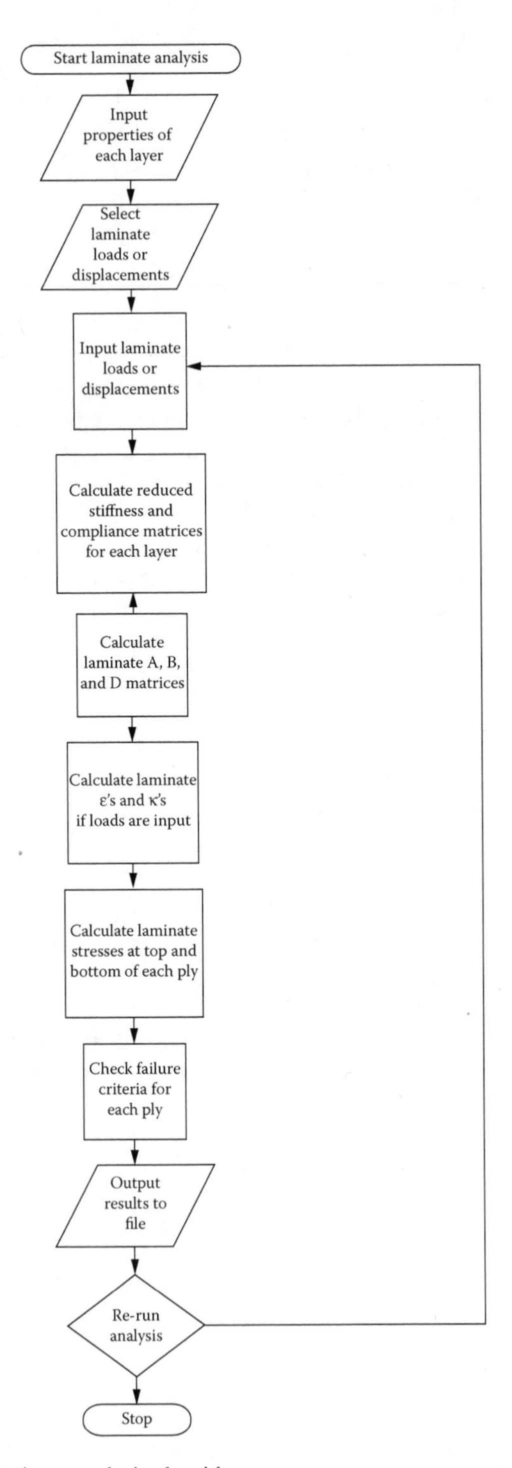

FIGURE 9.10 Laminate analysis algorithm.

From Equations 8.68 the following values of Q_{ij} are calculated:

$$Q_{11} = 1.727(10^6)\text{ psi},\quad Q_{22} = 0.656(10^6)\text{ psi},\quad Q_{12} = 0.690(10^6)\text{ psi},\quad Q_{66} = 0.160(10^6)\text{ psi}$$

From Equation 8.67 the following values of Q_{ij} are calculated for $\theta = +45°$

$$\bar{Q}_{11} = 1.10\,(10^6)\text{ psi}$$
$$\bar{Q}_{12} = 0.781\,(10^6)\text{ psi}$$
$$\bar{Q}_{22} = 1.10\,(10^6)\text{ psi}$$
$$\bar{Q}_{66} = 0.250\,(10^6)\text{ psi}$$
$$\bar{Q}_{16} = 0.267(10^6)\text{ psi}$$
$$\bar{Q}_{26} = 0.267\,(10^6)\text{ psi}$$

From Equation 8.67 the following values of Q_{ij} are calculated for $\theta = -45°$

$$\bar{Q}_{11} = 1.10\,(10^6)\text{ psi}$$
$$\bar{Q}_{12} = 0.781\,(10^6)\text{ psi}$$
$$\bar{Q}_{22} = 1.10\,(10^6)\text{ psi}$$
$$\bar{Q}_{66} = 0.250\,(10^6)\text{ psi}$$
$$\bar{Q}_{16} = -0.267\,(10^6)\text{ psi}$$
$$\bar{Q}_{26} = -0.267\,(10^6)\text{ psi}$$

Next the bending stiffness matrix D_{ij} values are calculated from Equation 9.40c:
$D_{11} = 11.47(10^3)$ lb-in., $D_{22} = 11.47(10^3)$ lb-in., $D_{12} = 8.14(10^3)$ lb-in., $D_{16} = 0.00$, $D_{26} = 0.00$, and $D_{66} = 2.61(10^3)$ lb-in. The zero values for D_{16} and D_{26} occur because the plies are antisymmetric about $z = 0$. This means there is no coupling in this case between bending moments and twist of the composite plate. For a symmetric $[45/-45/-45/45]$ layup, these terms would be non-0.

Example 9.4

Assume a rectangular piece of plywood with the properties of Example 9.3. Assume that the width a and length b of the sheet are 48.0 in. and 24.0 in., respectively. Assume the sheet is simply supported around the edges. Find the deflection at the center of the plate under a uniform load of 1.0 psi.

SOLUTION

From Timoshenko and Woinowsky-Drieger,[8] the deflection w is given by $w = \alpha q_o b^4 / D_{22}$ where $q_o = 1.0$ psi, $b = 24$ in., $D_{22} = 11.47(10^3)$ lb-in., and where α is defined in terms of the parameter $\varepsilon = (a/b)(D_{22}/D_{11})^{0.25}$. For our case, $\varepsilon = (48/24)(1)^{0.25} = 2.0$, and from Table 80 in Timoshenko and Woinowsky-Drieger[8] $\alpha = 0.01013$, $w = 0.0103(1.0)(24)^4/11.47(10^3) = 0.29$ in.

Comment:

This value of deflection is the same as that of an isotropic plate with $D = 11.47(10^3)$ lb-in. The laminate layup in this case has produced a balanced composite plate with $D_x = D_y$, where $D_x = D_1$ and $D_y = D_{22}$ which reacts to a uniform load like an equivalent isotropic plate.

If the layup had $\varepsilon = 0$ for all 4 layers, the D_x value would be higher, but the D_y value would be lower, that is, $D_y/D_x = E_T/E_L = 0.38/1.0$. The plate would not be very stiff in the y-direction.

9.3 THERMAL STRESSES

The thermal stresses from cooling down from the curing temperature must also be included in the analysis to determine ply stresses and subsequent failure. Thermal stresses are handled by calculating equivalent thermal forces and moments due to the temperature change as

$$N^T = \Delta T \sum_{k=1}^{N} \left[\bar{Q}\right]_k \alpha_k (z_k - z_{k-1}) \tag{9.45}$$

and

$$M^T = \frac{\Delta T}{2} \sum_{k=1}^{N} \left[\bar{Q}\right]_k \alpha_k (z_k^2 - z_{k-1}^2). \tag{9.46}$$

Equivalent loads and moments for the laminate are then calculated from

$$N^E = N + N^T \tag{9.47}$$

and

$$M^E = M + M^T. \tag{9.48}$$

The equivalent loads and moments are used to calculate the net mid-plane strains and curvatures as before. The ply stresses are then calculated from the strains and curvatures.

9.4 SUMMARY

The material presented in Chapters 8 and 9 should provide the reader with the fundamentals needed to understand the performance characteristics of composite materials and lamina, and laminates used in the design of engineering structures. The presentation concentrates on the mechanics of a unidirectional composite lamina, with emphasis on its elastic and strength response. A brief development of the micromechanics of a lamina is also presented to provide a basis for understanding how constituent material elastic and strength properties of the fiber and resin contribute to the macroscopic response of the lamina.

The unidirectional (orthotropic) lamina is the primary building block for developing the theory for mechanics of composite laminates. The elastic and strength properties of the orthotropic lamina provide the engineering data needed for design of composite structures. There are many excellent introductory texts available which deal exclusively with composite materials some of which are listed in the bibliography at the end of the chapter. The reader, with the background provided in Chapters 8 and 9, should be able to select from these texts to achieve the next level of understanding on a self-study basis.

HOMEWORK PROBLEMS

9.1. A unidirectional lamina of glass/epoxy composite shows the following strength properties:

$$X_L = 500\,\text{MN/m}^2 \quad X_T = 5\,\text{MN/m}^2$$
$$X_L' = 350\,\text{MN/m}^2 \quad X_T' = 75\,\text{MN/m}^2$$
$$S = 35\,\text{MN/m}^2$$

Estimate the off-axis shear strength of the lamina for fiber orientations of 15° using the Tsai–Hill Failure Theory, when τ_{xy} is the only nonzero stress component.

Solve for

1. Positive shear stress
2. Negative shear stress

Problems 9.2 and 9.3 are suitable for design problems.

9.2. Design an airplane wing for an elliptical lift/unit span distribution (this implies an elliptical profile to the wing as a function of distance from the plane). The area under the lift/unit span curve should be

$$L = K\frac{W}{2}\eta,$$

where

W = weight

K = load factor typically 4, and

η = factor of safety, typically 1.5.

The length of the wing is 3.65 m. The weight of the fuselage is 4000 N. There should be 9.3 m² of lift area on the wing. The design goal is to optimize both cost and weight. Assume that weight is 1.5 times as important as initial cost.

Explore two options. Option one has a cantilevered spar running from the root of the wing to the tip. The skin provides additional strength. When the spar is located 25% of the way back from the leading edge it takes out the twist. Option two is to eliminate the spar and just carry the loads with the skin. Since there is no located spar to take out the twist for this system, you must account for the twisting moment in your analysis.

Hint: You must assume that the wing can be approximated by a laminate with parallel layers separated by a low modulus and strength core. This is analogous to a wing wound around a removable bladder. Additionally make your calculations based on stations from the root of the wing to the tip.

9.3. Design a snowboard using the laminate analysis program. We will determine the loadings as a group in class. The snowboard should have factor of safety of 3 against static/impact failure. The size should be 170 cm × 30 cm. Your design goal should be to minimize cost with the given factor of safety.

9.4. Consider a laminate that is required to act as shear panel predominately (x–y loading). In addition it is required to carry a compressive load in the x-direction, but retain flexibility due to a moment M_x around the x-axis. Assume that the laminate should be balanced as much as possible in its out of plane deformation response to load. What is the minimum number of plies? How should they be oriented? Sketch the layup.

How would incorporation of a foam spacer change this?

For the Problem 9.5 through Problem 9.8 use these material properties or those assigned by your instructor:

$$E_1 = 21.0(10^6)\text{ psi},\quad E_2 = 1.4(10^6)\text{ psi},\quad G_{12} = 0.85(10^6)\text{ psi},\quad \mu_{12} = 0.25.$$

9.5. A standard modulus carbon lamina is loaded with a stress in the longitudinal direction of 10,000 psi. What are the strains and rotations of this lamina?

9.6. A standard modulus carbon lamina is loaded with a stress in the longitudinal direction of 20,000 psi and in the transverse direction with a stress of 15,000 psi. What are the strains and rotations of this lamina?

9.7. A composite laminate must resist a twisting torque around the x-axis. Assume that this laminate is made up of only four plies with a foam spacer. What is the most logical choice for the orientation of the four plies? What advantage is there to four instead of two plies, other than strength?

If a moment is added on the y-face around the x-axis, but its effect is less than the twisting torque, what plies should be added (location, orientation and why)? Sketch the loading on the laminate.

9.8. Consider a helicopter blade that is designed to change its angle of attack as it speeds up. Obviously, the blade will reduce its angle of attack with an increasing speed. Assume that the blade is analogous to the airplane wing of the design project.

Draw the loading on the blade. What lamina would lead to this behavior? Why?

How do you engineer in twist?

Hint: Think about fundamentals of both lamina and laminate analysis in your result. Actually, it may be best to think how you would not engineer in twist.

BIBLIOGRAPHY

Agrawal, B. D. and L. J. Broutman, *Analysis and Performance of Fiber Composites*, J. Wiley & Sons, New York, NY, 1980.

Calsson, L. A. and R. B. Pipes, *Experimental Characterization of Advance Composite Materials*, Prentice-Hall, New York, NY, 1987.

Gibson, R. F., *Principles of Composite Material Mechanics*, McGraw Hill, New York, NY, 1994.

Halpin, J. C., *Primer of Composite Materials*, Technomic Publishing Company, Lancaster, PA, 1984.

Hull, D., *An Introduction to Composite Materials*, Cambridge University Press, New York, NY, 1981.

Jones, J. M., *Mechanics of Composite Materials*, CRC Press, Boca Raton, FL, 1989.

Mallik, P. K., *Fiber-Reinforced Composites: Materials, Manufacturing, and Design*, Marcel Dekker, New York, NY, 1988.

Piggott, M. R., *Loading-Bearing Fiber Composites*, Pergamon Press, Elmsford, NY, 1980.

Tsai, S. W., *Composites Design*, Think Composites, Dayton, Ohio, 1986.

Tsai, S. W. and H. T. Hahn, *Introduction to Composite Materials*, Technomic Publishing Company, Lancaster, PA, 1980.

Vinson, J. R. and R. L. Sierakowski, *The Behavior of Structures Composed of Composite Materials*, Martinus Nijhoof Publishers, Dordrecht, Holland, 1986.

Whitney, J. M., I. M. Daniel, and R. B. Pipes, *Experimental Mechanics of Fiber Reinforced Composite Materials*, SEM Monograph No. 4, Society for Experimental Mechanics, Bethel, Connecticut, 1984.

REFERENCES

1. Jones, R. M., *Mechanics of Composite Materials*, Taylor & Francis, London, 1998.
2. Tsai, S.W. and E. M. Wu, A general theory of strength for anisotropic materials, *J Composite Matter*, 5, 58–60, 1971.
3. Goldenblat, I. and V. A. Koponov, Strength of glass-reinforced plastics in the complex stress state, *Polym Mech,* 1, 54–60, 1966.
4. Tsai, S. W., A survey of macroscopic failure criteria for composite materials, *J Reinforced Plast Composites,* 3(1), 40–62, 1984.
5. Hinton, M. J., A. S. Kaddour, and P. D. Soden, eds., *Failure Criteria in Fibre Reinforced Polymer Composites: The World-Wide Failure Exercise*, Elsevier, London, 2004.
6. Neal-Sturgess, C. E. ed., *Failure of Polymeric Compsite Structures: Mechanisms and Criteria for Prediction of Performance*, SERC/I. Mech.E Annual Expert Meeting, Sopwell House, St. Albans, UK, 1991.
7. Daniel, I. M., Composite materials, *Handbook of Experimental Mechanics*, Chap. 19, pp. 830–833, Prentice-Hall, New York, NY, 1987.
8. Timoshenko, S. and S. Woinowsky-Krieger, *Theory of Plates and Shells*, McGraw-Hill, New York, NY, 1959.
9. Bogetti, T. A., C. P. R. Hoppel, V. M. Harik, J. F. Newill, and B. P. Burns, Predicting the nonlinear response and progressive failure of composite laminates, *Compos Sci Technol*, 64, 329–342, 2004.

10. Bogetti, T. A., C. P. R. Hoppel, V. M. Harik, J. F. Newill, and B. P. Burns, Predicting the nonlinear response and progressive failure of composite laminates: Correlation with experimental results, *Compos Sci Technol*, 64, 477–485, 2004.
11. Cuntze, R. G and A. Freund, The predictive capability of failure mode concept—based strength criteria for multidirectional laminates, *Compos Sci Technol*, 64, 343–377, 2004.
12. Cuntze, R. G and A. Freund, The predictive capability of failure mode concept—based strength criteria for multidirectional laminates: Part B, *Compos Sci Technol*, 64, 487–516, 2004.
13. Puck, A. and H. Schurmann, Failure analysis of FRP laminates by means of physically based phenomenological models, *Compos Sci Technol*, 64, 1045–1068, 1998.
14. Puck, A. and H. Schurmann, Failure analysis of FRP laminates by means of physically based phenomenological models—Part B, *Compos Sci Technol*, 64, 1633–1672, 2002.
15. Liu, K.-S. and S. W. Tsai, A progressive quadratic failure criterion of a laminate, *Compos Sci Technol*, 64, 1023–1032, 1998.
16. Kuraishi, A., S. W. Tsai, and K.-S. Liu, A progressive quadratic failure criterion of a laminate, Part B, *Compos Sci Technol*, 62, 1682–1696, 1998.
17. Zinoviev, P., S. V. Grigoriev, O. V. Lebedeva, and L. R. Tairova, Strength of multilayered composites under plane stress, *Compos Sci Technol*, 58, 1209–1224, 1998.
18. Zinoviev, P., O. V. Lebedeva, and L. R. Tairova, Coupled analysis of experimental and theoretical results on the deformation and failure of laminated composites under plane state of stress, *Compos Sci Technol*, 62, 1711–1724, 2002.
19. Nielsen, L.E., *Mechanical Properties of Polymers and Composites*, Vol. 2, Marcel Decker, Inc., New York, pp. 469–470, 1974.

10 Polymer Processing

10.1 EXTRUSION

A single screw extruder is shown in Figure 10.1. Granules of a polymer are fed into the feed hopper. These granules become compacted, forwarded, heated, and mixed. (The study of the flow of the molten polymer through the extruder is a subject of research in graduate fluid mechanics. The flow is that of a non-Newtonian viscous fluid. One has to be concerned with thermomechanical coupling in the equation of state.) The molten polymer is forced through a die after which the polymer cools and solidifies as shown in Figure 10.2. The polymer expands radially and shrinks longitudinally. The die designer has to be concerned with these thermal changes in dimensions.

What comes out of the die? An anisotropic- or orthotropic- oriented-type structure. This orientation will affect the mechanical properties. The tensile strength will increase in the axial direction and decrease in the transverse direction. The elongation (ductility) will generally be increased in the axial direction for brittle amorphous polymers. However, for ductile crystalline polymers the elongation in the transverse direction can be greater than in the oriented axial direction because of the reorientation of molecular chains that occurs with application of force.

Twin screw extruder machines have been invented to replace the single-screw machines because of certain advantages. The twin screws rotate in the same direction but have opposite tangential velocities in the contact region as shown. The advantages claimed for the twin screw machine are positive transport of material with little backflow, less dependence of the residence time of the polymer in the screw upon the back pressure, easier control of heat input by oil heating of screws and heater bands on the barrel, and greater homogenization due to the shearing between the flights on the screws and due to the calendaring effect of one screw against the other.

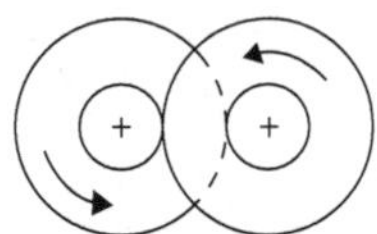

Twin screws

A good twin screw design has eight functions:

1. To take in a maximum amount of powder per screw revolution
2. To transport, melt, and meter material into die without hangup
3. To mix the material without creating too much friction by too small or too large a gap between the outer diameter of one screw and core of other
4. To homogenize and melt material by shearing it between flights

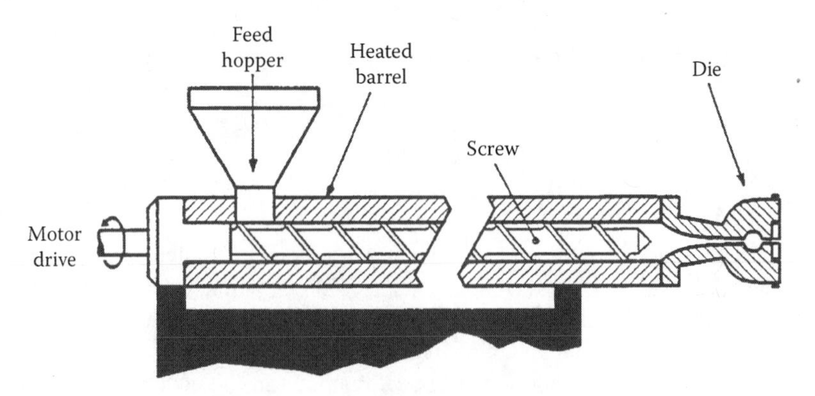

FIGURE 10.1 Schematic diagram of an extruder.

5. To homogenize material by creating slight backflow in certain zones
6. To vent material when it is agglomerate, not when it is still powder or already melt
7. To compress material before venting so that flights are full and so that a vacuum cannot act toward the hopper
8. To create frictional heat to plasticize efficiently, but keep the temperature within a range that is controllable by heating and cooling of screws and barrel

A typical twin-screw extruder is divided longitudinally into six zones with different flights and thread designs in each zone:

1. The *feed zone* has open flights to insure maximum material intake.
2. The *preheating zone* has large surface area multiple-thread design to assure good heat convection.
3. The *preplasticizing zone* should precompress materials and create no overheating.
4. The *throttle zone* should have a high compression to prevent drawing of powder into the vacuum zone.

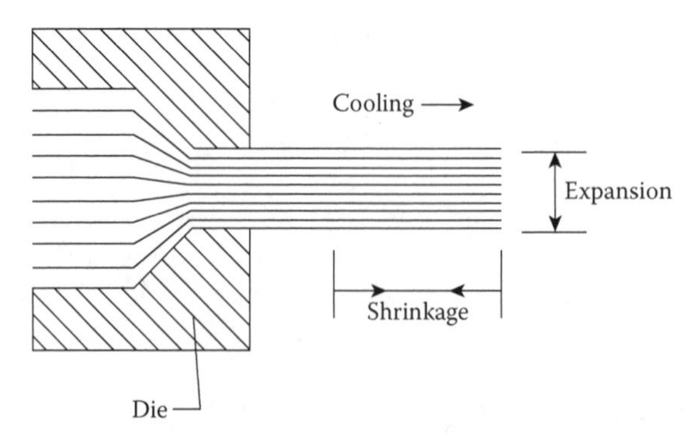

FIGURE 10.2 Cooling of the product on exit from die.

5. The *vacuum zone*—should have good metering action—has to be a decompression zone with open flights to prevent material from getting into the venting holes—should ensure effective degassing.
6. The *metering zone* should have an appropriate pressure buildup and balanced energy to insure good homogenization without overheating to make high output possible.

Very strong gearing is needed to run such a twin-screw machine.

10.2 MANUFACTURE OF PVC PIPE BY EXTRUSION

More than 90% of PVC pipe is produced on twin-screw extruders. Two types of extrusion dies for pipe are shown in Figure 10.3.[1] In the modern design (Figure 10.3a), a more uniform flow and less back pressure occurs. There are three important factors in extrusion die design: the pressure drop, the shear rate, and the residence time.

The pressure drop is calculated from

$$\Delta p = \frac{Q^{1/m}}{K^{1/n} G}, \tag{10.1}$$

where Q = volume flow rate, K = material coefficient in nonlinear stress/strain law: $\dot{\gamma} = K\tau^n$ where $\dot{\gamma}$ is the shear strain rate and τ is the shear stress, and where n and m are power law exponents. G is a die constant that is dependent on the geometry.

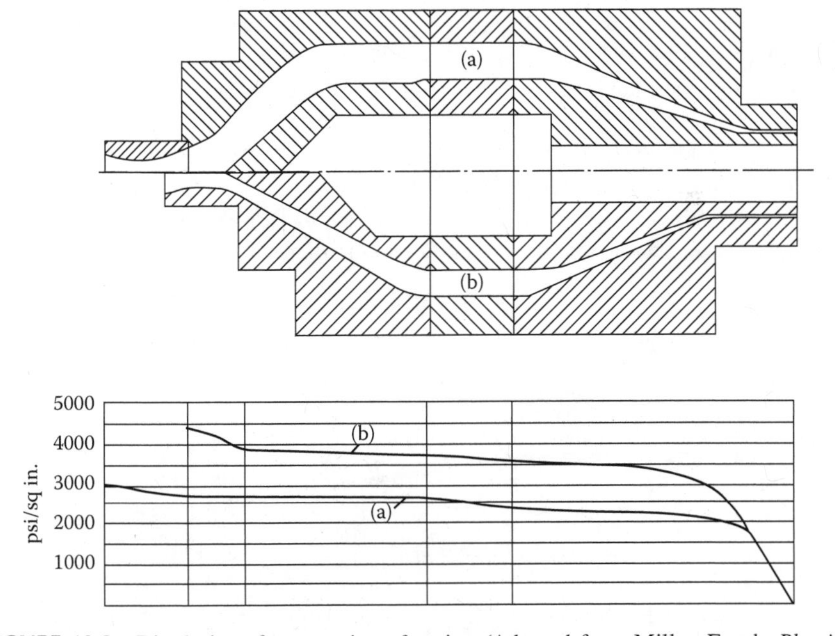

FIGURE 10.3 Die designs for extrusion of a pipe. (Adapted from Miller, E. ed., *Plastics Products Design Handbook, Part A, Materials and Components*, Marcel Dekker, New York, NY, 1981.)

For a thin annulus,

$$G = \frac{\pi}{6}^{1/n} \frac{(R_0 + R_i)^{1/n}(R_0 - R_i)^{2/n+1}}{2L}, \tag{10.2}$$

where L = length.

For

$$\frac{\pi(R_0 + R_i)}{(R_0 - R_i)} \geq 37,$$

the shear rate $\dot{\gamma}$ is given by

$$\dot{\gamma} = \frac{6Q}{\pi(R_0 + R_i)(R_0 - R_i)^2}. \tag{10.3}$$

The residence time τ is given by

$$\tau = \frac{AL}{Q} \tag{10.4}$$

where

$$A = \pi(R_0^2 - R_i^2).$$

Pipe can be coextruded with more than one layer, which results in a composite sandwich wall. Figure 10.4 shows a multimanifold die for extruding PVC or ABS pipe with three layers where the intermediate layer is a foam core. This results in a saving of weight; the foam core pipe is 75–80% of the weight of solid pipe. Such pipe is used in city sewer applications. The ASTM standards for foam core pipe are F-268 for ABS and F891 for PVC.

The main advantage of the sandwich pipe is the increased bending resistance. It is left for an exercise to show that there is little increase in hoop strength (see Homework Problem 10.2). The beam bending stiffness depends on the second moment of area. The second moment of area of a thin pipe is

$$I = \pi R^3 T, \tag{10.5}$$

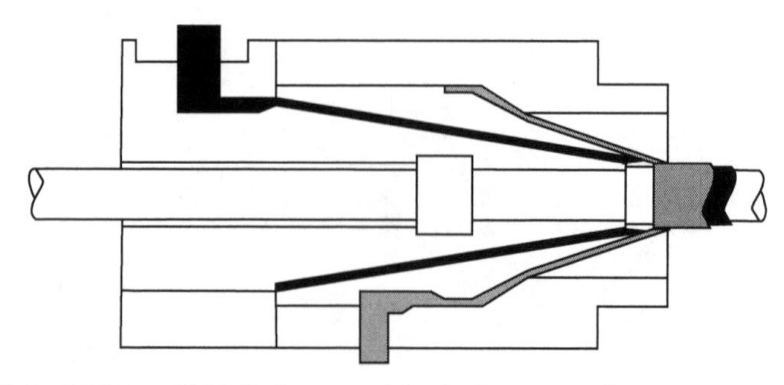

FIGURE 10.4 Multimanifold die for extrusion of a foam core pipe.

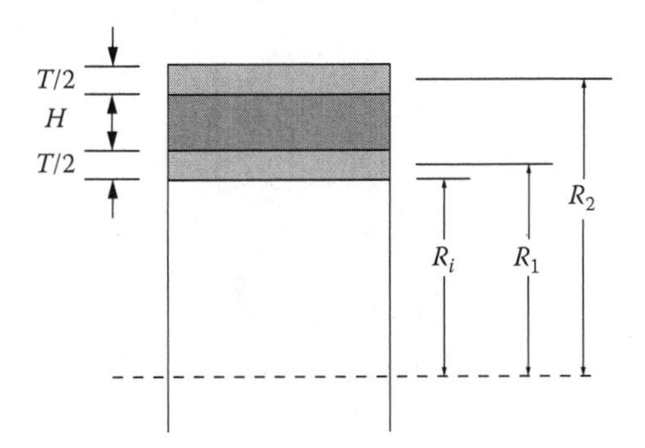

FIGURE 10.5 Dimensions of a foam core pipe.

where R is the mean radius and T is the thickness. Consider a foam core (sandwich pipe) with two layers of thickness $T/2$ separated by a core of thickness H as shown in Figure 10.5. The mean radii of the inner and outer layers are

$$R_1 = R_i + \frac{T}{4} \quad \text{and} \quad R_2 = R_i + \frac{T}{2} + H + \frac{T}{4}, \tag{10.6}$$

where as,

$$R = R_i + \frac{T}{2} \tag{10.7}$$

The increased stiffness is

$$\frac{EI_{\mathrm{f}}}{E_1 I} = \frac{(0.5E_1 R_1^3 T + E_2 R_{\mathrm{b}}^3 H + 0.5E_1 R_2^3 T)}{E_1 R^3 T} \tag{10.8}$$

if the modulus of the core is E_2 and R_{b} is the average radius. For example if $E_2 = 0.5\,E_1$, $H = T$, and $R_i/T = 5$, Equation 10.8 becomes

$$\frac{EI_{\mathrm{f}}}{E_1 I} = 2.008 \tag{10.9}$$

The stiffness is doubled by separating the layers. If the density of the foam cores is 50% of the solid material, then there is only a 50% weight gain for a gain in stiffness of 100%. In other words, the stiffness-to-weight gain is ~2/1.5 = 1.33 for this example.

Plastic pipe is also subject to denting. The denting resistance is dependent on the shell bending stiffness D, where for a solid wall thin pipe

$$D = \frac{ET^3}{12(1-\mu^2)} \,. \tag{10.10}$$

The result is equivalent to the bending stiffness EI of a unit width strip of the wall of the pipe with the addition of a biaxial effect represented by the term involving Poisson's ratio μ. For the equivalent rectangular beam section $I = bT^3/12$ and

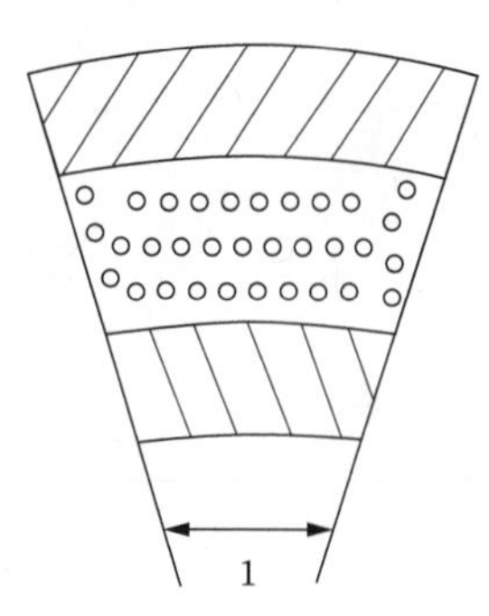

FIGURE 10.6 Unit strip of wall used to represent equivalence of shell bending stiffness with beam bending stiffness.

$b = 1$. See Figure 10.6. For the increase in shell bending stiffness of the foam core pipe, an analogy can be made with the sandwich beam in Chapter 9, Section 9.2.1.

10.3 INJECTION MOLDING

In injection molding, a reciprocating ram is used rather than a screw. Ram pressures up to 30,000 psi are developed in the chamber of the machine, which is heated to a range of 300–700°F. Injection molding is a repetitious process rather than a continuous process (like extrusion). Figure 10.7 shows an injection molding machine. There are three basic components that make up an injection mold: the cavity, the core, and the

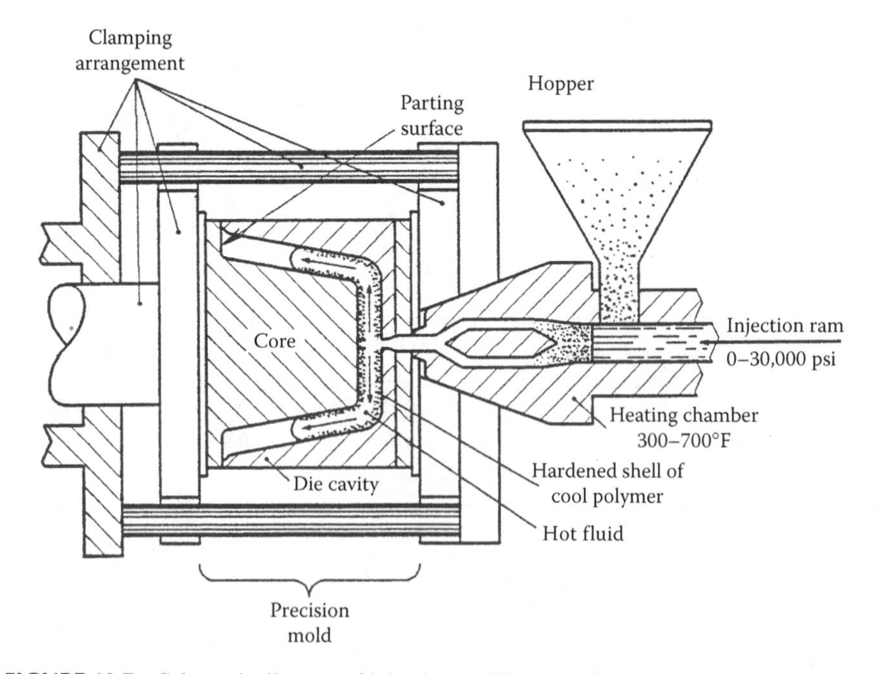

FIGURE 10.7 Schematic diagram of injection molding machine.

mounting plates that clamp the core and cavity together. As the polymer cools in the mold, the hot fluid core becomes surrounded by a hardened shell of cooled polymer at the mold interfaces. In order to reduce the effects of shrinkage upon the dimensions of the final part, the molten polymer must be packed into the mold under pressure.

Computer-aided design (CAD) and modern methods of tool design called "group technology" should be used in mold design. Some companies are designing and using a complete and separate die set for each part they make. This is a waste of money. Applying the group technology concept to the injection mold shown in Figure 10.8, three levels of mold parts can be defined:

Level I comprises the exterior base plates and clamps that never change for a family of parts with the same overall maximum dimensions.

Level II comprises the basic mold sections, core and die cavity main bodies that may or may not change for a subfamily of parts.

Level III comprises the core and die inserts that are specific to an individual manufactured part.

The external shape and sizes of the die inserts are standardized to fit all the basic mold bodies of the family. Likewise, the external screw holes and alignment pins on the mold bodies are standardized to fit the one configuration of base plates. Thus, the designer has the Level I and Level II designs always stored in a database, and one only has to design the interior of the cavity inserts and the exterior of the core inserts (Level III) to fit. This procedure saves design time, and manufacturing and assembly

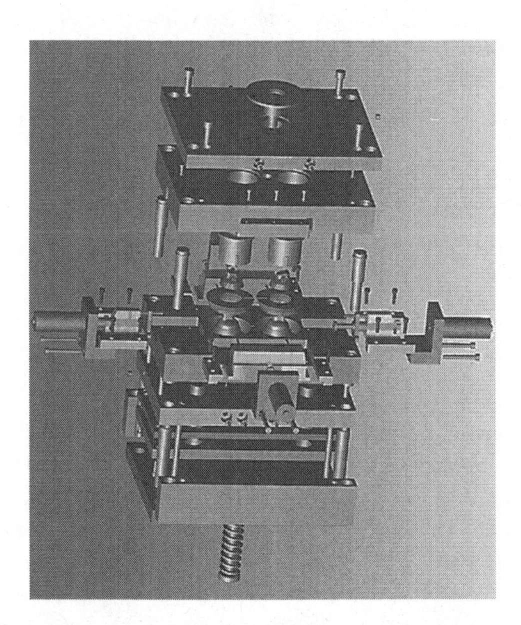

FIGURE 10.8 The level concept of mold design applied to group technology. (Adapted from R and B Mold & Die Solutions, www.rnbusa.com.)

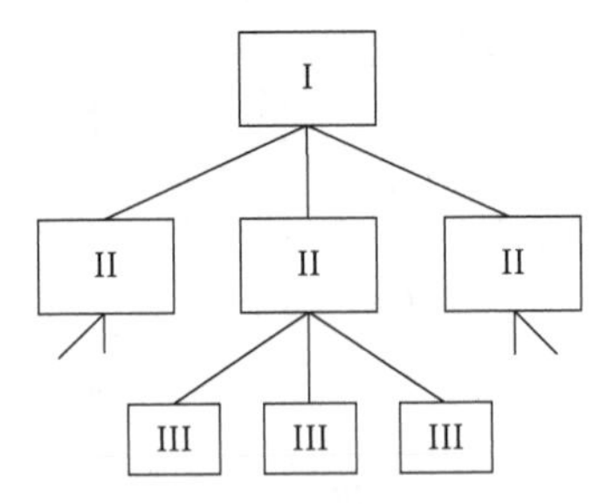

FIGURE 10.9 Family tree of mold components showing different levels.

costs. Figure 10.9 shows schematics of the assembly and design of molds that illustrate the concept of levels.

As in any manufacturing process, the cycle time is important for productivity. The geometry of the molded part affects the molding cycle time. In thick sections the time to remove the heat from the molded part is the limiting factor, whereas in thin sections, the problem is to fill the mold before solidification is complete.

Crystalline thermoplastics can be produced by injection molding, but sometimes a hot mold may need to be used if the polymer crystallizes slowly. There is a large volume change accompanying crystallization (Figure 1.13) and shrinkage during cooling can be quite large. Therefore, molds have to be designed oversized to compensate.

In order to design a mold for shrinkage, an ASTM test bar can be used to check the shrinkage of the material. This test bar is molded at a specific pressure, mold temperature, melt temperature, and cure time. The thickness of the test bar is about 1/8 in. Shrinkage rates of 0.008 in./in. are considered good and readily predictable. Rates greater than 0.010 in./in. are also easily accommodated if the part is symmetrical and can be center gated. A shrinkage ratio greater than 0.0150 in./in. is difficult to predict and control. Examples of the latter are nylon, polyethylene, and acetal polymers.

There are available CAD software programs that model the mold filling process. Many of the commercial finite-element analysis programs contain mold flow modules as well as programs that are specifically developed to perform mold flow analysis.

10.4 THERMOFORMING

Thin, flat sheets of thermoplastic commonly called sheet molding compounds (SMCs) can be formed into various thin shell shapes by various techniques. One is vacuum forming shown in Figure 10.10. The sheet is heat-softened and pulled by the vacuum onto the cold mold surface where it is cooled and hardened. A vacuum corresponding to a pressure difference of only one atmosphere is all that is usually required. Another process, opposite to vacuum forming, uses positive air pressure. The air, rather than the sheet, is preheated. However, air pressure up to 10 atm is required.

Thin sheets are also stamped similar to sheet metal stampings where a punch pushes the sheet into a matching die cavity. In this case, the sheet is usually softened first by preheating.

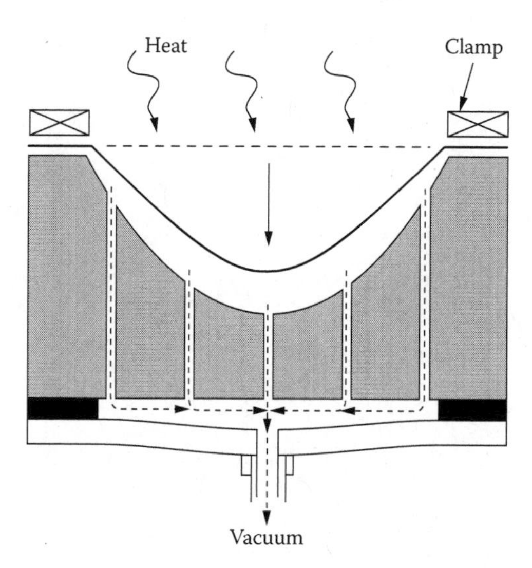

FIGURE 10.10 Vacuum forming of a thermoplastic sheet.

An interesting design application is the forming of inner door panels for automobiles. First a loose, coarse mat of wood fibers and phenolic resin is thermoformed—stamped in warm molds at 250–350°F. This wood/resin composite preform then serves as the mold for the next process—vacuum forming of a vinyl plastic skin over the wood liner. The composite preform is porous and permits pulling of the vacuum directly through it as shown in Figure 10.10.

10.5 BLOW MOLDING

To form hollow parts such as thin-walled bottles, the blow molding process (Figure 10.11) is used. Blow molding starts with a cylindrical hollow preform of smaller diameter called a "parison" which is first produced by extrusion in a continuous molding process, or by injection in an intermittent molding process.

The blow molding cycle consists of four steps:

1. The mold is open and the parison is dropped.
2. The mold is closed and the parison is pinched in two places.
3. Air is blown into the parison and the parison is inflated until it touches the mold whereupon it cools and solidifies.
4. The mold is opened and the part is stripped off.

One of the earlier design factors used in blow molding was the "blow ratio" which was defined as the depth of the cavity divided by the width of the cavity.[4,5] It was recommended that the blow ratio be kept within 1:1 for total biaxiality, which is equal stretching in both directions. However, the blow ratio can change during the process. This change occurs when there are multiple cavities to be filled. When the parison touches the walls of the first cavity, it now must stretch locally to fill

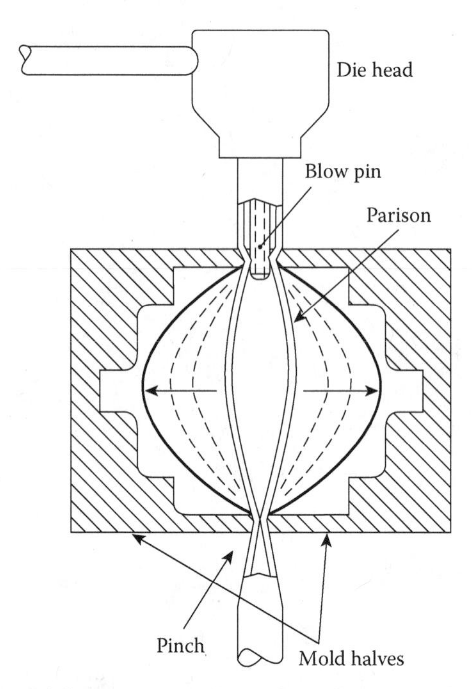

FIGURE 10.11 Blow molding process. (Adapted from Gerdeen, J. C. and M. Huang, *Trans. NAMRI/SME*, 21, 41–49, 1993. This abstract is used by permission of the Society of Manufacturing Engineers, Copyright 1993 from NAMRC XXI Proceedings.)

another subsequent cavity. This new cavity now has a new blow ratio. One of the main problems in design of the parison is to provide sufficient thickness in the right regions to stretch the parison into cavities without blowing holes in the wall of the stretched part.

Recently, other kinds of stretch ratios have been defined for blow molding to better define the limits of the process for different materials.[5] The hoop ratio H is defined as the ratio of the maximum outside dimension (diameter) of the finished molded part to the maximum outside dimension of the parison after emerging from the die (parison diameter):

$$H = \frac{D}{d}, \tag{10.11}$$

where D is the diameter of the blown part, and d is the diameter of the parison. Normally, a hoop ratio by this definition of 4:1 is considered a practical limit; however, values of 7:1 have been achieved in some cases.

Similarly, an axial ratio A can be defined as

$$A = \frac{L}{l}, \tag{10.12}$$

where L is the length of the blown part and l is the length of the parison. Axial ratios A range from 1.4 to 2.6.[5] A resultant blowup ratio BUR is defined as the product of the hoop ratio and the axial ratio:

$$\mathrm{BUR} = HA = \frac{DL}{dl} \tag{10.13}$$

Note that none of the definitions of blow ratio take the thickness into account. However, if the BUR ratio is applied to blow forming a cylinder from a cylindrical parison, and if volume constancy is assumed during the process, then the thickness ratio can be approximated by

$$\frac{t}{T} = \mathrm{BUR} \tag{10.14}$$

where T is the thickness of the blown part and t is the thickness of the parison. Different materials have their own stretch limits. Table 10.1 gives the stretch limits in terms of BUR for some materials.

Table 10.1 shows why PET is the material used to form 2-L soft-drink bottles. There is another reason in addition to the high BUR ratio, and which is that uniform stretch is possible like in blowing up a balloon. Whereas in other materials like PE and PVC, and local blow out failures occur because these materials are very sensitive to slight variations in heating which cause local weak spots to occur; that is, they have a narrow range (or window) of temperature. PET on the other hand which has a melting temperature of 480°F, can be blow molded at 205°F for optimum orientation.

Many structural parts are now being formed by blow molding from engineered glass-reinforced resins such as ABS, including automobile bumpers and dash boards, and hollow structural panels for appliances. Blow molding offers savings in tooling and assembly costs over multiple-pieced injection-molded parts which must be later joined or fastened together.

Gerdeen et al.,[4,6,7] have analyzed blow molding of spherical shapes. Two examples in the analysis of blow molding of cylinders will be considered here.

TABLE 10.1
Stretch Limits BUR for Various Polymers

Material	Stretch Limit, BUR	Temperature (°F)
PET	16	195–240
PP	6	260–280
PVC	7	210–230
PC	12	
PS	12	290–320

Source: Adapted from Belcher, S. L. *Plastic Blow Molding Handbook*, Van Nostrand Reinhold, New York, 1990.

10.5.1 Inflation

Example 10.1

The first problem is to determine the pressure as a function of time for the inflation phase of a certain design. If the material properties were known at elevated temperatures this could more easily be done. However, often these data are not known. Therefore, the following design procedure is postulated:

Step 1: Conduct experiments on a simulated problem.
Step 2: Measure the pressure as a function of time response, $P(t)$.
Step 3: Use a mathematical model to determine the mechanical properties modulus E and viscosity η at the operating temperature.
Step 4: If this model works then the determined properties can be used to assess the inflation phase of the proposed design.

For the simulated problem, assume the parison and the final shape are both cylindrical as shown in Figure 10.12. The dimensions of the original parison are a mean diameter of 1.00 in. and a thickness of 0.10 in. The final diameter of the product is 3.00 in. Consider "plane strain" such that the longitudinal strain $\varepsilon_x = 0$, and only the circumferential hoop strain ε_θ and the thickness strain ε_t are non-zero.

Some experiments were run,[4] and the following pressure/time response in Figure 10.13 was found for HDPE at a temperature of 180°C. Note that the pressure rises, reaches a maximum, and then drops off until contact with the wall at 15 s where there is a sharp rise in pressure. If one has experience blowing up a toy rubber balloon, then one has experience to confirm the fact that initially one has to exert a higher pressure to initiate the inflation phase. Thereafter, it is easier to continue the inflation. (This represents parts 1 and 2 of the prior procedure.)

To complete step 3, a Maxwell fluid model is assumed and E and η are determined using the following method. A constant strain-rate process is assumed, where the strain rate is estimated using the final radius at the contact time with the mold. Then different values of (P, t) and $(\dot{P}, t)$ are used to estimate E and η, taking into account the concepts of initial elastic response and retarded viscous response.

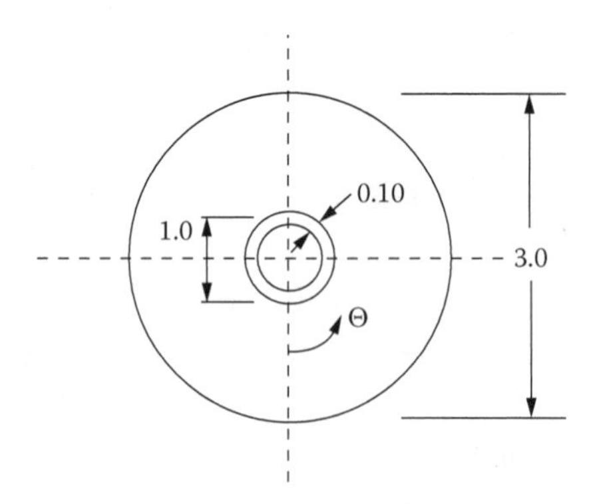

FIGURE 10.12 Geometry of blow-molded cylinder example. (Adapted from Gerdeen, J. C. and M. Huang, *Trans. NAMRI/SME*, 21, 41–49, 1993. This abstract is used by permission of the Society of Manufacturing Engineers, Copyright 1993 from NAMRC XXI Proceedings.)

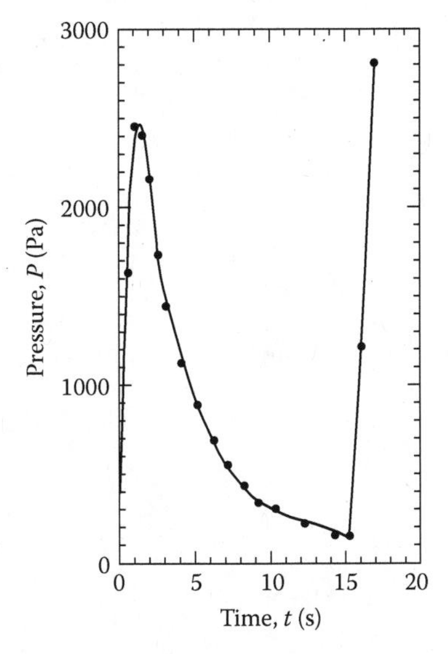

FIGURE 10.13 Pressure (time) response for blow molding HDPE cylinder. (Adapted from Gerdeen, J. C. and M. Huang, *Trans. NAMRI/SME*, 21, 41–49, 1993. This abstract is used by permission of the Society of Manufacturing Engineers, Copyright 1993 from NAMRC XXI Proceedings.)

The solution of the blow molding of the cylinder is outlined in the following.[4,6–8] The circumferential true strain and strain rate are given by

$$\varepsilon = \varepsilon_\theta = \ln\left(\frac{R}{R_0}\right) \tag{10.15}$$

and

$$\dot{\varepsilon} = \frac{1}{R}\frac{\mathrm{d}R}{\mathrm{d}t} \tag{10.16}$$

The differential equation for the Maxwell fluid (Equation 3.9) assuming true stress and true strain, is

$$\dot{\sigma} + \frac{E}{\eta}\sigma = E\,\dot{\varepsilon} \tag{10.17}$$

where $\sigma = \sigma_\theta$.

The equilibrium equation for the circumferential direction for the deformed position in terms of true stress is

$$\sigma = \frac{PR}{T}, \tag{10.18}$$

where σ, P, R, and T are all functions of time. Assume that the volume remains constant (for a dense polymer with no voids):

$$V_0 = 2\pi R_0 T_0 L_0 = V = 2\pi R T L_0.$$

Thus,

$$R_0 T_0 = RT. \tag{10.19}$$

For a constant strain-rate process, the solution of Equation 10.17 is

$$\sigma = Ce^{\left(-\frac{-Et}{\eta}\right)} + \eta\dot{\varepsilon}. \tag{10.20}$$

The initial condition is $\sigma = 0$ when $t = 0$, which gives

$$C = -\eta\dot{\varepsilon}. \tag{10.21}$$

Thus,

$$\sigma = \eta\,\dot{\varepsilon}\left(1 - e^{-\frac{Et}{\eta}}\right) = \frac{PR}{T} \tag{10.22}$$

and

$$\dot{\sigma} = \eta\dot{\varepsilon}\frac{E}{\eta}e^{-\frac{Et}{\eta}} \tag{10.23}$$

Solving for P and eliminating T gives

$$P = \sigma\frac{T}{R} = \sigma R_0 \frac{T_0}{R^2} \tag{10.24}$$

Then,

$$\dot{P} = \dot{\sigma} R_0 \frac{T_0}{R^2} - \sigma R_0 T_0 \frac{2\dot{R}}{R^3} \tag{10.25}$$

For the specific problem at hand

$$\dot{\varepsilon} = \frac{\ln\left(\frac{3}{1}\right)}{15} = 0.073\frac{1}{s}\ . \tag{10.26}$$

Thus,

$$\dot{\sigma}(0) = E\,\dot{\varepsilon}, \quad \dot{P}(0) = E\,\dot{\varepsilon}\,R_0\frac{T_0}{R_0^2} - 0. \tag{10.27}$$

Taking the initial elastic response as the slope of the initial slope of the curve (Figure 10.13) gives

$$\dot{P}(0) = \frac{1500}{0.5} \tag{10.28}$$

and

$$E = \frac{\left(\frac{1500}{0.5}\right)\left(\frac{0.5}{0.1}\right)}{0.073}$$

or

$$E = 205{,}500\,\text{Pa} = 29.7\,\text{psi} \tag{10.29}$$

Next match the maximum pressure at $t = 1.5$ s. First from Equations 10.15 and 10.16:

$$\frac{R}{R_0} = e^{0.073t} = 1.116 \quad \text{when } t = 1.5, \tag{10.30}$$

which gives

$$R = 0.5(1.116) \quad \text{at}\, t = 1.5\,\text{s}. \tag{10.31}$$

Assume that the time where P is maximum corresponds to the characteristic time for the material, that is try

$$\tau = \frac{\eta}{E} = 1.5 \text{ s}. \tag{10.32}$$

Then Equations 10.22 and 10.24 give

$$\text{P} = 2500 = \frac{\eta(0.073)(0.5)(0.1)(4)\left(1 - \frac{1}{e}\right)}{(1.116)^2} \tag{10.33}$$

which gives

$$\eta = 340{,}000 \text{ Pa} - s(49.5 \text{ psi} - s). \tag{10.34}$$

Substitution of η from Equation 10.34 and E from Equation 10.29 approximately confirms Equation 10.32.

Next check the pressure at $t = 15$ s.

$$P = \frac{340{,}000(0.073)(0.5)(0.1)(1 - e^{-10})}{(1.5)^2}$$
$$= 554 \text{ Pa}$$

which is a little high as shown in Figure 10.14. This indicates that we may need an additional dashpot in the model to model the longer time response. Also, the strain rate is not exactly constant over the whole range of response as shown in Jorgensen et al.[4] However, the analysis is useful for estimating the response of the process.

Analysis of blow molding of spherical shapes and experiments are described in Jorgensen et al.[4] Reasonably good results have been obtained.

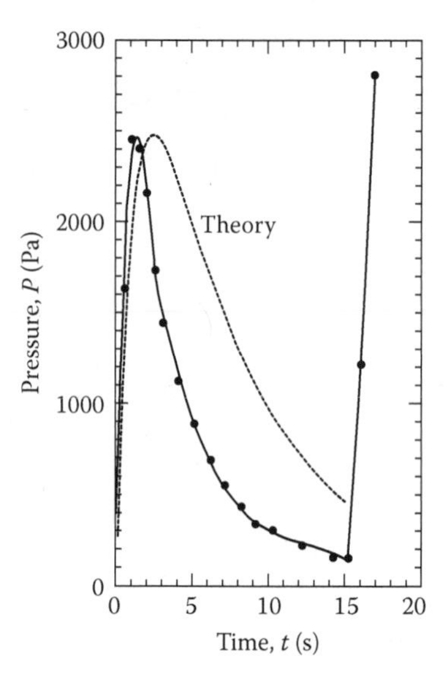

FIGURE 10.14 Theoretical versus experimental data for blow molding of a cylinder. (Adapted from Gerdeen, J. C. and M. Huang, *Trans. NAMRI/SME*, 21, 41–49, 1993. This abstract is used by permission of the Society of Manufacturing Engineers, Copyright 1993 from NAMRC XXI Proceedings.)

10.5.2 Cooling Phase

Example 10.2

The second example problem involves the cooling stage after the molten polymer hits the cool metal mold. After the polymer is solidified and has become sufficiently rigid it is removed from the mold. The cooling stage in the blow molding operation accounts for a large proportion of the production cycle. Therefore, the ability to estimate the cooling time is of practical interest.

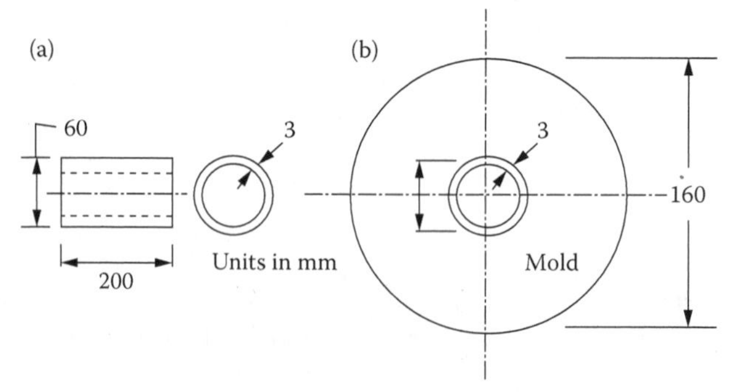

FIGURE 10.15 Dimensions of cylinder and mold for analysis of cooling stage. (a) Polymer cylinder, (b) mold. (Modified from Gerdeen, J. C. and M. Huang, *Trans. NAMRI/SME*, 21, 41–49, 1993. This abstract is used by permission of the Society of Manufacturing Engineers, Copyright 1993 from NAMRC XXI Proceedings, Figure 5.)

Consider again an idealized blow molding operation where a cylindrical container of dimensions shown in Figure 10.15. Also assume that the metal mold can be idealized as a cylindrical container with a wall thickness of 50 mm as shown in Figure 10.15.

For this idealized blow molding operation the objectives are the following:

1. Determine the temperature/time curves for both the polymer and the mold.
2. Determine the necessary cooling time in the mold before it can be removed.
3. Determine the effects of the following variables on the cooling time:
   a. Material properties of the polymer
   b. Thickness of the polymer
   c. Thickness of the mold
   d. Temperature of the mold
   e. Heat transfer coefficient between polymer and air

In order to proceed with the solution the following assumptions are made:

1. One-dimensional heat transfer in the x direction in Figure 10.16
2. Perfect thermal contact between the polymer and the mold
3. Convection between the polymer and air
4. Constant air temperature of $T_a = 20°C$
5. Any heat transfer from the outside wall of the mold is small enough to be neglected (Check dimension of mold to assure this.)
6. For the present analysis the material property data are constant with temperature, which is not necessarily true. For example, the specific heat of HDPE does vary with temperature as shown in Figure 10.17

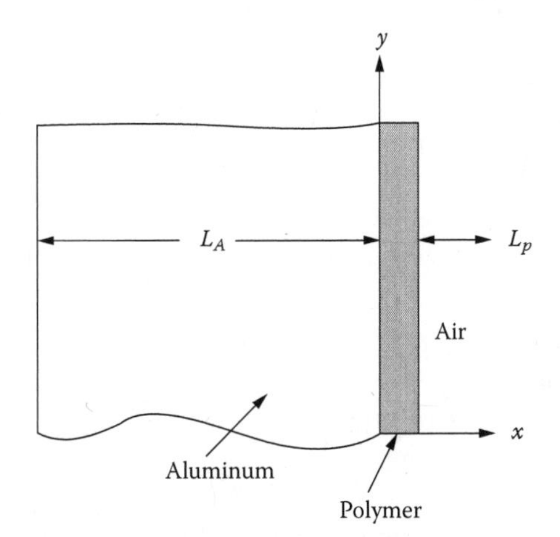

FIGURE 10.16 One-dimensional heat transfer model. (Adapted from Gerdeen, J. C. and M. Huang, *Trans. NAMRI/SME*, 21, 41–49, 1993. This abstract is used by permission of the Society of Manufacturing Engineers, Copyright 1993 from NAMRC XXI Proceedings.)

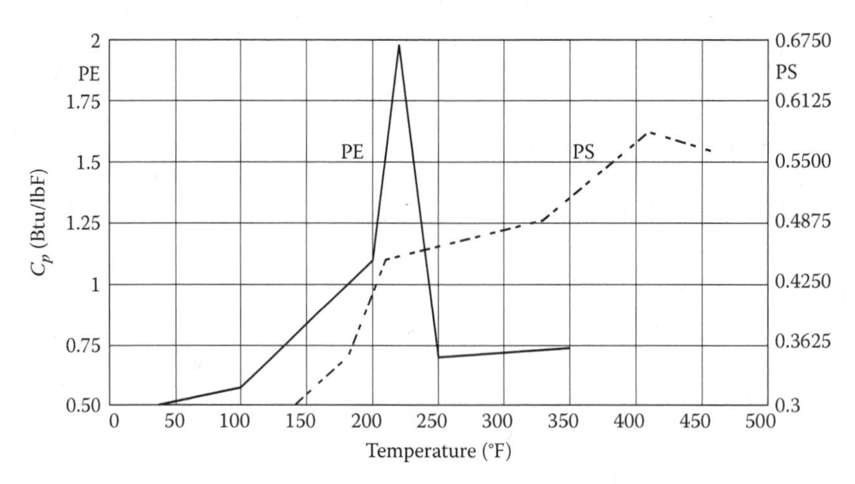

FIGURE 10.17 Variation of specific heat with temperature for HDPE. (Adapted from *The Effect of Temperature and Other Factors on Plastics*, Plastic Design Library Staff, New York, NY, 1991.)

Under assumption 5, a lumped analysis then is possible if the appropriate Biot modulus[10] for the polymer satisfies the following condition:

$$Bi = \frac{hL_p}{k} < 0.1. \tag{10.35}$$

For the data given in Table 10.2,

$$Bi = \frac{10(0.003)}{0.46} = 0.065,$$

and so this condition is satisfied.

To satisfy assumption 5, the dimension of the aluminum mold needs to be checked. We require that

$$h_A < h_p \tag{10.36}$$

TABLE 10.2
Assumed Material Data for Heat Transfer Analysis

Material Property	HDPE	Aluminum Mold
Density, kg/m³	$\rho_p = 940$	$\rho_A = 2700$
Specific heat, J/(kg °C)	$C_p = 2700$	$C_A = 890$
Thermal conductivity, J/(m s °C)	$k_p = 0.461$	$k_A = 170$
Initial temperature, °C	$T_i = 180$	$T_a = 20$

Source: Adapted from Gerdeen, J. C. and M. Huang, *Trans. NAMRI/SME*, 21, 41–49, 1993. This abstract is used by permission of the Society of Manufacturing Engineers, Copyright 1993 from NAMRC XXI Proceedings.

or in terms of the Biot number,

$$L_A < \frac{L_p k_A}{k_p} = \frac{0.003(170)}{0.461} = 1.1 \text{ m}, \tag{10.37}$$

which is satisfied for $L_A = 0.05$ m in Figure 10.15.

Perfect thermal contact between the polymer and the aluminum mold and satisfaction of Equation 10.36 means that the heat transfer rate by conduction can be represented approximately by a lumped system, such that

$$\frac{q}{A} = -\frac{\left(k_A(T_i - T_A)/L_A\right)}{2} = -\frac{\left(k_p(T_p - T_i)/L_p\right)}{2} \tag{10.38}$$

If the temperature at the interface T_i is eliminated, there results

$$\frac{q}{A} = -K(T_A - T_p) \tag{10.39}$$

where K is the effective or reduced thermal conductivity of the lumped system given by

$$K = 2\left(\frac{k_A}{L_A}\right)\frac{(k_p/L_p)}{(k_A/L_A) + (k_p/L_p)} \tag{10.40}$$

The energy balances for the aluminum mold and for the polymer are given by

$$KA_A(T_A - T_p) = -\rho_A C_A V_A \frac{dT_A}{dt} \tag{10.41}$$

$$KA_p(T_p - T_A) + hA_p(T_p - T_A) = -\rho_p C_p V_p \frac{dT_p}{dt} \tag{10.42}$$

These two first order differential equations can be solved for simultaneously numerically using a Runge–Kutta or other differential equation solver. Here, however, a closed-form solution is found by first solving for T_A from Equation 10.42, differentiating and substituting into Equation 10.41. This gives a second-order differential equation for T_p represented by

$$A_1 \frac{d^2T_p}{dt^2} + A_2 \frac{dT_p}{dt} + hT_p = hT_a \tag{10.43}$$

where,

$$A_1 = \frac{\rho_A C_A L_A \rho_p C_p L_p}{K} \tag{10.44}$$

$$A_2 = \rho_p C_p L_p + \rho_A C_A L_A\left(1 + \frac{h}{K}\right)$$

which, for the example data, become

$$A_1 = 3.111(10^6), \quad A_2 = 0.132(10^6) \tag{10.45}$$

The solution of Equation 10.43 is

$$T_p = C_1 e^{mt} + C_2 e^{nt} + T_a \tag{10.46}$$

where

$$m,n = \frac{(-A_2/A_1) \pm \sqrt{(A_2/A_1)^2 - 4h/A_1}}{2} \quad (10.47)$$

which for the above data become

$$m = -7.598(10^{-5}), \quad n = -0.04230. \quad (10.48)$$

Applying the initial conditions at $t = 0$:

$$T_p = T_i = C_1 + C_2 + T_a \quad (10.49)$$

$T_A = T_a$, or from Equation 10.42:

$$KA_p(T_i - T_a) + hA_p(T_i - T_a) = -p_p C_p V_p \frac{dT_p}{dt} \quad (10.50)$$

This results yields

$$C_2 = \left(\frac{K + h}{p_p C_p L_p} + m \right) \frac{T_i - T_a}{m - n} \quad (10.51)$$

where C_1 can be found from Equation 10.49. The solution for the constants is

$$C_2 = 151.016, \quad C_1 = 8.9845 \quad (10.52)$$

The solution for the temperatures T_p and T_A is plotted in Figure 10.18 where it is shown that the polymer temperature approaches the mold temperature after

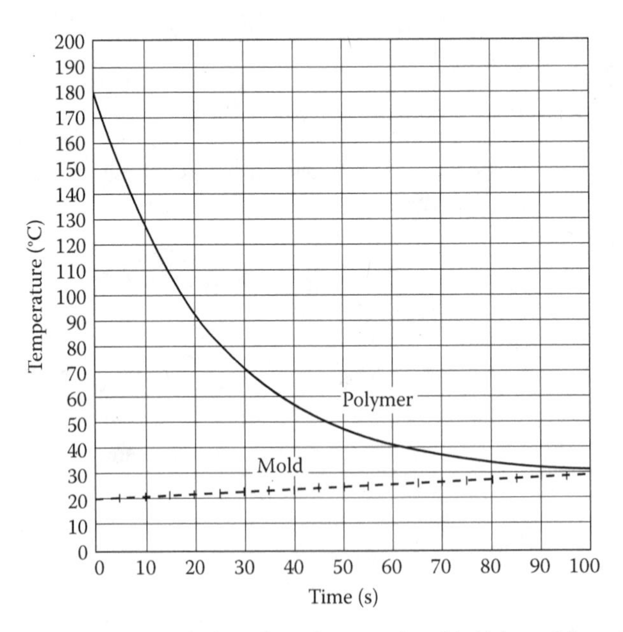

FIGURE 10.18 Temperature solutions for polymer and mold. (Adapted from Gerdeen, J. C. and M. Huang, *Trans. NAMRI/SME*, 21, 41–49, 1993. This abstract is used by permission of the Society of Manufacturing Engineers, Copyright 1993 from NAMRC XXI Proceedings.)

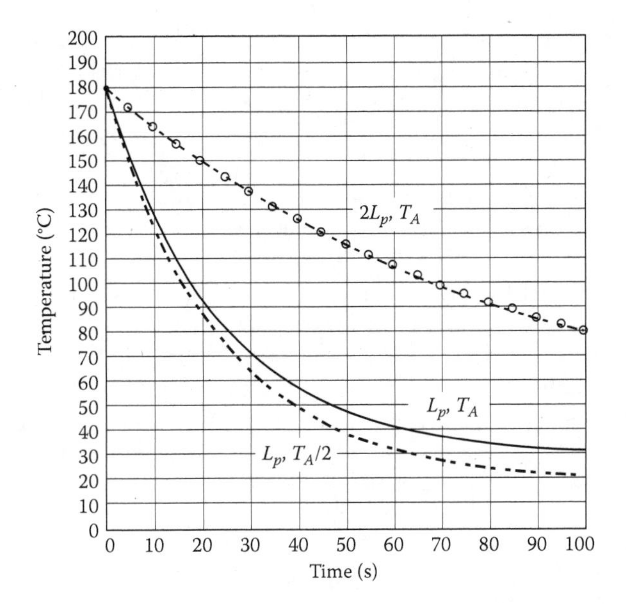

FIGURE 10.19 Effects of polymer thickness and mold temperature on a cooling curve. (Adapted from Gerdeen, J. C. and M. Huang, *Trans. NAMRI/SME*, 21, 41–49, 1993. This abstract is used by permission of the Society of Manufacturing Engineers, Copyright 1993 from NAMRC XXI Proceedings.)

30 s. The mold temperature only slightly rises to 22°C after 30 s. However, in mass production there would be a 2°C temperature rise of the mold per cycle without water cooling.

The effect of increased polymer thickness and decreased mold temperature on the cooling curve is shown in Figure 10.19.

HOMEWORK PROBLEMS

10.1. In blow molding operations, a tubular parison of molten polymer, usually produced by an extruder, is blown by gas pressure (p) against the walls of a metal mold, as illustrated in Figure 10.11.

The problem is to determine the pressure (time) response for the inflation phase. If material properties were known at elevated temperature this could be done. But they are not. Therefore, the following design procedure is postulated:

Step 1: Conduct experiments on a simulated problem.
Step 2: Measure the pressure (time) response $p(t)$.
Step 3: Use a mathematical model to determine the mechanical properties modulus E and viscosity η at temperature.
Step 4: If this model works then the determined properties can be used to assess the design of the inflation phase.

In an idealized blow molding operation consider that both the parison and final shape are cylindrical as shown in Figure 10.12. Assume the original parison has a mean diameter of 1.00 in., and at thickness of 0.10 in. Assume that the final diameter of the product is 3.00 in. Assume "plane strain" such that the longitudinal strain $\varepsilon_x = 0$. Consider only the circumferential hoop strain ε_h and the thickness strain ε_t.

Experiments were run and the pressure/time response (Figure 10.13) was found for HDPE at a temperature of 180°C. (This represents parts of steps 1 and 2 of the prior procedure.)

To complete part (3) of the problem, assume a Maxwell fluid model and determine E and η using the following method:

Assume a constant strain rate process during inflation, $\dot{\varepsilon} = $ constant. Estimate the strain rate using the final radius at the contact time with the mold. Then use different values of (p, t) or $(\mathrm{d}p/\mathrm{d}t, t)$ to estimate E and η. (Hint: Use the concepts of "elastic response" and "viscous response.") Then calculate a theoretical $p(t)$ curve and see how it fits the experimental data overall.

With the E and η values determined, what is the effect of changing the strain rate? Give any recommendations for an improved model if considered necessary.

10.2. Show that the little increase in hoop stress is a sandwich pipe with a foam core as shown in Figure 10.5. Assume that the thickness of the foam core H is equal to the thickness of each pipe layer $T/2$. Additionally, assume that $R_i = 10(T/2)$.

REFERENCES

1. Miller, E. ed., *Plastics Products Design Handbook, Part A, Materials and Components*, Marcel Dekker, New York, NY, 1981.
2. R and B Mold & Die Solutions, www.rnbusa.com.
3. *The Design Book, Design Concepts in Blow Molding Engineering Thermoplastic*, General electric, PBG-13, Pittsfield, MA.
4. Jorgensen, H. K., J. C. Gerdeen, K. J. Weinmann, and M. G. Hansen, An analytical and experimental study of the inflation stage in blow molding of polymers, *16th NAMRI Proc., SME*, May 25–27, 1988.
5. Belcher, S. L. *Plastic Blow Molding Handbook*, Van Nostrand Reinhold, New York, 1990.
6. Gerdeen, J. C. *A Theory for the Blow Molding of Polymers, Third International Conference on Computer-Aided Production Engineering*, University of Michigan, Ann Arbor, MI, June 1–3, 1988.
7. Gerdeen, J. C., The prediction of free equilibrium shapes in the forming of sheet materials, NUMIFORM 89, *Proceedings of 3rd International Conference on Numerical Methods in Forming Processes*, Ft. Collins, CO, June 26–30, 1989.
8. Gerdeen, J. C. and M. Huang, Polymer blow molding process design, *Trans. NAMRI/SME*, 21, 41–49, 1993. (This abstract is used by permission of the Society of Manufacturing Engineers, Copyright 1993 from NAMRC XXI Proceedings.)
9. *The Effect of Temperature and Other Factors on Plastics*, Plastic Design Library Staff, New York, NY, 1991.
10. Incropera, F. P. and D. P. Dewitt, *Fundamentals of Heat and Mass Transfer*, 4th ed., John Wiley & Sons, New York, NY, 1996.

11 Adhesion of Polymers and Composites

11.1 INTRODUCTION

An adhesive is a material (typically polymeric) which forms a structural bond between two materials. Adhesion is when two surfaces are held together. The two surfaces that are held together are called adherends. A discussion on the details of a simple adhesive joint will aid in visualization of the concepts presented in this chapter. Examination of an idealized and a realistic adhesive joint will elucidate the issues required when using adhesives, coatings, matrix materials, and reinforcements for composites. Figure 11.1 shows both the idealized and the realistic adhesive joints. In the idealized case there are three materials, which result in four bonding surfaces and two interfaces that create a simple joint. A generally held view of a surface is that it is a discrete boundary with no associated dimension. However, relative to adhesion, a surface has dimension. The first thing to consider is the surface roughness, or more appropriately, the surface topography. All surfaces, regardless of how smooth they appear, possess topography. Additionally, the material properties near the surface are not necessarily the same as in the bulk material. Thus, the interfacial zone between materials that are adhered cannot realistically be viewed as possessing zero thickness as in the idealized case. In fact, due to the surface roughness effects, varying material properties, and the possible mixing of materials, the interface is referred to as the interphase. The interphase is the zone where the material properties are not the same as the bulk materials. The interphase zone has an associated dimension dependent on the materials and their preparation.

11.2 FUNDAMENTALS OF ADHESION

11.2.1 Wetting and Work of Adhesion

A review of the fundamentals of surface science will establish both the technical basis for adhesion and provide methods to quantitatively determine a surface's ability to be bonded. From a physics point of view, objects and materials are either attracted or repulsed from one another. Since adhesives are typically used in the liquid state, the terms -phillic and -phobic are often used to describe the attraction or repulsion, respectively. For example, materials that attract water are called hydrophilic, while those that repulse water are hydrophobic. Water is a good example, since historically many adhesives have been solvent borne, but now with increasing environmental consciousness there is a large effort to replace the solvent-borne adhesives with aqueous or water-borne adhesives.

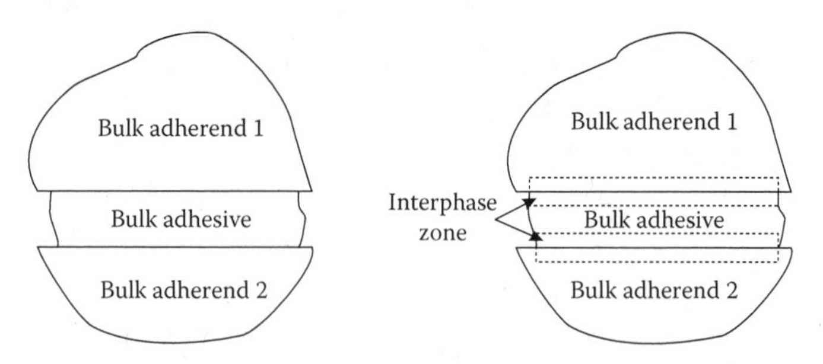

FIGURE 11.1 Adhesive joints.

It should be apparent that it is of interest to either know or rank the surface attraction of adherend materials to prospective adhesives. The surface attraction of a material is referred to as the surface energy, surface free energy, or surface tension. Therefore, measurement of the surface energy of a material will rank the attraction of that material to the prospective adhesive. If possible it is desired to measure the attraction of the adhesive to the surface materials that are to be used as adherends. Often, this is not possible and a more fundamental measurement is made to determine the surface energy.

A simple, yet powerful, method of determining or ranking surface energy of materials is through the use of contact angle analysis. Contact angle analysis measures the angle formed by a drop of liquid on a smooth, flat surface. The angle of contact of the liquid drop is indicative of the attraction or repulsion of the surface and liquid to each other. The higher the attraction, the more the liquid is pulled in contact with the surface and the lower the angle of contact. When this angle of contact approaches 0°, the liquid is said to wet the surface and the liquid will spread on the surface. For phobic materials, the liquid drop will bead, and the angle of contact can be greater than 90°. This method involves observing the liquid droplet and surface interface from the side and optically measuring the angle of contact as shown in Figure 11.2. It is not uncommon to be able to resolve the contact angle within ±1°. Another method to determine the surface energy is dynamic contact analysis, where fibers or other shapes are sequentially pushed into a liquid and then pulled out of the liquid. The greater the attraction or wetting of the liquid to the solid surface, the greater the measured force to pull out the solid material will be. Both of these methods can be used with the actual materials. In fact, a cursory examination of wettability can be made visually, without the use of laboratory equipment.

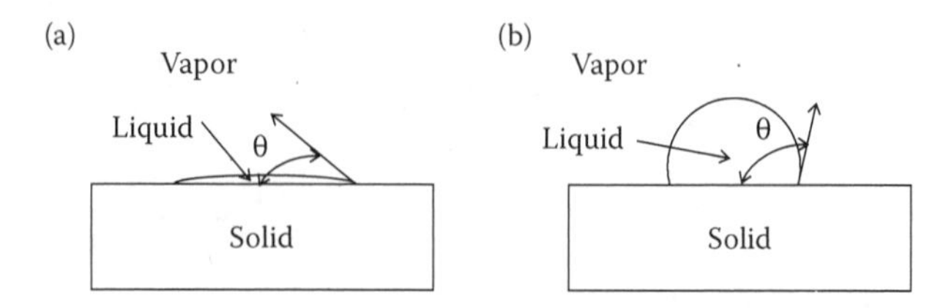

FIGURE 11.2 Contact angle measurement. (a) High surface energy substrate and (b) low surface energy substracte.

The preceding section has discussed the philosophy behind the surface measurements. Now the theory will be presented, which will allow for quantification of the attraction between the solid and liquid. Young's equation for the balance of forces when a liquid drop is placed on a surface is

$$\gamma_{sv} = \gamma_{sl} + \gamma_{lv}\cos\theta \tag{11.1}$$

where

γ_{sv} = surface tension, or surface free energy of the solid–vapor interface
γ_{sl} = surface tension, or surface free energy of the solid–liquid interface
γ_{lv} = surface tension, or surface free energy of the liquid–vapor interface
θ = the angle of the liquid drop on the surface.

The vapor refers to the gaseous environment in which the measurement is made. The term of interest in Equation 11.1 is the surface energy γ_{sl}, of the solid–liquid interface, which relates to the surface tension for the solid γ_s. Even though γ_{sl} and thus γ_s cannot be calculated directly from the contact angle measurement, the angle is strongly indicative of the value of γ_s.

This value is relevant to adhesion through the Young–Dupré equation which relates the surface tension to the free energy of adhesion.

$$\Delta G_{sl} = \gamma_{sl} - \gamma_s - \gamma_l = -W_{sl} \tag{11.2}$$

where

W_{sl} = work of adhesion
γ_s = surface tension, or surface free energy of the solid interface
γ_l = surface tension, or surface free energy of the liquid interface

Where is all this going? Basically, the measurement of the contact angle that a liquid drop makes on a surface gives an indication of the surface free energy of that surface. Through the use of a variety of known liquids the actual surface energy of a material can be obtained. A ranking of the surface energy of a variety of surfaces can be performed by the sole use of water. Most of us have experience with waxing a car and watching water bead on the surface. When waxed, the surface has a lower surface energy ($\theta \approx 110°$) and the water droplet "stands off of the surface." Water will spread on a clean unwaxed car. This is an excellent practical example of the type of measurements that are made to determine the surface energy of solids.

Application of this to the use of adhesives is actually quite simple and direct. First the surface should be as clean as possible, without the presence of contaminants which will usually lower the surface energy. The higher the surface energy of the substrate is, the greater the wetting or spreading of the adhesive on the substrate. Spreading enhances macro- and microscopic contact between the adhesive and substrate. Additionally, for the same area of contact between adhesive and substrate, the greater the surface energy of the substrate the greater the work of adhesion and thus the macroscopically observed force required to separate the two materials.

11.2.2 Measurement of Adhesion

Adhesives are used in shear, tensile, and mixed loadings. Therefore, it is of interest to attempt to perform measurements in the fundamental loading modes. Additionally, a qualitative description of the failure mode of an adhesive joint is of paramount importance. There are two extremes of failure modes that are used to describe the adhesion failure. Cohesive failure is when either one of the adherends or the adhesive fails in the bulk of the material away from the interface. Adhesive failure is referred to when the interface, or more appropriately, the interphase fails. In one sense it is virtually impossible to have true adhesive failure, since on the microscopic scale there is always residual adhesive left on the adherend surface. However, the adherend surface appears devoid of adhesive under macroscopic, or visual, observation. In many cases, the adhesive strength of a joint is limited by the bulk properties of either the adherends or the adhesive. Indeed, in cases where the bulk strength of the adhesive and the interfacial strength of the joint exceed the strength of the adherends the joint will fail in one of the adherends. Often the qualitative description of cohesive failure (or stock tear) in the adherends is considered to be the upper limit on the joint strength. However, this criterion can be misapplied. For example, for polymeric materials, the locus of failure can be dependent on the specific rate and temperatures of the loading. When testing is performed at rates or temperatures that are not representative of the application conditions, failure can be forced into failure modes, which will not occur in the application. Interpretation that the adhesive and adhesion strength have exceeded that of the adherends due to the cohesive failure of the adherends, would only apply at those conditions tested. Consider the time–temperature superposition principle. Similar to any other property it is desirable to test adhesion in a laboratory test at a rate and temperature that would be an equivalent state to that seen in application conditions. For example, one uninstrumented "test" that has been used to qualitatively measure the adhesion of tensile cord in an elastomeric matrix was that of manually pulling the cord out of the matrix and adhesion gum. The cord would then be examined for adhesive or cohesive failure. However, for many combinations of cord/adhesive/elastomer matrix it was possible to force the failure mode between adhesive and cohesive failure dependent on how fast the cord was manually pulled. A simple solution to this situation was to test the adhesion of cord in the laboratory over a range of rates, if not temperatures.

Measurement of the adhesion between two surfaces is often performed by a tensile test of a lap shear or double-lap shear specimen. The configurations of the tests are shown in Figure 11.3. Historically, due to its ease of sample construction and

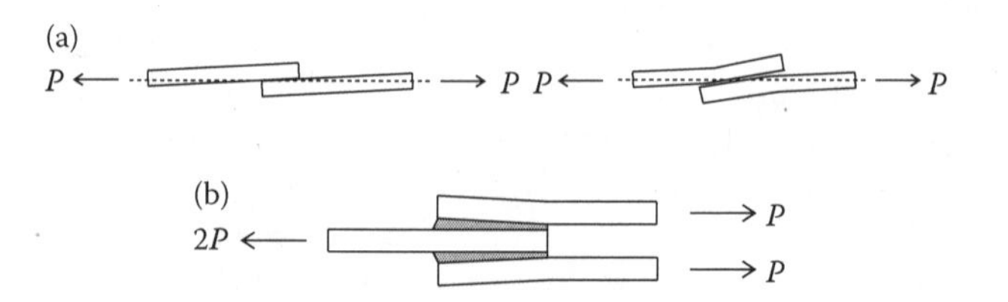

FIGURE 11.3 Lap shear specimens. (a) Single lap shear and (b) double lap shear.

testing, the lap shear test has been used quite extensively. Lately, this test has fallen into less widespread use due to the complexity of the stress state that exists in the specimen during testing. Specifically, at the ends of the bonds the samples are under combined tensile and shear loading. Adhesives are best used in shear where they can withstand much larger forces than in tension. The combined stress state is exacerbated by deformation in the samples under testing. The deformation problems are largely mitigated by the double-lap shear specimen, which, due to symmetry, deforms symmetrically with minimal curvature during testing. Both lap shear tests can be used with metal and polymer samples.

The single lap shear specimens actually undergo both tensile and shear stresses at the adherend/adhesive interfaces. Ignoring the complex state of stress and deformation which is set up in the specimen, the simplest analysis possible results in the following expression for average shear stress:

$$\tau = \frac{P}{bl}, \tag{11.3}$$

where P = load (N), b = width (mm), and l = length (mm). This result can be extended to the double-lap shear by realizing that each joint carries half of the applied load.

The peel test is as the name implies a test which measures the peeling of an adhesive joint. The test can be performed at a 90°, 180°, or at an arbitrary angle peel test as shown in Figures 11.4a, b. The 90° peel test is often used to evaluate the performance of tapes and other flexible media adhered to a rigid substrate. The 180° peel test (or T test) is used to evaluate the adhesion of two flexible media, such as elastomers. These tests have the advantage of being easy to perform and interpret.

The single-fiber fragmentation test is often used to determine the interfacial shear strength of fibers embedded in matrix materials. For this test to be valid the matrix materials should have an elongation to break which is at least three times the elongation to break the fiber. This test is especially applicable to fiber-reinforced composites such as carbon/expoy. In the test a single fiber is embedded in the

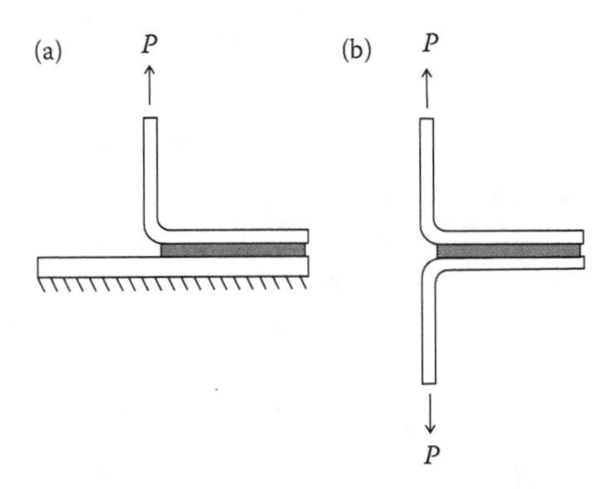

FIGURE 11.4 Peel adhesion tests. (a) 90° peel test and (b) 180° peel test.

matrix and a tensile load is applied to the composite test specimen. The fiber will break at the weakest section. After the initial break, the tensile load is transferred into the fiber sections by shear. The fiber continues to break until a critical length is reached where the shear stress cannot transfer a stress above the facture strength of the fiber. This occurs because the shear stress distribution necessarily goes to zero at the fracture ends of the fiber and increases with the distance away from the broken end as

$$\tau = \frac{\sigma_f d}{2l_c}, \tag{11.4}$$

where σ_f = tensile strength of the fiber at the critical length, d = average fiber diameter, and l_c = critical length. In practice the broken lengths are described by the Weibull distribution and Equation 11.4 becomes

$$\tau = \frac{\sigma_f}{2\beta}\Gamma\left(1 - \frac{1}{\alpha}\right), \tag{11.5}$$

where α = Weibull shape factor, β = Weibull scale factor, and Γ = the Gamma function. The Weibull distribution is used when there is a wide distribution to the strengths and it is more meaningful to describe the probability distribution than the mean and standard deviation.

The wedge test is used to evaluate the adhesion of rigid substrates, such as aluminum and steel. Often this test is used for observation of crack propagation rates through the adhesive. A wedge is driven between the rigid substrates (see Figure 11.5) and the crack propagation is measured as a function of time. This test has been used to evaluate the effect of environment (humidity, saline solution, and so on) on adhesion. The test specimen becomes a self-contained test device and thus is an inexpensive method to test specimens over extended periods of environmental exposure. This test is not intended to provide quantitative values of adhesion, but qualitative rankings.

The blister adhesion test (see Figure 11.6) is used to measure the adhesion of coatings and applied films. This is a nontrivial task when it is considered that it is often impossible to grab the coating and pull it from the substrate. Other methods to measure adhesion of coatings are scratch tests and indentation tests.

The measurement of adhesion between fibers and matrix in a fibrous composite is much more difficult than the measurement of the systems described above. Two methods that have been investigated in the past are the fiber pullout and short beam fragmentation test. The fiber pullout test consists of a drop or disk of adhesive or matrix material placed on the fiber. The fiber is then pulled through an opening which strips the adhesive drop from the fiber.

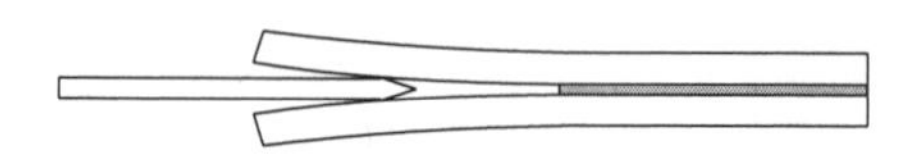

FIGURE 11.5 Wedge adhesion test.

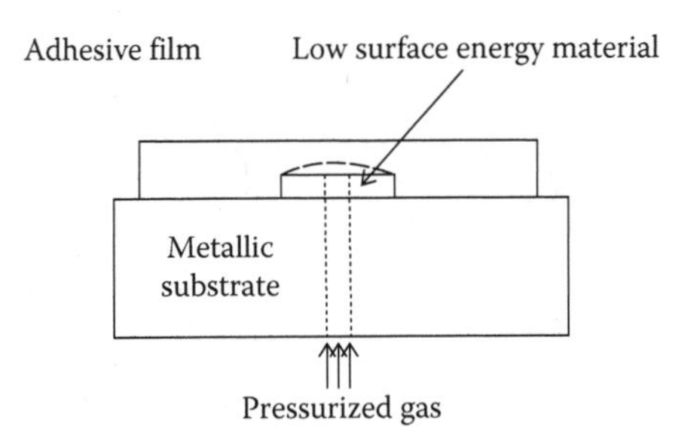

FIGURE 11.6 Blister adhesion test.

11.2.3 Viscoelasticity of Adhesion

Adhesion is not an intrinsic property of a materials system, but is dependent on many factors. By now it should come as no surprise that the measurement of adhesion is sensitive to both rate and temperature as all of the other mechanical properties have been. In fact, adhesion can often be transformed by the WLF (Williams, Landel, Ferry) equation or Arrhenius transformation in the same manner as modulus and other properties. Figure 11.7 shows the transformation of isothermal peel data transformed into a master curve along with the polyester adhesive's shear and tensile strength properties. In another study investigating the effect of temperature and surface treatment on the adhesion of carbon fiber/epoxy system, five epoxy systems were found to fit an overall master curve when corrected for the material T_g. This result is quite remarkable and is shown in Figure 11.8.

Another consideration in the use of adhesives in structural members is creep. In very rigid structures composed of steel, aluminum, or fiber-reinforced composites,

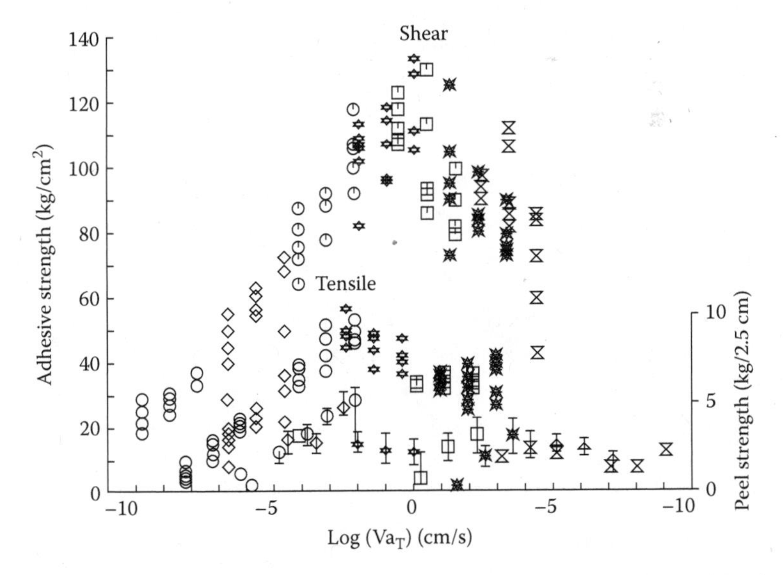

FIGURE 11.7 Time–temperature transformation of adhesion data. (Adapted from Turreda, L. D., Hatano, Y., and Mixumachi, H., *Holzforschung*, 45, 371–375, 1991.)

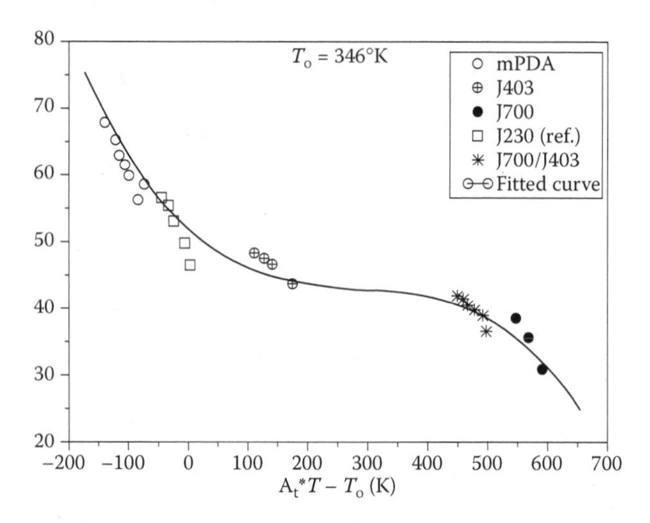

FIGURE 11.8 Master curve for various carbon fiber/epoxy systems. (Adapted from Rao, V. and Drzal, L. T., *Journal of Adhesion*, 37, 83–95, 1992.)

the adhesive can be the component most prone to creep. The same principles that apply to structural composites relative to creep also apply to adhesives and adhesive joints. Foremost, adhesives will be either thermoplastics or thermosets. If an adhesive joint is expected to withstand a load for extended periods of time, the adhesive should be a thermoset and not a thermoplastic. As in structural components, a thermoplastic will eventually behave as a fluid and the joint will eventually flow depending on the load and the temperature.

While adhesion is typically rate and temperature dependent, many adhesive systems do not behave in a manner that is amenable to time–temperature transformation.

11.3 ADHESIVES

11.3.1 Common Polymeric Adhesives

As with all other polymeric applications, a large number of polymeric adhesive systems exist. However, for the sake of brevity, only the generic families of adhesives will be discussed.

Epoxy: Epoxy adhesives have demonstrated a wide applicability as adhesive systems. Their use ranges from 1-part, 2-part adhesive packages to the matrix materials for composite laminates. Epoxies are based upon the epoxide ring shown below:

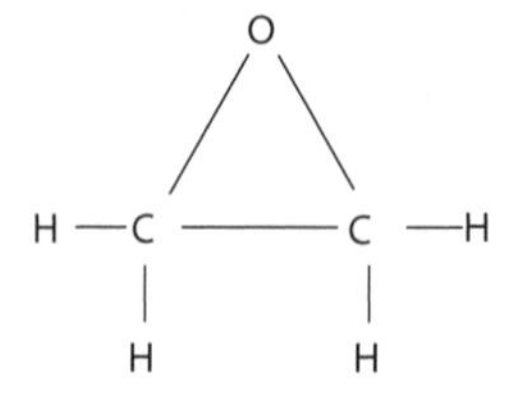

This ring opens and crosslinks with other functional chains to form a thermoset material. The cross-linking can be in the presence of hardeners, catalysts, or high temperatures. The various epoxy materials lead to a range of cure conditions and performance. Epoxies are compatible with a wide range of materials such as aluminum, steel, wood, rubber, and so on. The compatibility with other materials makes epoxy a useful intermediary for the bonding of dissimilar materials. Typical properties include low creep, high stiffness, low shrinkage, high tensile shear strength, with poor peel or cleavage strength, and moderate-impact strength. The poor peel, cleavage strength, and impact resistance can be improved by compounding with rubber fillers, resulting in a rubber-modified epoxy. Epoxy is also mixed with polyamides (nylons) for both better cure and to increase flexibility. High-temperature epoxy adhesives can be created by mixing phenolic with the base epoxy material. The epoxy–phenolic adhesives have operating temperatures from 150°C to 250°C. While epoxies are generally cross-linked and thus thermosets, there are epoxies that are cured by chain extensions of low-molecular-weight precursors and result in a thermoplastic material.

Polyester: Polyester adhesives can be both thermoplastic and thermoset materials. Thermoplastic polyester is best known for its use in fiberglass composites, such as boat and car bodies.

Cyanoacrylates: Cyanoacrylates polymerize at room temperature, and the presence of moisture in air or surface promotes polymerization. Curing does not require heat or pressure and is quick. It is generally thought that the curing reaction occurs due to either moisture in the atmosphere or moisture on the surfaces that are being bonded. Curing is very quick, often of the order of 30 s.

Polyurethane: Polyurethane adhesives are created from isocyanates. Often used for bonding polymers (especially elastomers) to metal. A specific example is the bonding of SMC that is used in automotive body panels to steel.

Silicone: Silicone is used as both a sealant and an adhesive, often functioning simultaneously in both capacities. Silicones are based on polydimethyl siloxane (PDMS) as shown:

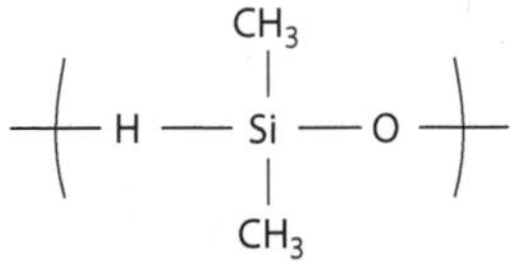

$$\left(H - \underset{CH_3}{\overset{CH_3}{\underset{|}{\overset{|}{Si}}}} - O \right)$$

PDMS has the distinction of having the lowest glass transition temperature ($T_g \approx -135°C$) of any polymer. While silicone materials are relatively weak when compared with hydrocarbon-based polymers, their low-temperature flexibility allows them to be used in many applications where strength is not a primary concern. Silicones are often referred to as room-temperature vulcanizing rubbers. Typical uses include gaskets and seals.

Anaerobic: Anaerobic adhesives cure in the absence of oxygen. One of the most common applications is that of thread-locking compound. Beyond the actual use, there are two interesting things about thread-locking compounds. First, why do not anaerobic adhesives cure in the container? It is because the containers are porous to air and thus the material will not cure. Second, if you observe a bolted connection that has utilized thread-locking compound, you will notice that the adhesive exposed to air looks and feels wet. This is because the adhesive on the surface has not cured.

11.3.2 Polymers as Matrix Materials (*In Situ* Adhesives) in Polymeric Composites

The material properties of the polymeric adhesives presented here are summarized in Table 11.1. The tensile properties of various adhesives are shown in Figure 11.9.

TABLE 11.1
Mechanical Properties of Adhesives

Adhesive	T_g (°C)	Adhesive Shear Strength[a] (MPa)	Adhesive Peel Strength[a] (kN/m)	$T_{service}$ (°C)
Epoxy	100–180	12–14		−20 to 180
Cyanoacrylate		>17		−85 to 350
Polyurethane	80–120	30–35		−30 to 175
Silicone	−125		2.5–3.5	−65 to 200

[a] Typical values for steel/adhesive/steel systems.

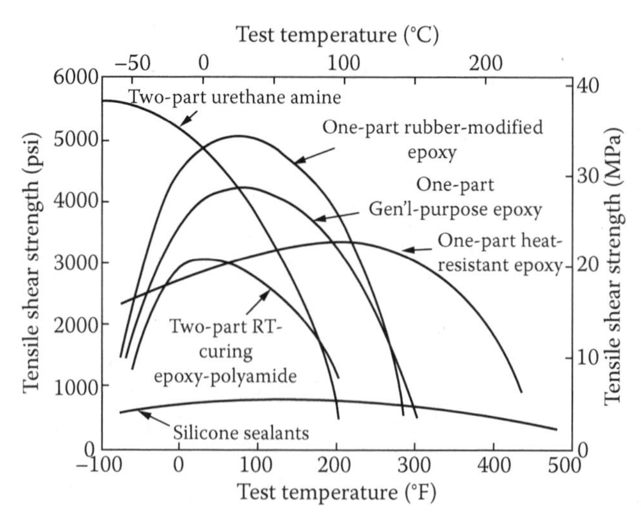

FIGURE 11.9 Tensile properties of various adhesive materials. (Adapted from Patric, R. L., ed., *Structural Adhesives with Emphasis on Aerospace Applications, Treatise on Adhesion and Adhesives*, Vol. 4, Marcel Dekker, Inc., New York, 1976.)

11.4 ENHANCEMENT OF ADHESION IN COMPOSITES

Polymeric adhesives function by providing a bond between the two adherends. This bond is obtained by the adhesive wetting both surfaces. Wetting occurs when the adhesive has both a low-enough viscosity to flow over the surface and an attraction to the surface which causes it to be pulled into intimate contact with the surface topography. While in theory it is desirable to obtain chemical adhesion (covalent bonding), in practice this rarely occurs. Usually, mechanical bonding and attraction forces dominate between the adherend and the adhesive.

In order to enhance wetting of the surfaces, and subsequent attraction, they are often cleaned with corona discharge or plasma gas. Both of these methods remove contaminants from the surfaces which can reduce the adhesive bonding. The cleaning methods can also functionalize chemical groups on the surface of polymeric materials yielding potential bonding sites for the adhesive. Covalent bonding is desired since it requires the greatest amount of energy or force to break when compared to weaker interaction forces, such as Van der Waals.

Silation of glass fibers is used to create a surface which is much more acceptable for bonding with polymeric adhesives. Glass fiber or fiberglass is known by S-glass (silica—which has the highest concentration of SiO_2), E-glass (electrical—the most commonly used glass fiber reinforcement), and A-glass (alkali). The silating agents work the best with S-glass, then E-glass, and finally A-glass. Silane coupling occurs at the silica sites in the fiberglass or filler material. Coupling agents allow covalent bonding with silica- or oxide-containing polymers as shown in Figure 11.10. The adhesion effect of silation is shown in Figure 11.11 for a fracture of a glass fiber in polypropylene composite.

Various surface pretreatments, often referred to as primers, are put on fibers and other textiles by the manufacturers to enhance subsequent bonding. Depending on the subsequent use of the textiles, the change in adhesion can be negative, nonexistent, or positive. In interlaminar shear strength tests of untreated and oxidative surface-treated polyacrylonitrile-based carbon fiber/epoxy composites[1] the shear stress went from 14.9 to 22.1 MPa.

Most discussions and analyses of composite materials assume that the bonding between the matrix and reinforcement is perfect (i.e., adhesion strength greater than either of the two constituents). However, in reality, adhesion is imperfect. This results in reductions in stiffness and strength of the composite. Additionally, imperfect adhesion results in an increase in hysteretic losses at the interface.

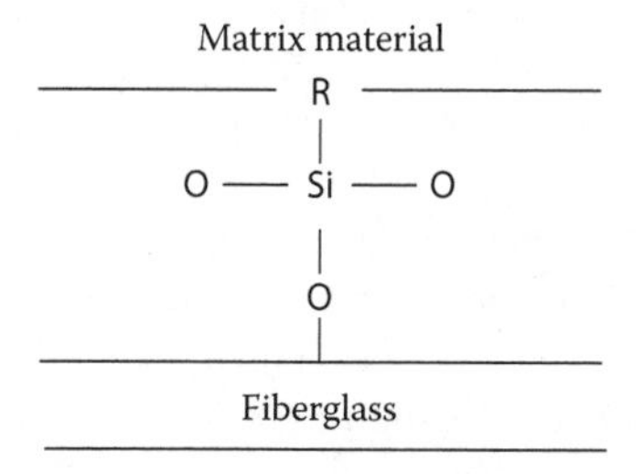

FIGURE 11.10 Silation of fiberglass.

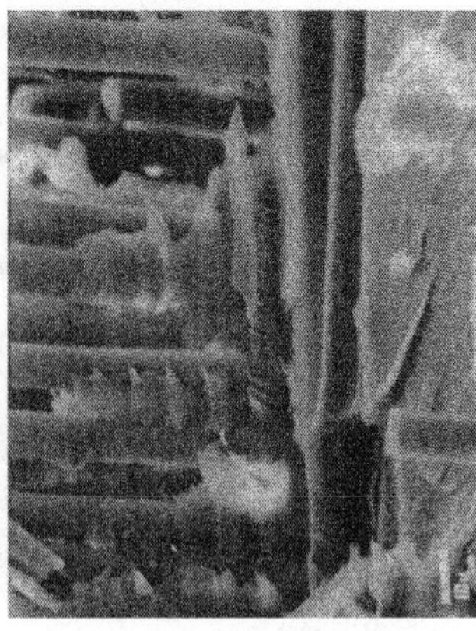

FIGURE 11.11 Tensile test of silation of fiberglass. (Images courtesy of Dow Corning Corporation.)

11.5 CURING OF ADHESIVES

Curing of adhesives and polymer systems depends on the characteristics of the particular material. A few of the more common curing modes are presented as follows:

1. Thermoplastic materials are typically cured by evaporation of solvent or by cooling below their melt temperature.
2. Thermosets are cured by chemical cross-linking, which is enhanced by temperature or catalysts.
3. *In situ* polymerization occurs with monomeric materials, where the monomer polymerizes in the presence of some catalyst (e.g., cyanoacrylate in the presence of water).

A discussion on the curing of adhesives aids in an understanding of their use and final properties. Thermoplastics are applied either in a solvent carrier or above their melt temperature. Thus, the cure associated with thermoplastic adhesives is the time required for either the majority of solvent to evaporate or the adhesive to cool below its melt temperature. Thermosets are applied either as solvent-borne or low-molecular-weight materials. In the case of thermosets, sufficient time and temperature must be applied to cure the material through chemical cross-linking. A phase diagram of cure temperature and cure time of a simple thermoset adhesive is shown in Figure 11.12.

For a thermoset adhesive there are at least four regimes of phases of final cure state. The four regimes are as follows:

1. The liquid regime occurs when sufficient crosslinking has not occurred to form a three-dimensional network of the polymer. Bulk liquid motion is still possible and the adhesive responds as a fluid.
2. In the rubbery regime the adhesive has been crosslinked beyond the minimal cure state of a three-dimensional network (this point is called the gelation point). While the three-dimensional network of the polymer has

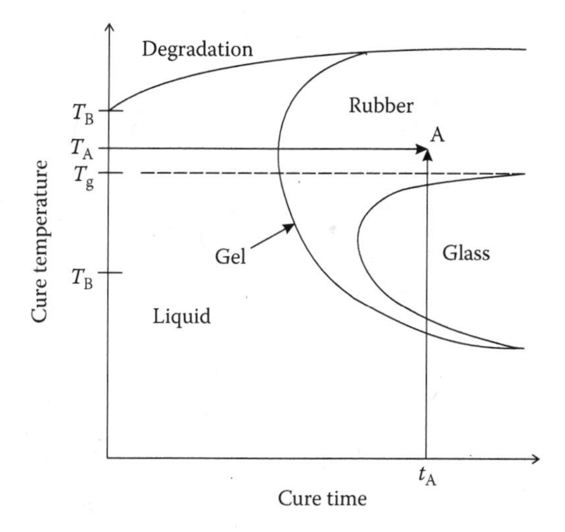

FIGURE 11.12 Phase diagram of thermoset adhesive.

been established, large motions of the polymer are still possible due to the liquid or liquid-like nature of the bulk of the adhesive. The rubbery state of polymers is often defined as the state of a lightly cross-linked liquid.

3. The glassy regime occurs when the material has been cross-linked to sufficient degree to eliminate large-scale polymer motion due to the minimal liquid polymer volume.
4. Degradation occurs when the cure state is such that there is a reversal or reduction in polymer properties due to either an increasing cure time or temperature. With additional cure, polymeric chains break down faster than they are crosslinked. This results in a reduction of the physical and mechanical properties of the polymer.

The phase diagram shows the state of the adhesive at the end of the cure cycle at a given cure temperature. For example, for a cure time of t_A and a cure temperature of T_A, the adhesive is in the rubbery state at the intersection point labeled A. However, if this material is cooled to an operating temperature below T_g, then this material would be in a glassy state. Do not confuse this glassy state with the glassy state from the thermomechanical spectrum, where the material is "glassy" or often brittle.

The amount or percent cure is obtained by curing a sample with a given cure profile. In the case of an adhesive used as a matrix material there may be various cure stages that are composed of time, temperature, and pressure. After curing, the sample is analyzed via various techniques such as differential scanning calorimetry or thermogravimetric analysis to obtain the percentage of cured and uncured material. Torque rheometry is also used to monitor the cure of cross-linked materials including adhesives. An uncrosslinked material is placed into an oscillating rheometer at elevated temperature. The rheometer is oscillated isothermally at constant strain amplitude. The torque is monitored as a function of time as shown in Figure 11.13.

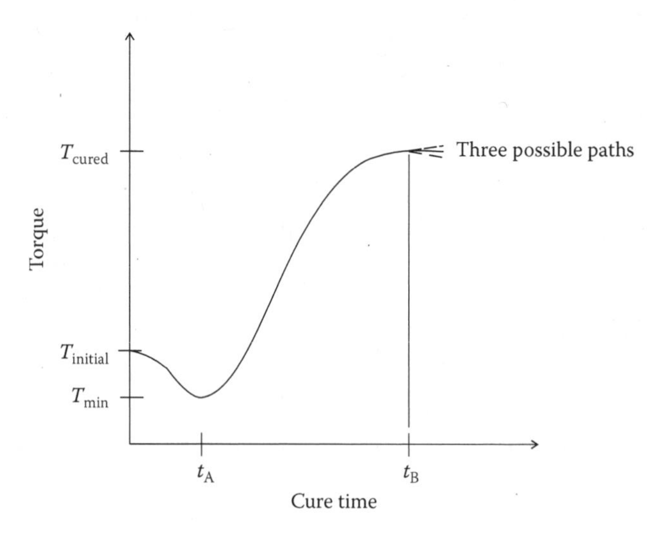

FIGURE 11.13 Cure of cross-linked adhesive in a torque rheometer.

As the material heats up from room temperature, the torque falls from its initial value due to a decrease in viscosity caused by the temperature increase and lack of significant cross-linking. Viscosity is a function of polymer chain length, cross-linking, and temperature. Cross-linking increases the size or molecular weight of the resin components as a function of time. Eventually, the increasing molecular weight overwhelms the decrease in viscosity due to temperature and a minimum torque is reached at t_A. Cure times beyond t_A result in a formation of the three-dimensional network of the adhesive. At time t_B the state of cure is approximately 100% dependent on the specific adhesive or adhesive system, additional cure time beyond t_A will result in one of the following:

1. A slight increase in cure state
2. No perceptible change of cure state
3. Degradation

There are many cases of both polymers and adhesive that are not cured to a complete state of cure. This can occur with polymeric materials that are not just used as adhesives. Most commonly, this will occur with elastomeric materials. The reasons for this incomplete or undercure are as follows:

1. Accommodate additional thermal exposure and thus cure in the application. Final cure of some polymer (especially elastomer) products can be expected to occur in the application. This can be beneficial in some cases and extremely dangerous in others.
2. At full state of cure, the material may be too brittle and possess minimal damping to absorb shock and impact loads.
3. The life of the product may be longer due to increased fatigue resistance at a lower state of cure.

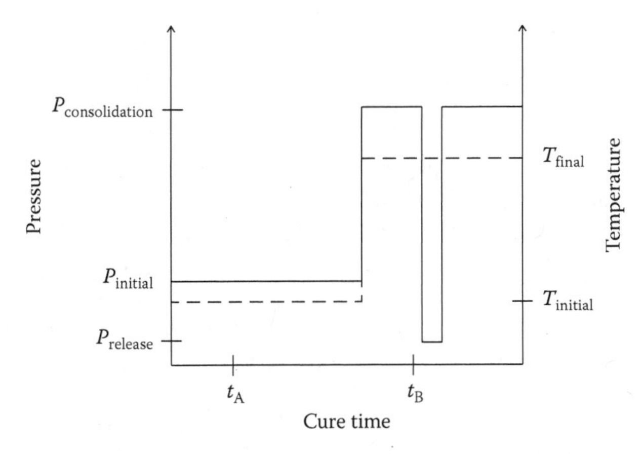

FIGURE 11.14 Step cure of composites.

Pressure is the other processing variable that is of great importance to the processing of adhesives, particularly when the adhesive is the matrix of a composite material. The application of pressure must be high enough to consolidate the composite or adhesive joint, yet not too high to cause the polymeric materials to flow out of the mold. Consolidation of the composite reduces the voids or trapped gases, consequently increasing the density and typically material properties. Revisiting Figure 11.13, it is seen that prior to time t_A, the material will flow due to a lowering of viscosity and significant consolidation pressure is not needed. Many materials and composites have a cure cycle that incorporates various stages. Initially a low temperature and pressure stage (the temperature is still greater than room temperature) is used to allow the materials to flow and consolidate with minimal cross-linking. This flow of adhesive or matrix material is extremely important to the consolidation process. Subsequently, an increase in pressure and temperature results in cross-linking and densification. There can also be bumps where the pressure is reduced to allow solvent escape from the product. A representative cure cycle is shown in Figure 11.14. For large-volume manufacture the temperature will also be approximately constant when it is provided by an isothermal heating source such as steam.

11.6 SUMMARY

For the engineer, *a priori* prediction of the value of adhesion that will occur in a structural situation is difficult. Typically, it is necessary to test the adhesives under consideration via the manufacturing processes that will be employed, on the materials that will be used, and under equivalent rates and temperatures that will be exhibited. Often the selection of adhesives can be facilitated by working with adhesive suppliers.

HOMEWORK PROBLEMS

As the discussion on adhesion has been predominately qualitative, the following homework requires research outside of the text.

The following homework will require research beyond the scope of the material provided in the text. This will be facilitated by use of both the internet and your local library.

11.1. What adhesives should be used to bond a carbon fiber/epoxy composite laminate to an aluminum framework? Why? Comment on the advantages and disadvantages of various adhesives under potential application conditions.
11.2. What issues are present when bonding steel or aluminum under high humidity or wet conditions? What adhesion test presented in the text is applicable to rank adhesive materials relative to long-term use under these conditions?
11.3. If you are joining to composite parts of the same constituent materials together what adhesive should you use? Does it matter if the parts are fiberglass/polyester resin or carbon/fiber epoxy?
11.4. What are the base polymer(s) of construction adhesives such as Liquid Nail™ and other such adhesives? Are they thermoplastic or thermoset? If they are thermoplastic, what ramifications does this have in use?

REFERENCES

1. Tai, D. W. and Penn, L. S., Interfacial adhesion to residual stress in carbon fiber/epoxy composites, *Journal of Composite Materials*, 25, 1445–1458, 1991.
2. Turreda, L. D., Hatano, Y., and Mixumachi, H., Rate-temperature dependence of bond strength properties of polyester adhesive, *Holzforschung*, 45, 371–375, 1991.
3. Rao, V. and Drzal, L. T., The temperature dependence of interfacial shear strength for various polymeric matrices reinforced with carbon fibers, *Journal of Adhesion*, 37, 83–95, 1992.
4. Patric, R. L., ed., *Structural Adhesives with Emphasis on Aerospace Applications, Treatise on Adhesion and Adhesives*, Vol. 4, Marcel Dekker, Inc., New York, 1976.

12 Polymer Fusing and Other Assembly Techniques

12.1 INTRODUCTION

Obviously, the fundamental difference between most polymer fusing of two components and adhesion bonding is that for fusing there is no adhesive present. While most fusion processes are used to fuse the subcomponents of the same material to create a final component of a single material, it is possible to fuse components comprised of two different, albeit typically very similar, polymers. In addition, some of the fusion techniques discussed will use a gasket with either particles that heat due to an oscillating magnetic field or an actual AC or DC heating element.

Why use some of the fusion techniques presented here instead of adhesion bonding? In general, adhesion bonding a few parts can be accomplished far more inexpensively than the fusion techniques that will be discussed in this chapter, usually due to the capital cost of the welding or fusing equipment involved. Fusion techniques can create joints that exceed the strength of adhesion joints. Some, not all, fusion techniques also have the advantage of creating assemblies of one single polymeric material without the presence of a second material, the adhesive. This is important when polymeric components have to be selected, for example, to minimize or eliminate contamination in a gas or fluid flow system. One field requiring minimization or elimination of contaminants is medical applications. The fusion techniques in this section are often used to create a hermetic seal for the products that allow for the flow and storage of gases and fluids.

The purpose of this chapter is to expose the design engineer to a variety of fusing and assembly techniques that are available. For many engineers, not intimately familiar with these techniques, it is surprising as to not only the range of techniques available, but also the range of materials and components that are fused. For a broader discussion on many of the techniques covered here and a few more the reader is referred to Troughton[1] and Grewell et al.[2] In addition, it should be acknowledged that much of the work over the last 20 years on the topic of the various fusion methods and polymer weld performance has been performed by V. K. Stokes.[3] One will not only see that reflected in many of the references used, but for the reader needing or interested in learning more, a search of his relevant publications is highly recommended.

As design engineers it is necessary to be able to estimate the strength of the joints or assemblies that are created. The equipment manufacturers and raw material suppliers work closely with the design engineer when selecting and implementing most of the processes presented here. In fact there is often a dearth of published literature to provide a reasonably objective guide to joint performance for the designer and analyst. If for no other reason than the variability of processing methodologies and equipment, most of the assembly techniques discussed here requires some degree of experimentation to determine the assembly strength. However, despite the aforementioned caveats, strength values will be here reported for as many material systems as possible to guide the reader by both example and analogy in the design and analysis process. As to be expected, strength values are usually referenced to either the weaker of the two materials in an assembly or an idealized joint created for reference in the particular study. While it should be obvious that the fusion process can change the material properties (especially for semicrystalline polymers), due to the slow cool down and the generation of flash, the first estimation of joint strength is to assume the strength properties of the weaker material.

The polymer fusion techniques rely on mixing of the materials from two mating counterfaces or in the case of welding with a third component of the same or similar material. Note that, while in general, the two components are typically of the same material, the materials do not have to be the same. Dependent on the specifics of the individual technique, this can be driven thermally or by a combination of mechanical and thermal input. The utilization of a design of experiments to find the optimal processing conditions is a common theme in many of the studies related to the use of the fusion bonding techniques. As will be seen in the various method sections, the compatibility of polymers for the various techniques is typically summarized by a square block diagram or matrix with the materials that can be bonded by the technique on the left hand and top side of the diagram. Since all the materials are compatible with themselves, the diagonal always shows excellent compatibility and the off-diagonal entries will indicate the level of compatibility, usually marginal or occasionally dependent on specific formulations.

12.2 HEATED TOOL WELDING

Heated tool welding uses a heated platen that is placed between two polymer components. This is one of the simplest techniques to fuse two thermoplastic materials together. There are two similar methodologies used. For both paths, one or both of the components are heated against a platen until the surface starts to melt and then the components are pressed together. The mating configurations are typically either a lap joint for sheet materials or a butt joint for plates or pipes. One method is to apply one or both of the components to be fused on a heated platen and then remove when melting starts to occur and press together as shown in Figure 12.1. The initial heating phase has two pressures applied sequentially, P_1 to promote heat transfer and then a reduction in pressure to P_2 as the material begins to melt.[4] The cooling phase has a third pressure, P_3 that holds the assembly during the cooling phase. For two different polymers, separate platens are used to melt the materials as shown in Figure 12.2. Even though Figures 12.1 and 12.2 show flat platens, in practice the platens can be shaped in three dimensions.

When the methods shown in Figures 12.1 and 12.2 are controlled by only by pressure and not penetration depth the final overall length of this process is not tightly controlled. In order to control the overall length, stops can be used that control the

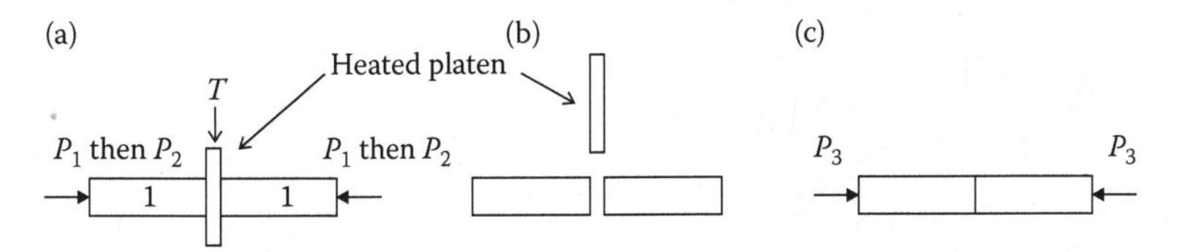

FIGURE 12.1 Heated tool welding of same or similar polymer. (a) Heating stage, (b) platen removal, and (c) welding pressure.

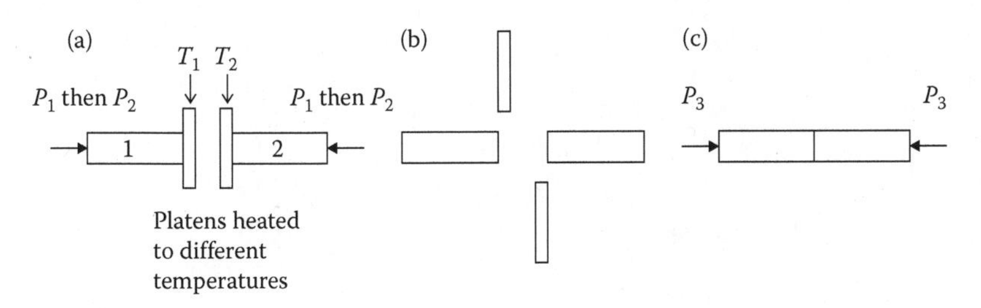

FIGURE 12.2 Heated tool welding of different polymers ($T_{m1} \neq T_{m2}$). (a) Heating stage, (b) platens removed, and (c) welding pressure.

displacement, otherwise the final dimension of the assembled component is variable. One disadvantage of heated tool welding compared to some of the other fusion techniques is that the cycle time is of the order of 20–30 s. Another is that there is flash that occurs during this process, which while not necessarily functionally detrimental, is often considered aesthetically displeasing.

Heated tool welding has been used to weld parts such as complicated 3-D joints of tail light assemblies of ABS to PMMA and also both PP and PE pipe. In addition to butt welding pipe, fittings can also be welded onto the pipe with this technique. Plastic flash, in the form of a bead, is created through this technique. Dependent on the aesthetic quality required of the weld, weld traps can be designed into the part to allow a path for flash flow to not be visible. Strength of hot plate welds are reported based on the original area. While the actual strength of the material, inclusive of the flash, will typically be less than the original parent material, the increase in material area can result in weld strengths in terms of load that approaches the parent material or the weaker of the two parent materials. As an example of this Stokes[4] found that welding dissimilar materials at optimized conditions when controlling the weld with stops, PC-PBT joints had strengths that were equivalent to the weaker PBT material. For PC-polyetherimide (PC-PEI), the maximum strength was 83% of the PEI material.

12.3 ULTRASONIC WELDING

Ultrasonic welding is used to assemble amorphous and semicrystalline thermoplastic components that have joints that are butt, lap, tongue and groove, and shear. The typical ultrasonic normal vibration in the range of 15–70 kHz and 10–50 μm oscillates two mating components to create the weld. The amplitude of vibration generates the heat required to melt and flow the joint and is thus polymer dependent. Ultrasonic

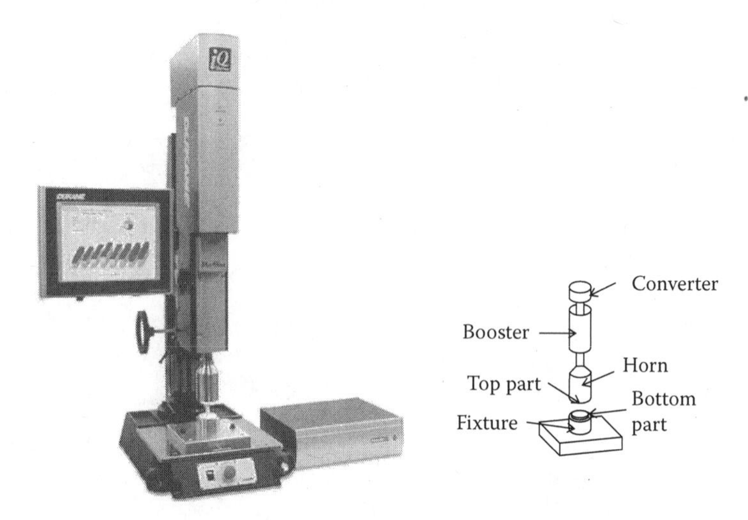

FIGURE 12.3 Ultrasonic equipment. (Courtesy of Dukane.)

welding has the advantage of being a rapid assembly technique on the order of 1–2 s total time, 0.1–1 s weld time, with a subsequent hold time of 0.1–1 s at pressures of 0.1–10 MPa. The low loads and short times required to create a bond make ultrasonic welding ideal for automation and multistation processing. A typical benchtop ultrasonic welder is shown in Figure 12.3. Table 12.1 shows the compatibility of various polymer combinations that can be used in ultrasonic welding.

12.3.1 Joint Design

There are a variety of joints that can be designed for ultrasonic welding. The simplest is the butt joint as shown in Figure 12.4a. Figure 12.4b shows the joint after welding such that there is no external flash. This is a familiar aesthetic design feature that most are familiar with, even if the primary reason for flash control was not known. The no external flash condition is established for this design by controlling the approach of the upper component. In other joint designs it can be beneficial to create flash traps. While it is recommended that joints be welded in the near field (6.35 mm or less from horn) one advantage of ultrasonic welding is that it can weld in the far field (>6.35 mm).

For butt and lap joints, energy directors are used on the joints, the most common historically being the triangular director shown in Figure 12.4. Other joints are used such as the tongue and groove and the circular cap shown in Figure 12.5a,b. However, if it is desired to control the location more so than the slip fit of butt joint or the tongue and groove, it is necessary to use a shear joint as shown in Figure 12.6.

There are a variety of methods to create hermetic seals with ultrasonic welding. The first most common is to create triangular energy directors perpendicular to a main energy director as shown in Figure 12.7a. However, a textured surface[7] can also be used on one surface in conjunction with the triangular energy director on the

TABLE 12.1
Ultrasonic Welding Compatibility Matrix

	ABS	ABS/PC	ASA	HIP S	PC	PC/PBT	PEI	PES	PET	PMMA	PPEO	PPO	PS	P S Rubber Modified	PSU	PVC	SAN	SAN-NAS-ASA	SBR
ABS	x	x																	
ABS/PC	x	x																	
ASA			x																
HIP S				x															
PC		A			x														
PC/PBT		D			A	x													
PEI					B		x												
PES								x											
PET									x										
PMMA	x	B			D					x									
PPEO											x								
PPO						D				B		x							
PS												x	x						
P S Rubber Modified														x					
PSU						B									x				
PVC	D															x			
SAN	D			D						D		D					x		x
SAN-NAS-ASA	B									B		B	B					x	
SBR										B							x		x

	SBC	Semicrystalline	Cellulo sics	Fluoropolymers	LCP	PA	PBT	PC/PET	PE	PEEK	PET	PMP	POM	POM	PP	PP	PPS	PPS	TEO	UH M WP E
ABS																				
ABS/PC																				
ASA																				
HIP S																				
PC																				
PC/PBT																				
PEI																				
PES																				
PET																				
PMMA																				
PPEO																				
PPO																				
PS																				
P S Rubber Modified																				
PSU																				
PVC																				
SAN																				
SAN-NAS-ASA																				
SBR																				

TABLE 12.1 (continued)
Ultrasonic Welding Compatibility Matrix

SBC	D			D						D			D							x																			
Semicrystalline																																							
Cellulosics																						x																	
Fluoropolymers																							x																
LCP																								x															
PA																									x														
PBT						A																				x													
PC/PET																											x												
PE																												x											
PEEK																													x										
PET																														x									
PMP																															x								
POM																																x							
POM																																	x						
PP																																		x					
PP																																			x				
PPS																																				x			
PPS																																					x		
TEO																																						x	
UH M WP E																																							x

Source: Adapted from Anonymous, *Thermoplastic Compatibility Guide*, Form No. 11549-F-02, Dukane Corporation, IL, 2002; Anonymous, *Polymers: Characteristics and Compatibility for Ultrasonic Assembly*, Technical Information PW-1, Branson Ultrasonics Corporation, Danbury, 1971.

Note: x, excellent; D, Dukane; B, Branson; A, All.

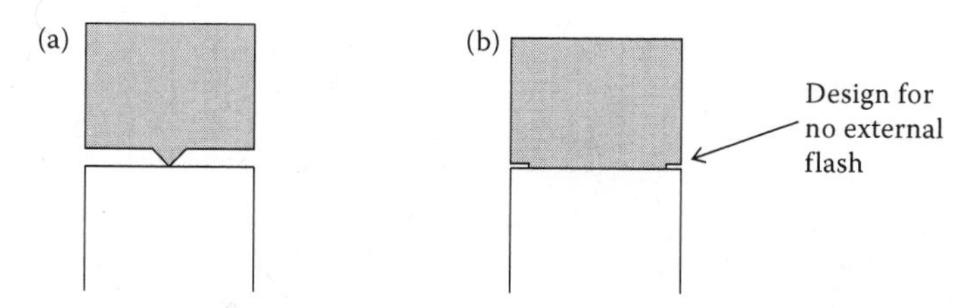

FIGURE 12.4 Simple energy director butt joint before (a) and after ultrasonic welding (b).

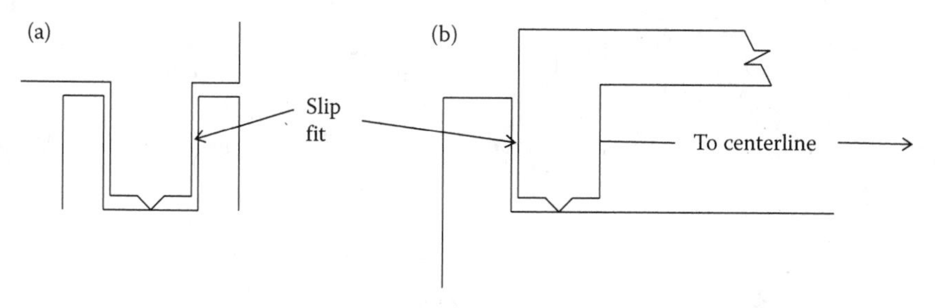

FIGURE 12.5 Other ultrasonic joints. (a) Tongue and groove and (b) circular cap.

other. Additionally, an o-ring seal can be used with ultrasonic welding, where the o-ring is permanently confined and compressed by the weld.

The main parameters that are varied in ultrasonic welding are the normal pressure, amplitude of vibration, and time of welding. Liu et al.[8] performed a Taguchi Design of Experiments on polypropylene and polystyrene to determine the effect of

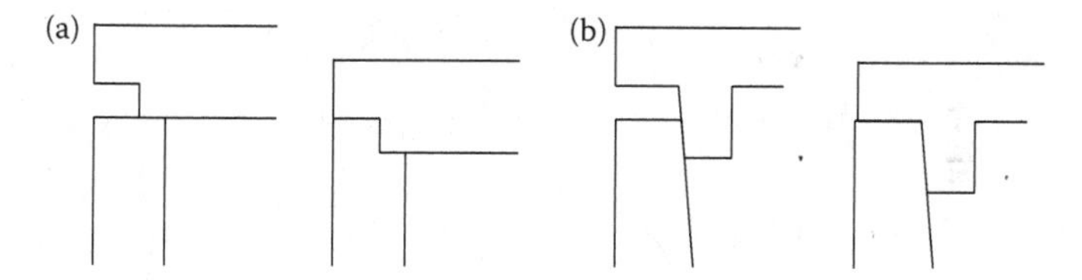

FIGURE 12.6 Ultrasonic shear joints. (a) Simple shear joint and (b) tapered shear joint.

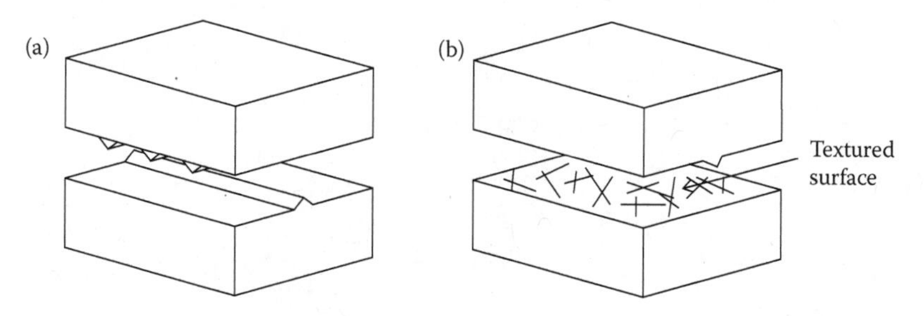

FIGURE 12.7 Ultrasonic hermetic sealing configurations. (a) Crossed energy directors and (b) textured surface (Branson).

weld time, weld pressure, amplitude, hold time, and hold pressure on the weld strength of the butt-mated cylindrical tubes. This study is instructive for a few reasons. As stated previously, it is usually necessary to perform a design of experiments to determine the effect of the process parameters on the weld strength. Importantly, the test geometry of mating cylinders is applicable to actual geometries used in practice and not simply a lap joint. In addition to determining the optimal conditions of the test variables, it was shown that weld strength correlated with input energy. The optimal weld strength was 8.28 MPa for the polystyrene and 26.36 MPa for the polypropylene. The authors also created a weldability diagram, which did not catch on with the research community. However, it is useful for the understanding of individuals new to the topic. The essential result of the weldability diagram is that it identifies that at low weld times and amplitudes the weld joint is not effective or not bonded and at high weld times and amplitudes excessive flash is generated. The area for effective ultrasonic welding is in the zone of moderate weld times and amplitudes.

12.3.2 Staking

Staking refers to the melting of a polymer protuberance, typically cylindrical, which secures another material to the original polymer material. The polymer that is used in the staking process is usually an amorphous thermoplastic and the material being staked is either a semicrystalline or crystalline polymer or even a metal. The reason for which some polymers or an amorphous polymer is not also staked is that the ultrasonic method has a tendency to flow those materials as well as the stake. This process is usually used to connect sheets of material to each other. As can be seen by the various staking configurations shown in Figure 12.8, this process is an extension of ultrasonic welding, with the exception being that the materials are not fused together. In fact, it is the exception, to stake two similar melting point, amorphous thermoplastic materials, since it is undesirable to melt the material not being deformed.

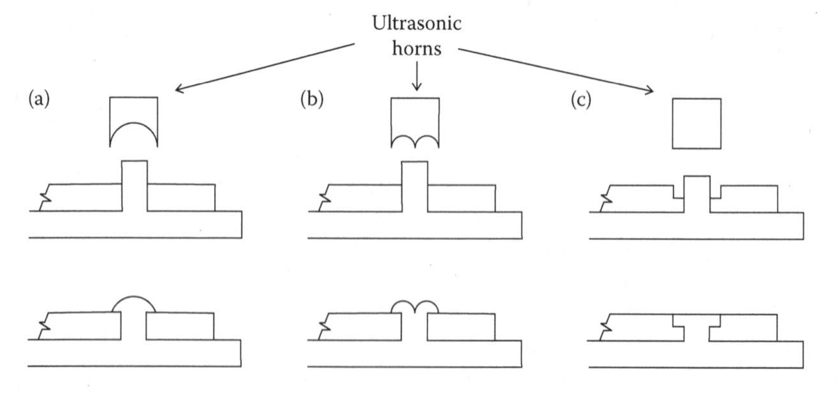

FIGURE 12.8 Various staking geometries. (a) Button, (b) rosette, and (c) flat.

12.4 FRICTION WELDING

Friction welding techniques encompasses: linear vibration, orbital welding, angular oscillation, and spin welding. All four techniques use the frictional heating at the interface to melt the subcomponents under pressure for mixing. One advantage that friction welding techniques offer over ultrasonic welding is the ability to weld amorphous and semicrystalline thermoplastics to each other. In addition, plastic materials that are difficult to weld ultrasonically, due to fillers can be welded by vibration welding. The compatibility matrix for friction welding is shown in Table 12.2.

12.4.1 Linear Vibration and Orbital Welding

Linear vibration and orbital welding are essentially the same technique with the obvious difference being that one oscillates back and forth in a linear motion and the other oscillates in an orbital path. These techniques are applicable to parts that are larger (200–1500 mm) than those handled by ultrasonic welding. Large automotive under the hood air boxes and manifolds are vibration welded from PA (nylon). Recently, high-temperature polyamide oil sumps have been welded with this technique.

For linear vibration welding the two components are pressed together with a pressure of 0.5–14 MPa oscillated at a frequency of 100–500 Hz (120 and 240 Hz are industry standard frequencies) and an amplitude of 0.5–5.0 mm for a times of the order of 10 s. The welding process is described as four stages:

1. The initial solid friction stage where there is frictional heating, but not penetration or approach of the two components.
2. The transition stage where the surfaces start to melt and flow, penetration increases nonlinearly as a function of time.
3. The steady-state phase where penetration joint is approximately a linear function of time.
4. The fourth and final phase is the cooling, where the normal load is removed and the joint solidifies.

Linear and orbital friction processes primarily work best with joints that are in the same plane. Secondarily, joints that have a slight amount of curvature can be welded. There are special cases where the joints can be nonplanar but parallel. Different thermoplastics can be welded together, but since the bond is driven by the temperature nominally at the interface, one requirement is that the melt temperatures of the two plastics are within 20°C of each other. The parts must be rigid enough, due to both geometry and material properties, to transmit the normal and tangential loads without plastic deformation of the part due to the combination of stress and temperature. The bottom part is held by a fixture to the stationary platen. The upper vibratory part is then pressed against the bottom part. One interesting alignment feature is to design in transverse shear pins and mating holes that align the parts prior to application of the normal force and relative motion.

An interesting study of polyamide 66 was carried out by Potente and Uebbing.[13] They performed three separate sets of experiments on vibration and spin welding. The

TABLE 12.2
Vibration Welding Polymer Compatibility Matrix

	ABS	ABS/PC	ASA	HIP S	PC	PC/PBT	PEI	PES	PMMA	PPEO	PPO	PS	PSU	PVC	SAN	SBC	Semicry stalline	LCP	PA	PBT	PE	PEEK	PET	PMP	POM	PP	PPS	TEO	UHMWPE
ABS	x	x			D		x		x		x	D		D	x	x				x			x						
ABS/PC	x	x			x	D			x																x				
ASA			x																										
HIP S				x								D			D	D													
PC	D	x			x	D	D		x		D		D							x			x						
PC/PBT		D			D	x			D											D									
PEI	x				D		x													x			x						
PES								x					D																
PMMA	x	x			x	D			x						x	D													
PPEO										x	x																		
PPO	x				D					x	x	x			x														
PS	D			D							x	x			D	D													
PSU					D			D					x																
PVC	D													x															
SAN	x			D					x		x	D			x														
SBC	x			D					D			D				x													
Semicry stalline																													
LCP																		x											
PA																			x										
PBT	x				x	D	x													x									
PE																					x								
PEEK																						x							
PET	x				x		x																x						
PMP																								x					
POM		x																							x				
PP																										x			
PPS																											x		
TEO																												x	
UHMWPE																													x

Source: Adapted from Anonymous, *Thermoplastic Compatibility Guide*, Form No. 11549-F-02, Dukane Corporation, IL, 2002.
Note: x, excellent; D, Dukane.

first set of samples were vibration welded, 3.3 mm sheets of standard, heat stabilized, 40% short glass fiber filled, and 40% talc-filled PA 66. The samples were tested by longitudinal butt welds as well as transverse welds ("T" shaped). The most significant results from this testing showed that the talc filled specimens had both longitudinal and transverse weld strengths of ≈90% of the 70 MPa sheet strength, while the weld strength of the glass fiber-reinforced materials were approximately the same as the standard material in the range of 45–50 MPa. While the talc was able to provide reinforcement to the welded area after the welding, the glass fibers were not. A second set of experiments on 2 and 4 mm test samples confirmed that the glass fiber welds only possessed the strength of the unfilled material welds. In attempt to correlate the results from the sheet testing, two complex parts were made. The first part was a hollow hexagonal container that was vibration welded, while the second part was a cylindrical container that was spin welded. Unfortunately, with the variability in both the weld sample tests and the complex part tests, a correlation between the sample and part tests was not established. Perhaps the most important conclusion from the work was that the use of the glass-filled material was not justified over that of the unfilled material for these joints. Vibration welding of this and a series of other thermoplastics is shown in Table 12.3 for estimating the strength of various material combinations.

12.4.2 Spin Welding

Spin welding is used to weld axisymmetric joints. The lower component of the assembly is held stationary in fixtures, while the upper component is both rotated and pressed normally against the fixed component. The upper component typically has clocking features that allow it to be driven by the drive spindle. The simplest way to think about spin welding and indeed the simplest way to accomplish it is to consider a drill press spinning a part which is subsequently pressed down on a stationary part fixed to the press table. Spin welding can not only create hermetic seals, but it can also encapsulate other components within the volume created by the assembly.

12.5 LASER WELDING

Laser welding is predominately used to weld two polymers together, by transmitting laser light through the polymer that is transparent. Note that transparent means that it transmits light at the laser frequency, not that it is transparent to the human eye. When the light strikes the nontransparent or absorbing polymer, heat is created in the volume of the absorbing polymer at the interface as shown in Figure 12.9. Carbon black is the most common absorber that is added to materials. In addition, there is a patented[14] material, Clearweld® that can be applied as a coating or incorporated into transparent plastics that absorbs in the near IR spectrum. This allows laser welding to create optically clear assemblies. Comparison of the use of this material to vibration and laser welding has demonstrated comparable weld strengths.[15] One of the most common applications of this technique is the welding of automotive tail lights. The process can be controlled such that there is neither visible distortion of the welded assembly nor flash as is generated in other fusion processes.

TABLE 12.3
Vibration Weld Strengths of Various Thermoplastic Pairs

Material 1	Material 2	Reference Material and Strength (MPa)	Percentage of Strength	Notes	Literature Reference
ABS	ABS	ABS (39.8)	85–90	120 Hz	Stokes[9]
ABS	ABS	ABS (39.8)	80–85	250 Hz	”
ABS	ABS	ABS (39.8)	85–90	14% Rubber	”
ABS	ABS	ABS (39.8)	85–95	20% Rubber	”
ABS	PC	ABS (39.8)	70–38		”
ABS	PEI	ABS (39.8)	45–65		”
ABS	M-PPO	ABS (39.8)	70–75		”
ABS	GF-PPO	ABS (39.8)	50–65		”
PPO-PPS	PPO-PPS	PPO-PPS (57.7)	86	50%/50%	Stokes[10]
PPO-PPS-Rubber	PPO-PPS-Rubber	PPO-PPS-Rubber (47.4)	72	50%/30%/20% Rubber Mod.-1	”
PPO-PPS-Rubber	PPO-PPS-Rubber	PPO-PPS-Rubber (52.6)	89	50%/30%/20% Rubber Mod.-2	”
PC-ABS-1	PC-ABS-1	PC-ABS-1(56.5)	83		Stokes[11]
PC-ABS-2	PC-ABS-2	PC-ABS-2(64)	68	Flame retardant	”
PC-ABS-1	PC	PC-ABS-1 (56.5)	85		”
PC-ABS-2	PC	PC-ABS-2(64)	72		”

PC-ABS-2	PBT	PBT (44.1)	25		”
PC-ABS-2	PC-PBT	PC-PBT (50.1)	26		”
PC-ABS-2	M-PPO	M-PPO (45.5)	76		”
PC-ABS-2	PPO-PA	PPO-PA (55.4)	20		”
M-PPO	M-PPO	M-PPO (33.8)	83–101		Stokes[12]
PMMA	PMMA	PMMA (62.6)	85–105		”
PMMA	PC	PMMA (62.6)	60–100		”
PMMA	PBT	PBT (59.8)	13–21		”
PMMA	M-PPO	M-PPO (33.8)	61–94		”
PA	PA	PA (55)	90		Potente and Uebbing[13]
PA-GF	PA-GF	PA-GF (79)	50–60	40% Glass fiber	”
PA-talc	PA-talc	PA-talc (70)	90	40% Talc	”

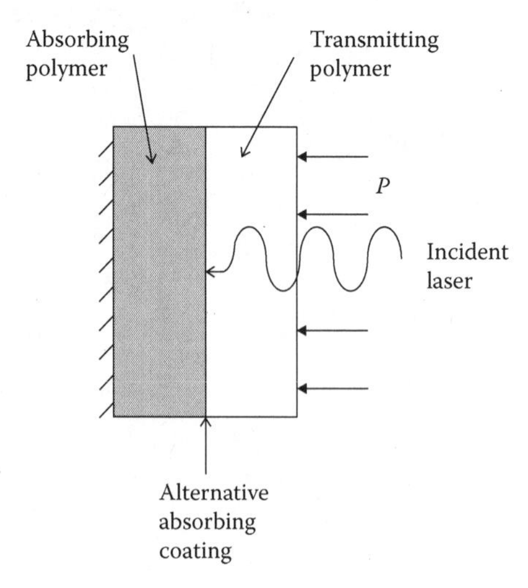

FIGURE 12.9 Laser welding.

12.6 HOT GAS

Hot-gas welding is the process that is the most similar to welding metals in that it uses a welding rod. Components are butted against one another with a groove that allows welding rod material to be melted along with the two components. This equipment for this process is typically handheld and thus extremely portable. Since the tooling is portable, there is no limitation on the size of part that can be welded. One application of this technique is to weld large thermoplastic tank sections that used in truck liquid transport. The most common application is the welding of roof sheet material. The technique is illustrated in Figure 12.10.

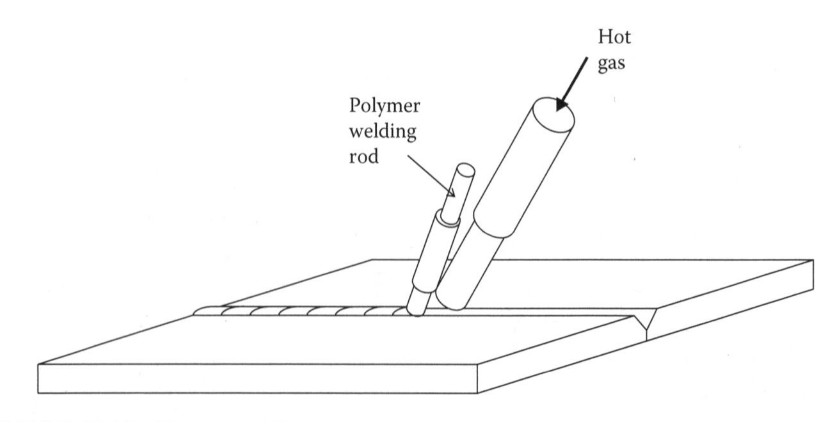

FIGURE 12.10 Hot-gas welding

12.7 RESISTANCE WELDING

Resistance welding utilizes implantable heating elements, which stay in the final assembly, to heat the two components under pressure until fusion occurs. The heating elements, which form a continuous electrical path during fusing, stay in final assembly. In this regard, resistance welding is similar to induction welding. Resistance welding was initially created to bond carbon fiber composites, where carbon fiber formed the electrical path and heating element. The resistance welding configuration is shown in Figure 12.11. Since the initial work, continuous carbon fiber, carbon fiber fabric, and stainless steel have all been used to provide the resistance heating element when bonding plastics and fiber-reinforced composites.[16]

Hou et al.[17] investigated resistance welding carbon fiber fabric in a PEI matrix. They fused lap shear specimens of 10 ply laminates with a single heating element comprised of carbon fiber prepreg with PEI films (no carbon fiber) molded to both sides. The various test specimens were compared to a compression molded lap shear sample for reference (≈32 MPa.) The test samples reached the lap shear strength of the compression-molded reference sample with molding pressures in the range of 0.10–0.40 MPa and welding energies of 2722–3111 kJ/m^2. They also ran the carbon fiber heating element without the PEI films and found that the lap shear strengths were 30% lower than those with the PEI film molded onto the fabric heating element. Ageorges et al.[18] investigated a similar system of carbon fiber in a PEI matrix. There were essentially three different test heating elements to bond the laminates: carbon fiber fabric, 0° unidirectional prepreg, and 90° unidirectional prepreg between the laminates. Instead of compression molding separate laminates to create a reference lap shear sample, this study compression molded a single laminate and then ground away the ends to create lap shear reference samples. This created as close to ideal lap shear sample as possible for reference based on the layup. In this study optimal welding occurred at pressures of 0.4–0.8 MPa and molding weld times of 90–120 s. In this study, the unidirectional samples had an optimum lap

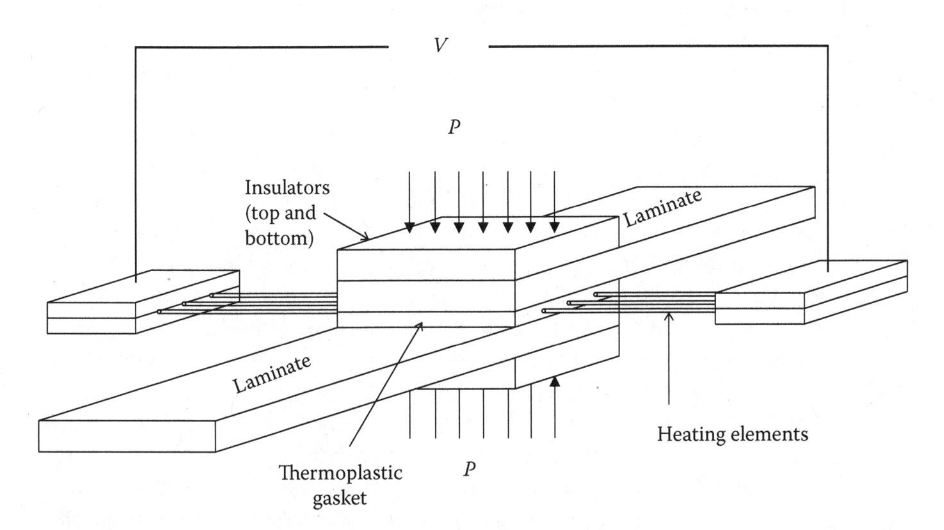

FIGURE 12.11 Resistance welding configuration.

shear strength of ≈27 MPa compared to the ground reference sample of ≈29 MPa and the fabric samples had an optimum lap shear strength of ≈30 MPa compared to the reference of ≈33 MPa. Unlike the Hou et al.[17] study, the Ageorges et al.[18] study was not able to equal the reference sample. This is not surprising, since the Ageorges et al.[18] study created what could arguably be considered a superior reference sample that would have a higher strength relative to the other test samples.

Other examples of this welding methodology follow. Hou and Fredrich[19] constructed lap shear specimens from continuous unidirectional glass fibers in a polypropylene matrix with a carbon fiber/polypropylene heating element. Dubé et al.[20] made stacked stringer samples (long laminate/heating element/short laminate stacked one on top of the other) to simulate aerospace composite joints, which were tested in bending. The unique aspect of this testing was that the heating element was a stainless-steel mesh between AS4 (polyacrylonitrile based) carbon fiber/APC2 (PEEK) laminates. The stainless-steel mesh had neat PEEK film placed on each side. The purpose of the film is to provide a resin-rich layer for bonding to the laminates. Bonding of dissimilar laminates was demonstrated by Ageorges and Ye,[21] when they bonded a thermoplastic carbon fiber/PEI laminate to a thermoset carbon fiber/epoxy laminate. The heating element was carbon fiber with PEI on both sides. There was a hybrid interlayer between the heating element and carbon fiber/epoxy laminate comprised of glass fiber fabric with PEI on the heating element side and epoxy on the carbon fiber/epoxy side to promote bonding.

12.8 INDUCTION WELDING

Induction welding utilizes an oscillating magnetic field that vibrates magnetic materials that are imbedded in a polymer. The magnetic material is either continuous wire or particles. Due to hysteretic losses, heat is generated and the thermoplastic material surrounding the magnetic material flows and upon solidification creates a bond. Initially this technique was used with particles that were imbedded in a polymer gasket. When bonding two components made of the same material, the gasket is ideally made of this material. However, the only requirement is that the gasket thermoplastic material be compatible with the polymeric materials of the two components. Table 12.4 shows a compatibility matrix for induction welding. Comparison of Table 12.4 with Table 12.1 evidences similarity between the amorphous polymers that can be welded by the two different methods. When bonding differing materials, it is possible to use a gasket material that is a blend of the two materials. This enhances the bond between the two components. Induction welding has been extended to composite materials that contain components that will heat when exposed to the oscillating magnetic field eliminating the need for the gasket. In addition, since the magnetic materials are still present in the joint after bonding, these materials allow for disassembly of the bond joint at a later date. Examples of induction welding are shown in Figure 12.12.

Similar to vibration welding, induction welding allows the bonding of larger parts (on the size scale of a meter) than most of the fusion bonding techniques. The size scale means that both liquid tanks and pressurized gas reservoirs are welded with this process. This welding technique also accommodates polymer materials that contain fillers.

TABLE 12.4
Induction Welding Polymer Compatibility Matrix

	ABS	ABS/PC	LCP	PA	PBT	PC	PBT	PE	PEEK	PEI	PEO	PET	PMMA	PMMA-Multi	POM	PP	PPO	PPS	PS	PSU	PVC	SAN/NAS	SBC
ABS	x	x											E	E					E			E	E
ABS/PC	x	x				x	E						E	E									
LCP			x																				
PA				x																			
PBT					x																		
PC		x				x	E						x										
PC/PET		E				E	x						E										
PE								x															
PEEK									x														
PEI										x													
PEO											x												
PET												x											
PMMA	E	E				x	E						x									E	E
PMMA-Multi	E	E												x								E	
POM															x								
PP																x							
PPO																	x		x			E	
PPS																		x					
PS	E																x		x			E	E
PSU																				x			
PVC																					x		
SAN/NAS	E												E	E			E		E			x	
SBC	E												E						E				x

Source: Data from http://www.emabond.com

Note: x, commonly accepted by industry; E, recommended by Emabond.

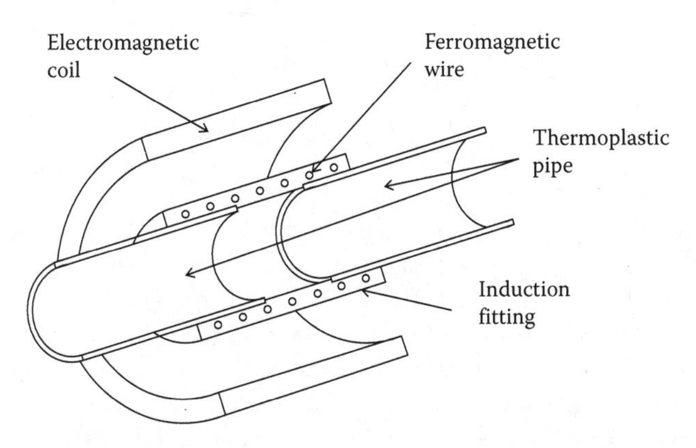

FIGURE 12.12 Induction welding.

TABLE 12.5
Butt Joint Strength of Thermoplastics from Induction Welding

Material	Percentage of Strength (%)	Reference Material and Strength (MPa)
PC	46	PC (66.5)
PBT	37	PBT (59.8)
PP	53	PP (31.9)
PC-PBT (PBT Gasket)	20	PBT (59.8)

Source: Dubé, M. et al. *Composites, Part A*, 38, 2541–2552, 2007.
Note: Failure stress is based on original material thickness, not inclusive of weld bead thickness that is formed.

Stokes[22] examined the strength of butt welds in PC, PBT, PP, and PC-PBT joints. The gasket material was the respective lap material for the PC, PBT, and PP joints. For the PC-PBT specimen, the gasket material was PBT. Table 12.5 shows the relative strength of the butt joints.

Suwanwatana et al.[23] compared the strength of lap shear joints of S2 glass fiber polyphenylene sulfide (PPS) laminates bonded in an autoclave and via induction welding. They found that 90% of the strength of 1 h autoclaved results could be obtained within 3–10 min via the induction welding process. The gasket was polysulfone with nickel particles ranging from 0.08 to 22 μm.

12.9 MECHANICAL FASTENER CONNECTIONS

Mechanical fastener connections are used when there is the need to disassemble the components. The most common reason for disassembly is for repair of the internal components of a device, which can be electronic, mechanical, or both. There are two main types of mechanical connections for assembling polymer components that will be considered. One type is the threaded inserts and bolts. The other common situation is when screws are threaded directly into the unthreaded polymer for mechanical fastening of another component. Threads can also be molded into the polymer component. Molding in threads increases the cost of the molding operation.

12.9.1 Screws

Direct threading of a screw into polymers is typically used for components that are expected to be disassembled only a few times for repair. The screw is typically self-tapping. A mechanical connection of this sort is not intended to carry significant load and is often used in conjunction with other fastening methods such as the snap connection discussed previously in the text. However, it is one of the quickest and least inexpensive methods to assemble components that do not experience significant loads and stresses. There are many general industrial guidelines about mechanical fastening in polymers, which are seemingly independent of material

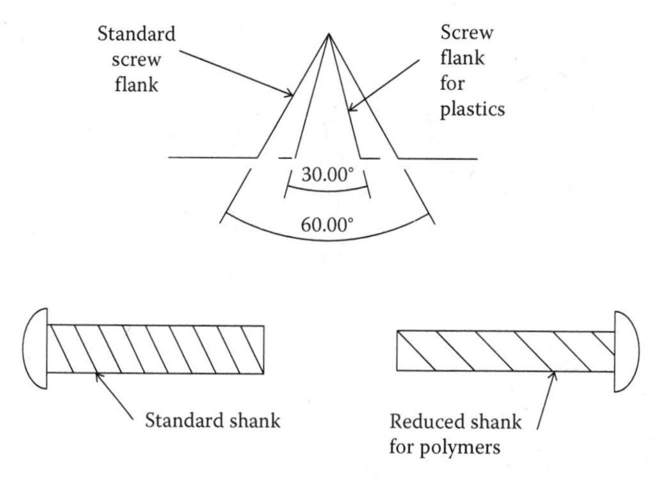

FIGURE 12.13 Thread forming and thread tapping screws.

properties. However, it should be understood that these guidelines are just guidelines and a place to start. One of the best examples of this is the selection criteria based on modulus for choosing between a self-tapping and thread forming screw. From a mechanics point of view, a thread forming screw is analogous to a radial press fit with an additional applied torsional stress. Failure occurs when the combined stress state exceeds either the yield or ultimate strength of the material.

There are two general types of self-tapping screws, thread forming and thread tapping as shown in Figure 12.13. Thread tapping screws, form the threads by displacing material away from the screw. Thread cutting screws, actually cut the threads removing material from the path of the screw, in a process analogous to a metal tap. Since thread forming screws displace material, they are typically only used for low modulus materials. While the critical stress can be calculated based on the modulus and strength of the material, a general rule, thread forming screws are used when the modulus is <1380 MPa (200,000 psi).[24] Above this modulus, thread cutting screws are used. An exception to this recommendation is the three-lobed thread forming screws made by various manufacturers as shown in Figure 12.14. The three-lobed screw reduces the stress by requiring less deformation of the polymeric material during installation, as well as providing volume for the material to flow after installation.

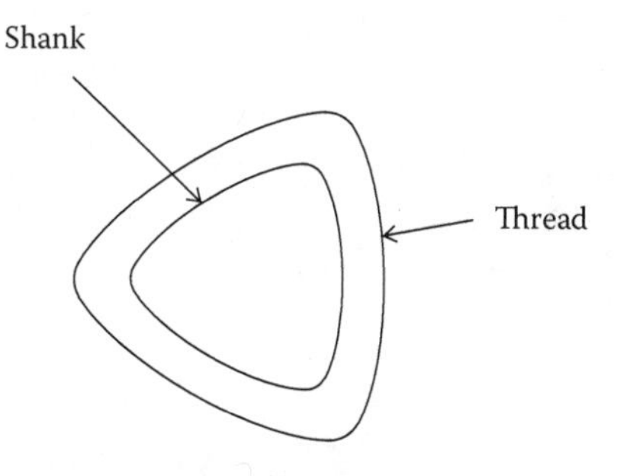

FIGURE 12.14 Three-lobed thread forming screw.

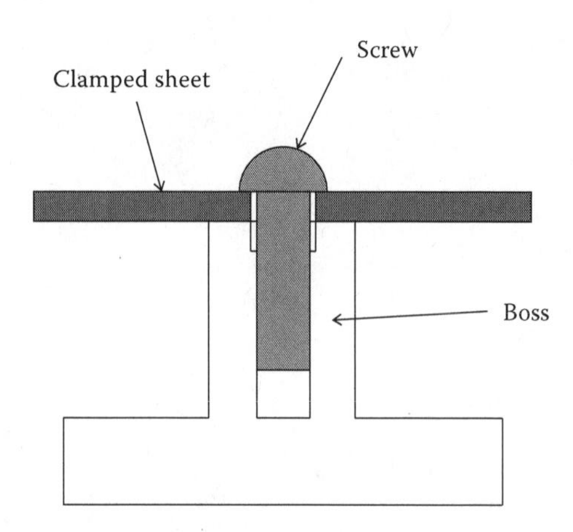

FIGURE 12.15 Boss design for self-tapping screws.

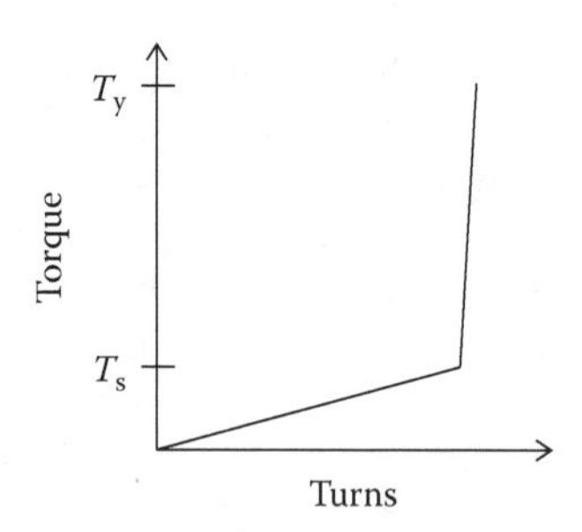

FIGURE 12.16 Screw torque when threading into plastic.

Typically a molded part will have bosses for the screws to engage as shown in Figure 12.15. It is generally recommended that the boss be 2.5–3 times the screw pitch diameter. The length of engagement should be based on the require pullout force, but is generally 2.5–3 times the screw diameter. The counterbore is present to reduce stresses in the boss material. Clearance for material displaced by the thread cutting is accommodated below the threaded screw. There is large variety of screw designs available for use from suppliers. Most screw manufacturers have recommended dimensions of the boss features based the specifics of a particular fastener.

Most people are familiar with overdriving a screw into a plastic boss and stripping the screw. The torque required to drive a screw increases (shown as a linear increase in Figure 12.16) as the thread engagement increases up until the screw head is seated at Ts. If the additional torque is applied, there will be a small amount of elastic torsional deflection until the screw starts to strip the plastic.

12.9.2 Inserts

The chief advantage of threaded metal inserts in a polymer component is the ability for multiple disassembly and assembly of a multicomponent assembly. The obvious

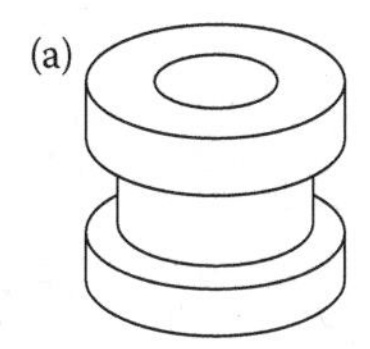

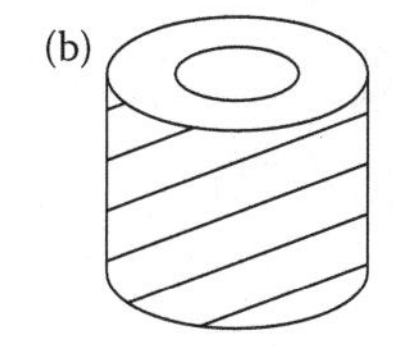

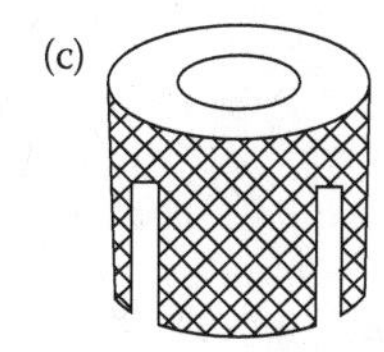

FIGURE 12.17 Inserts. (a) Injection molding, (b) threaded, and (c) knurled-expanding.

disadvantage is that the cost of the plastic component is increased due to the threaded metal inserts.

Threaded inserts are inserted via a variety of methods; molded in during injection molding process, ultrasonic insertion, threaded, and thermal insertion. There are essentially two types of threaded inserts, those that maintain their shape and those that expand as the screw is driven into the insert.

An ultrasonic insert is perhaps the best compromise between cost of manufacture and strength of fastener connection. Ultrasonic insertion is similar to ultrasonic welding. The difference is that the metal insert is driven into the premolded or drilled hole. The strength benefits of ultrasonic insertion are twofold. First the material is displaced while heated, such that it flows around the features of the insert, reducing postinsertion stresses compared to a threaded in place insert. Second, since the insert outer diameter is much greater than a typical screw fastener connection, the stresses transferred to the plastic component are lower. While the ultrasonic horn can be placed directly on the insert, it is recommended that the insertion configuration is such that, if possible, the horn contacts the plastic to reduce horn wear. There are essentially two types of mechanically inserted inserts. The first is the solid insert, which is typically threaded into the plastic boss. The second is an expanding insert, where the screw expands the insert, such that the serrated surface locks into the polymer. The holes for the inserts can either be drilled or preferably molded into the plastic. The three general types of inserts are shown in Figure 12.17.

HOMEWORK PROBLEMS

12.1. Consider a steel thread forming screw with a minor diameter of 2 mm. This screw is threaded into nylon and PC. Assume that the stresses from the polymer expanding radially from the screw centerline can be modeled as an interference fit. What are the stress components in the polymer at the screw? Will the material fail?

12.2. For a fusion process consider that the polymer material is heated above the melting point and then allowed to cool slowly to room temperature. What are the strengths of unfilled nylon, 20% glass fiber-filled nylon, and PC for this process? Base your answers on data from the polymer database or reference search.

REFERENCES

1. Troughton, M. J. ed., *Handbook of Plastics Joining: A Practical Guide,* 2nd ed., William Andrew Inc., Norwich, New York, 2008.

2. Grewell, D. A., A. J. Benator, B. Park, eds., *Plastics and Composites Welding Handbook*, Hanser Gardner Publications, Inc., Cincinatti, OH, 2003.
3. Potente, H. and J. Natrop, Computer-aided optimization of the parameters of heated-tool butt welding, *Polym Eng Sci*, 29, 1649–1654, 1989.
4. Stokes, V. K., Experiments on the hot-tool welding of three dissimilar thermoplastics, *Polymer*, 39(12), 2469–2477, 1998.
5. Anonymous, *Thermoplastic Compatibility Guide*, Form No. 11549-F-02, Dukane Corporation, IL, 2002.
6. Anonymous, *Polymers: Characteristics and Compatibility for Ultrasonic Assembly, Technical Information PW-1*, Branson Ultrasonics Corporation, Danbury, 1971.
7. Sager, T. B., *Ultrasonic Welding of Thermoplastic Workpieces*, U.S. Patent No. 4,618,516, 1986.
8. Liu, S.-J., W.-F. Lin, B.-C. Chang, and G.-M. Wu, Optimizing the joint strength of ultrasonically welded thermoplastics, *Adv Polym Technol*, 18(2), 125–135, 1999.
9. Stokes, V. K. and S. Y. Hobbs, Vibration welding of ABS to itself and to polycarbonate, poly(butylene terephthalate), poly(ether imide) and modified poly(phenylene oxide), *Polymer* 34(6), 1222–1231, 1993.
10. Stokes, V. K., The vibration welding of poly(phenylene oxide)/poly(phenylene sulfide) blend, *Polym Eng Sci*, 38(12), 2046–2054, 1998.
11. Stokes, V. K., The vibration welding of polycarbonate/acrylonitrile-butadiene-styrene blends to themselves and to other resins and blends, *Polym Eng Sci*, 40(10), 2175–2181, 2000.
12. Stokes, V. K., The vibration welding of poly(methyl methacrylate) to itself and to polycarbonate, poly(butylene terephthalate), and modified poly(phenylene oxide), *J Adhes Sci Technol*, 15(4), 457–466, 2001.
13. Potente, H. and M. Uebbing, Friction welding of polyamides, *Polym Eng Sci*, 37(4), 726–737, 1997.
14. Jones, I. A. and R. J. *Wise, Welding Method*, European Patent EP 1117502 B1, 2003.
15. Kagan, V. A. and N. M. Woosman, Efficiency of clearwelding technology for polyamides, *J Reinf Plast Comp*, 23(4), 351–359, 2004.
16. Stavrov, D. and H. E. N. Bersee, Resistance welding of thermoplastic composites—An overview, *Composites, Part A*, 36, 39–54, 2005.
17. Hou, M., L. Ye, and Y.-W. Mai, An experimental study of resistance welding of carbon fibre fabric reinforced polyetherimide (CF Fabric/PEI) composite material, *Appl Compos Mater*, 6, 35–49, 1999.
18. Ageorges, C., L. Ye, and M. Hou, Experimental investigation of the resistance welding of thermoplastic-matrix composites. Part II: Optimum processing window and mechanical performance, *Compos Sci Technol*, 60, 1191–1202, 2000.
19. M. Hou and K. Friedrich, Resistance welding of continuous glass fibre-reinforced polypropylene composites, *Compos Manuf*, 3(3), 153–163, 1992.
20. Dubé, M., P. Hubert, A. Yousefpour, and J. Denault, Resistance welding of thermoplastic composites skin/stringer joints, *Composites, Part A*, 38, 2541–2552, 2007.
21. Ageorges, C. and L. Ye, Resistance welding of thermosetting composite/thermoplastic composite joints, *Composites, Part A*, 32, 1603–1612, 2001.
22. Stokes, V. K., Experiments on the induction welding of plastics, *Polym Eng Sci*, 43(9), 1523–1541, 2003.
23. Suwanwatana, W., S. Yarlagadda, and J. W. Gillespie Jr., Hysteresis heating based induction bonding of thermoplastic composites, *Compos Sci Technol*, 66, 1713–1723, 2006.
24. Keller, C. T., *Self-Tapping Screws: How To Choose and Use The Right One*, Engineering Design, Dupont, 1983.

13 Tribology of Polymers and Composites

13.1 INTRODUCTION

Tribology is loosely translated from Greek as "the science of rubbing." In a lighthearted view, the "tri" can be considered to be the three areas of friction, wear, and lubrication. Much work has been historically performed in the sliding of metallic systems, both dry and lubricated. From this great body of work virtually all of the current "laws" of friction have evolved. Unfortunately, the subtopic of polymer-on-metal or polymer-on-polymer sliding systems violates most of the so-called "laws" thus invalidating them relative to polymers. However, examination of the popular misconceptions (typically those taught in high-school or undergraduate physics) is instructive for an understanding of polymeric systems. This chapter concentrates on dry and nonliquid lubricated polymer-on-metal or polymer-on-polymer sliding. The only lubrication that will be discussed is self-lubrication or dry lubricant films.

Examples of polymers in tribological applications are

1. Self-lubricated guides for slide assemblies
2. Gears
3. For low-speed lightly loaded shafts
4. Pumps
5. Conveyer chains
6. Fluid meters
7. Seals
8. Tires and wheels
9. Belts

Obviously, the potential application list is much more extensive. Polymers are used in tribological applications for many of the same reasons as they are used in structural applications. In many instances, polymers will perform a dual function as both a structural and a tribological material. An example of dual functionality is self-lubricated polymers that are injection molded into structural housings for low-speed lightly loaded rotating shafts. Concerns such as weight, cost, and manufacturability are some of the same issues with the use of polymers and polymeric composites for sliding applications as in structural applications.

The following is a compilation of general comments about the use of polymers in industrial tribological applications. Polymer failure in tribological applications is often catastrophic, occurring in a very short timescale of the order of minutes or hours.

Thermophysical properties are very important relative to polymer-on-metal sliding. Polymers often have orders of magnitude less ability to transfer heat away from the sliding interface compared to metals. There is often more potential for chemical reactions to occur in polymer-metal sliding than in metal-on-metal sliding. However, under light loads and low speeds, there is the potential to exceed the performance of metallic systems with extremely low friction ($\mu = 0.2$) and little or no wear debris. Hardness of a polymer is usually more important to tribological performance than modulus, even though the two can often be directly related. Surface roughness of the polymer material is rarely considered due to the fact that the initial polymer surface topography is worn away quickly during the break-in period, especially by an opposing metallic counterface. Typically the polymer component is at worst sacrificial relative to the other counterface and at best forms a transfer film during the sliding process that reduces subsequent wear. Due to both melt and degradation behavior, most polymers are used in sliding applications below 125°C.

13.2 CONTACT MECHANICS

Before entering into the discussion on friction and wear of polymers, a brief introduction to contact mechanics will be useful to the ensuing discussion. In this section, the Hertzian theory of elastic contact will be presented to calculate the theoretical area of contact assuming perfectly smooth surfaces. Without going through the proof of the Hertzian theory, general contact geometries will be examined. There are two basic geometries that will suffice for estimating most situations. The geometries are cylinder-on-cylinder and sphere-on-sphere and are shown in Figure 13.1. These default, in the special case when one surface is a flat plane (with an infinite radius of curvature) to a cylinder-on-plane and a sphere-on-plane. In addition, conformal contact situations can be estimated by considering the conformal radius as a negative quantity.

The macroscopic or apparent area of contact is calculated from the formulas shown in Table 13.1. However, for very hard materials, such as metals and some

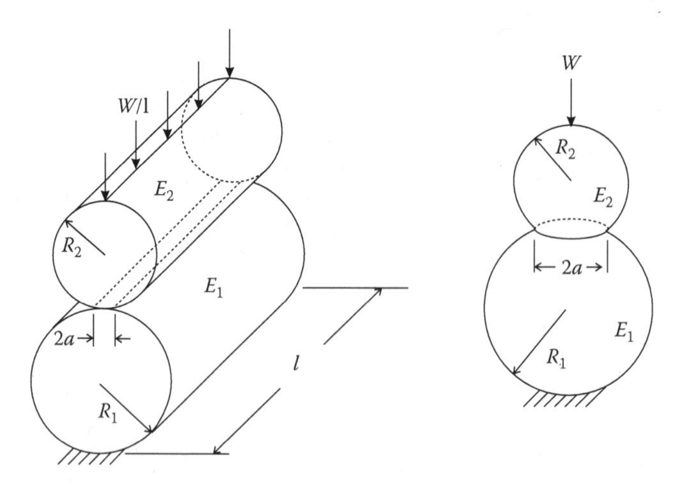

FIGURE 13.1 Contact geometries.

TABLE 13.1
Hertzian Contact Equations for Non-Conformal Geometries[a,b]

Private	Cylinder-on-Cylinder	Sphere-on-Sphere
a (mm) Contact halfwidth or radius	$a = 2\left(\frac{W'R}{4E'}\right)^{\frac{1}{2}}$	$a = 2\left(\frac{W'R}{4E'}\right)^{\frac{1}{2}}$
p (MPa) Normal surface pressure distribution	$p(x) = 2\frac{W'}{\pi a}\left(1 - \frac{x^2}{a^2}\right)^{\frac{1}{2}}$	$p(r) = \frac{3W'}{2\pi a^2}\left(1 - \frac{r^2}{a^2}\right)^{\frac{1}{2}}$
p_{mean} (MPa) Mean surface pressure	$p_{mean} = \frac{W'}{2a}$	$p_{mean} = \frac{W'}{\pi a^2}$
p_{max} (MPa) Maximum surface pressure	$p_{max} = 2\left(\frac{W'E'}{\pi R}\right)^{\frac{1}{2}}$	$a = \frac{1}{\pi}\left(\frac{6W'E'^2}{R^2}\right)^{\frac{1}{3}}$

[a] These formulae are approximately true when the mating geometries are similar (i.e., $R_1 \approx R_2$).
[b] Use $R = -R$ for conformal geometries.

polymers and composites the real area of contact is much smaller than the apparent and is determined by taking into account the surface topography.
where:

E' = composite modulus (MPa),

$$\frac{1}{E'} = \frac{1-\nu_1^2}{E_1} + \frac{1-\nu_2^2}{E_2}. \tag{13.1}$$

R = composite radius (mm),

$$\frac{1}{R} = \frac{1}{R_1} + \frac{1}{R_2}. \tag{13.2}$$

W' = Weight (force) per unit length for cylindrical contact,

$$W' = \frac{W}{l}. \tag{13.3}$$

Consideration of the Hertzian area of contact leads to a discussion on the real versus apparent area of contact. The apparent area of contact is the area that appears to be in contact. For example, if a brick is placed on a concrete sidewalk it appears that the face are of the brick is the same as the contact area. On closer inspection and consideration it is clear that this is not the case due to surface roughness of both surfaces. The real area of contact is much smaller than the apparent area. If instead, a rubber ball of given diameter is pressed against an optically smooth glass plate, one can observe the Hertzian contact area. Interestingly, the original Hertz work was performed with curved glass plates and has found great applicability in the contact and subsequent friction of elastomeric sliding. If, however, the glass is roughened the true or real contact area will be reduced from the Hertzian contact area.

13.3 SURFACE TOPOGRAPHY

The previous discussion of Hertzian contact assumed that the mating surfaces were perfectly smooth, thus ignoring the fact that all surfaces have topography or roughness. Under magnification surfaces have topographies that result from the manufacturing processes that have been used to create the surface. At high magnification a surface will appear as a distribution of peaks and valleys, much like a mountain range. An example of such a surface is shown in Figure 13.2. The peaks or protrusions which resemble mountains are called asperites. The size, shape, and distribution of the aspirates have a large influence on both friction and wear. A direct effect of surface roughness is to concentrate the distributed Hertzian contact into distinct contact regions. Thus, both the normal load and friction load are carried over a smaller area than predicted by the Hertzian equations.

Surface topography is often thought of in terms of surface roughness. While the surface roughness is often indicative of the topography, it must be complemented by another surface parameter to reasonably describe the surface. Surface parameters, such as R_a, and R_q or R_{rms} are widely used to describe the surface roughness:

$$R_a = \frac{l}{l}\int_0^1 |y|\,dx \approx \frac{1}{N}\sum_{i=1}^{N}|y_i|, \tag{13.4}$$

$$R_q = \sigma = Rms = \sqrt{\frac{1}{l}\int_0^1 \frac{y^2}{2}\,dx} \approx \sqrt{\frac{1}{N}\sum_{i=1}^{N} y_i^2}. \tag{13.5}$$

As can be seen from Equations 13.4 and 13.5, R_a and R_q describe how a surface deviates from its mean line. While R_a is often utilized in the United States as the only parameter to describe a surface, R_q is similar and is the standard deviation of a surface from the mean value. However, as can be seen from Figure 13.2, these two surfaces have the same R_a and R_q. It is obvious that the surface in Figure 13.2a is superior for systems that need to reduce wear due to the large bearing area. The surface shown in

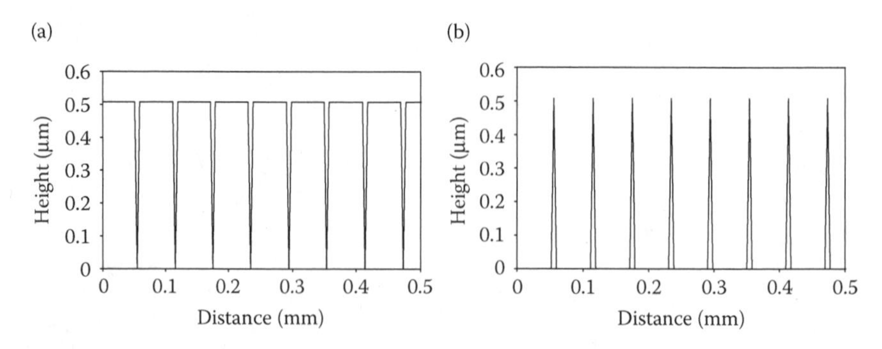

FIGURE 13.2 Surfaces with identical R_a and R_q. (a) Plateau and (b) peak.

TABLE 13.2
Comparison of Surface Parameters

	Plateau Surface	Peak Surface
R_a	0.0566	0.0566
R_q	0.100	0.100
Skewness	–3.330	3.330

Figure 13.2b would be more applicable to a situation where the polymer sliding against it was expected to be worn away. In this case an additional parameter such as skewness may be used to differentiate between the surfaces:

$$\text{Skewness} \approx \frac{1}{\sigma^3}\frac{1}{N}\sum_{i=1}^{N} y_i^3. \tag{13.6}$$

The hypothetical surfaces presented are useful for illustrative purposes. The surface parameters are shown in Table 13.2. However, when the same manufacturing technique is used to create wear specimens or actual machined components for use often, the simple use of R_a and R_q are adequate for process control of an ongoing process. There are a myriad of other surface parameters that can be used to describe a surface for tribological applications. For example, parameters such as the radius of curvature of the asperities can also be used. A metrologist would contend that three parameters are required to adequately describe a surface, whereas a tribologist would counter that two parameters will often suffice. Note, that R_a and R_q essentially quantify the second moment around the mean value and those do not constitute to independent measures of the surface.

13.4 FRICTION

What is the definition of friction? Simply put, friction is the tangential resistance to sliding of relative surfaces. A normal load holding the surfaces together is typically implied in this situation. This results in the following definition for coefficient of friction

$$\mu \equiv \frac{F}{W}, \tag{13.7}$$

where μ = coefficient of friction, F = tangential resistive force (N), and W = normal load or force (N).

Although controversy exists over whether or not the coefficient of friction is a fundamental property, it is obvious that if both F and W can be measured or estimated this ratio can be formed. An extremely important point must be made at this juncture. A single material does not possess a coefficient of friction. A coefficient of friction can only exist between two materials. This is not as trivial a point as it may first seem. While the coefficient of friction of a polymer against two materials may be similar or the same, in general it is not. Why? Because the coefficient of friction depends on many parameters of the mating surfaces, such as geometry, surface finish of both surfaces, material properties, and so on. For most discussion, both the text and practitioners will use friction and coefficient of friction interchangeably. The friction force, and thus the coefficient of friction, has many functional dependencies as shown below:

$$\mu = \mu(W, T, V, RH, R, L), \tag{13.8}$$

where T = temperature, V = velocity, RH = relative humidity, R = roughness of both surfaces, and L = presence of lubricants.

The functional dependence of coefficient of friction can be viewed as a statistical model of the main effects and interactions. The following are the "three friction laws," which should be more appropriately referred to as the "three metallic friction laws." After stating the laws, they will be examined for why they are applicable for metals and where polymers differ. Amoton's Laws (ca. 1699) state the following:

1. The coefficient of friction is independent of apparent area of contact.
2. The coefficient of friction is independent of the normal load.

Coulomb's Law (ca. 1785) states the following:

1. Kinetic friction is independent of speed.

The first law can be demonstrated experimentally by taking a rectangular steel block sliding on a steel surface, where both surfaces are nominally flat and smooth (It was originally demonstrated with a brick). If the block is slid on any of the three faces the friction force required to move the block is approximately the same. Why does this occur? If the three surfaces are of different macroscopic dimension we have experimentally shown that the coefficient of friction is independent of the apparent area. Placing another block of equal weight on top of the first block and pulling both, the tangential sliding force is twice that of the single block experiment. However, since we have doubled the normal load, the coefficient of friction is the same as in the first experiment. Why? This occurs because the real area of contact between the two surfaces is the same regardless of the apparent area of contact. There are two arguments with regard to why the real area of contact remains the same. Consider that both surfaces possess some degree of roughness.

The first theory is that when the surfaces are pressed together the outermost peaks of the surfaces are plastically deformed. Plasticity states that the flow pressure is equal to the applied stress.

$$P_f = \frac{W}{A_r} \text{ (MPa)}, \tag{13.9}$$

where A_r = real area of contact at the asperity peaks (mm^2).

The second theory considers the contact between surface asperities to be elastic and when the Gaussian distribution of surface heights of the asperites of the mating counterfaces is taken into account the real area of contact is once again linearly dependent on normal load. In the simplest view of friction it is assumed that the tangential friction force is proportional to the real area of contact,[1] such that,

$$F = A_r s(N), \tag{13.10}$$

where s = shear strength of the mating junction or softer material, whichever is weaker (MPa).

This result leads to the following relationship for the coefficient of friction under fully plastic conditions. The only functional dependencies are on the shear strength of the junction and the flow pressure of the softer material. For nominally flat surfaces the geometry of contact does not enter into the formulation.

$$\begin{aligned} \mu &= \frac{F}{W} \\ &= A_r \frac{s}{W} \\ &= \left(\frac{W}{P_f}\right)\frac{s}{W} \\ &= \frac{s}{P_f} \end{aligned} \tag{13.11a–d}$$

As a heuristic (without proof) rationalization of the third law, based on the development of the first two laws, it can be proposed that if neither s nor P_f depend on velocity or strain rate, then neither will the coefficient of friction. In dry metallic sliding the three laws hold over a wide range of normal loads and velocities. The laws do break down at extremely low and high loads as well as extremely high velocities, where the material properties become strain rate dependent.

How are sliding polymeric systems different from sliding metallic systems with respect to friction? First let us examine the similarities. In both sliding systems, the level of friction is usually determined or limited by the weaker of the two materials. Thus, even for a polymer-on-metal sliding system, the friction will be largely controlled by the polymer. The effect of surface roughness of the metal on friction will obviously play a role in the measured friction force.

Examination of the coefficient of friction of polymeric systems in the same manner as for metallic systems shows the functional differences. The primary difference between the two sliding systems is that even for polymeric systems that undergo initial plastic deformation, there will be a significant portion of the real area of contact that is in elastic contact. For elastomeric sliding systems virtually all of the real area of contact is loaded elastically. Thus, for the sake of discussion, assume that the contact is a sphere-on-flat configuration with completely elastic loading. One case where this occurs is for a soft elastomer on a hard, nominally flat counterface. In this situation, the real contact area is given by Hertzian contact theory radius calculation as

$$a = \left(\frac{3WR}{4E}\right)^{\frac{1}{3}}, \tag{13.12}$$

where a = contact radius (mm), R = radius of sphere (mm), and E = composite modulus (MPa). Thus, the real area of contact is

$$A_r = \pi a^2 = \pi \left(\frac{3WR}{4E'} \right)^{\frac{2}{3}} \tag{13.13a,b}$$

This condition results in the following development for the coefficient of friction

$$\mu = \frac{F}{W} = \frac{A_r s}{W} = \frac{\pi \left(\frac{3WR}{4E'} \right)^{\frac{2}{3}}}{W} = \pi \left(\frac{3R}{4E} \right)^{\frac{2}{3}} W^{-\frac{1}{3}} s \tag{13.14a–d}$$

From this simplified example, it is seen that for elastic contacts there is dependence on the normal load, which, in general, invalidates Amoton's second law. Additionally, since the real area of contact is dependent on the geometry, Amoton's first law is also invalid for polymeric sliding. The final issue is the strain rate dependence of material properties. Previously, in the text, it has been shown that the material properties of polymers are very strain rate dependent and, not surprisingly, this dependence holds in the friction measurements. Therefore, even Coulomb's law is generally not valid for polymeric sliding. As a final note, by inserting the appropriate geometry through the Hertzian calculations for the real area of contact, a first approximation can be made for designs where dimensional changes are being evaluated. If the test geometry is known for tabulated coefficient of friction data, the interfacial shear strength can be estimated. This value can then be used to calculate the coefficient of friction for various other geometries.

13.4.1 Static and Dynamic Coefficients of Friction

The static friction force is the tangential force required to initiate bulk sliding. The kinetic, or dynamic, friction force is the tangential force required to maintain sliding at nonzero (typically constant) velocity. Except on rare occasions the static and kinetic friction forces and hence coefficients of friction are not equal. The kinetic friction force can be higher or lower than the static friction force. For example, the static friction force can possess time dependence due to viscoelastic effects experience during contact. As the dwell time between two surfaces that are brought in contact increases, the area of contact (due to creep) can increase and the strength of this contact from attractive forces can increase. Either or both of these effects result in a dwell-time-dependent static friction. This can lead to the situation shown in Figure 13.3, where the static coefficient of friction increases with dwell time and the kinetic coefficient of friction remains essentially constant.

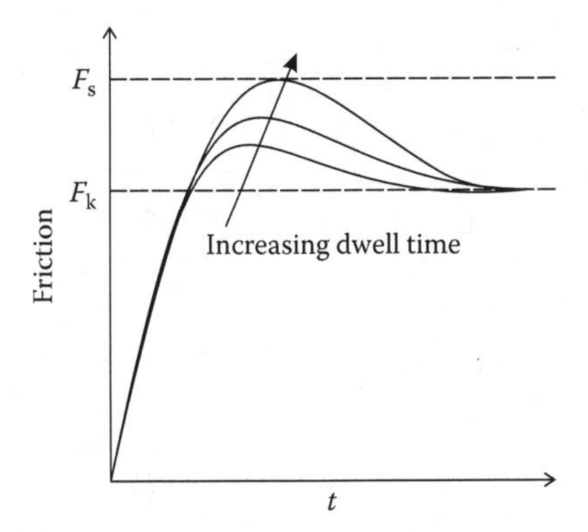

FIGURE 13.3 Dwell-time dependence of the static coefficient of friction.

The kinetic coefficient of friction can be greater than the static when there are heating or other effects occurring. For example, at the start of sliding there is no frictional heat generation and the counterfaces are at the surrounding environmental temperatures. As sliding ensues, frictional heat is generated, softening one or both surfaces resulting in a larger real contact area and thus coefficient of friction. In one sense this situation is not ideal, the friction at any time, t, after sliding has commenced is dependent on the temperature which is transient until steady state has been reached. Additionally, even after steady state has been reached the temperature is obviously greater than that at the beginning of the test ($t = 0$) when the static coefficient is determined.

13.4.2 Adhesive and Abrasive Friction

Dry sliding friction is usually broken down into two components, adhesive and abrasive friction. In general, both of these components or effects occur simultaneously. Adhesive friction occurs between when one of the surfaces is very smooth (typically $R_a < 25$ μ in.) or at the real contact areas where the materials have been flattened due to the extremely high pressures. Adhesive friction can be thought of as a result of the net energy loss required to slide to perfectly smooth contacts past one another. Conversely, abrasive friction occurs when one, or more of the surfaces, is extremely rough and hard. This allows the aspirates on the rough surface to penetrate into the softer surface, regardless of the softer surface's topography. When relative sliding ensues, plowing of the softer material resists the relative tangential motion of the harder material asperities.

13.5 WEAR

Wear is defined as the removal of material from one or both surfaces. This may or may not result in a net system weight loss. How can this happen? Material may be transferred from one counterface to the other, which means weight loss from one surface is gained by the other. Material may also be plowed out of the sliding track which does not result in appreciable weight loss. The modes of wear follow the fundamental modes of friction as well as additional effects, which occur due to the frictional stresses on the area of contact. There is adhesive, abrasive, fatigue, and delamination wear. Additionally, there are phenomena such as impact wear, rolling wear, and fretting wear.

Adhesive wear occurs due to very smooth surfaces creating an interfacial bond which is stronger than one or both of the mating materials. When bulk sliding forces this junction to slide or roll past the adhered material a volume of material is removed. Abrasive wear is caused by plowing of asperities of the harder surface through the softer surface. Both adhesive and abrasive wear start immediately at the beginning of the sliding motion.

Fatigue wear occurs due to the cyclic stressing of the surface by repeated passes of the opposing counterface. Fatigue wear is indicated when there is a finite number of cycles that occur prior to noticeable wear formation. The delamination theory of wear is a subset of fatigue wear, where a crack propagates parallel to the surface until it reaches a critical length at which point a platelet of material is separated from the surface. Delamination wear is indicated by the presence of the plate-like wear debris around the sliding path. Fatigue-dominated wear mechanisms are indicated by the presence of an incubation time, where no discernable wear occurs. The incubation time is followed by the generation of wear debris.

13.5.1 Archard Wear Law

The Archard Wear Law is the simplest functional form (linear) that can be construed for the wear of mating surfaces:

$$V = KWD\ (\text{mm}^3), \tag{13.15}$$

where V = volume of wear (mm^3), K = wear factor (mm^3/(MPa m/s s)), W = normal pressure (MPa), and D = distance traveled (m).

Equation 13.15 can also be rewritten in terms of time as

$$V = KWvt, \tag{13.16}$$

where v = velocity (m/s) and t = time (s).

Manufacturers will have the wear factor, K tabulated for comparison between materials. The Archard wear law, while developed for dry sliding of metals, often holds for polymeric materials. Archard's law implies that the volume of wear will increase linearly for all time. However, for most polymers against metals, wear volume will often deviate from linear behavior, by slowing as a function of time, and occasionally reaching a maximum value. This can result in a parabolic form to the volume loss as a function of time as shown in Equation 13.17.

$$V = KWvt^n, \tag{13.17}$$

where n = functional time dependence ($n < 1$).

Often, the wear factor will be determined by performing laboratory tests for a fixed duration of time and measuring the weight or volume loss at the end of that time to determine what can be best described as a secant wear factor. Laboratory wear tests are usually designed to be very severe, in order to accelerate the wear rates to draw a comparison in a reasonable amount of time. Typically, hours or days, compared to applications where the life may be of the order of years or decades. This can result in errors for estimates of K. Additional sources of confounding in estimates of wear rate are the different morphologies that can occur from the manufacturing processes, such as extrusion or molding. As discussed in previous chapters, molding creates varying surface properties due to crystallinity. For filled materials the surface layers will be resin rich, with the bulk of the filler (e.g., fibers or spheres) below the surface.

13.6 PV LIMIT

Polymeric materials have a strong dependence on PV values and limit. PV is the product of pressure and velocity. The interdependence of pressure and velocity on friction and wear is explored in an empirical manner by obtaining friction and wear data usually at specified pressures over a range of velocities. One of the primary purposes of acquiring these data is to determine the PV limit. The PV limit is the critical value above which the sliding system no longer performs adequately. Figure 13.4 shows a representative PV curve for polymeric materials. This curve can be rationalized by considering the limits of pressure and velocity. At virtually zero velocity there is a pressure value at which the material flows under the normal load (cold flow). At virtually zero load there is a velocity at which the material melts due to frictional heat generation. Thus, at combinations away from the extrema, a combination of the two effects leads to inadequate performance. While this rationalization does not completely explain what happens at given catastrophic values of PV, it begins to give physical understanding of the underlying phenomena. The PV limit is also dependent on the specific mating configuration, especially as it relates to heat transfer away from the sliding interface.

Historically, the PV limit was determined for a polymer against metal by utilizing a shaft rotating in a half-bearing as shown in Figure 13.5. This was done to simulate a polymeric bushing. Currently, PV limit is often determined by the use of a thrust washer configuration. This configuration is used due to the availability of test devices for this specimen geometry. The test procedure is relatively simple (see Figure 13.6). First a constant velocity is imposed. Load is then stepped to an increased level and held for some time (typically 30–60 min) while the temperature of the bearing, friction, and wear rate are monitored. If temperature, friction, and wear rate are within acceptable bounds, the normal load is again increased. While, historically, temperature and friction were the primary quantities monitored for stability, wear rate is often the more practical and direct evidence that the PV limit has been reached. The load is increased until conditions are deemed unacceptable. The product of pressure and velocity at

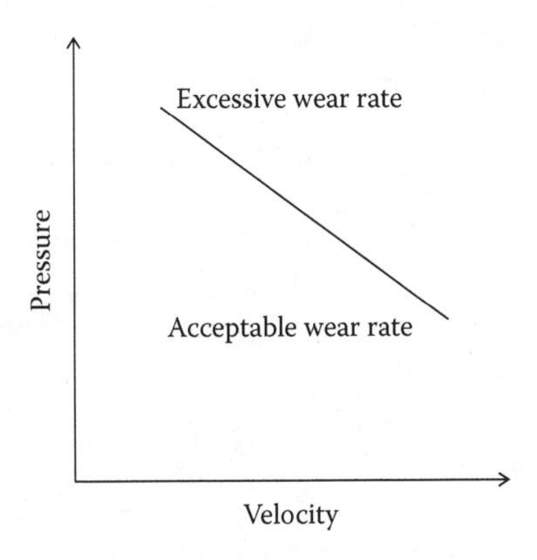

FIGURE 13.4 Representative PV curve for a polymeric material.

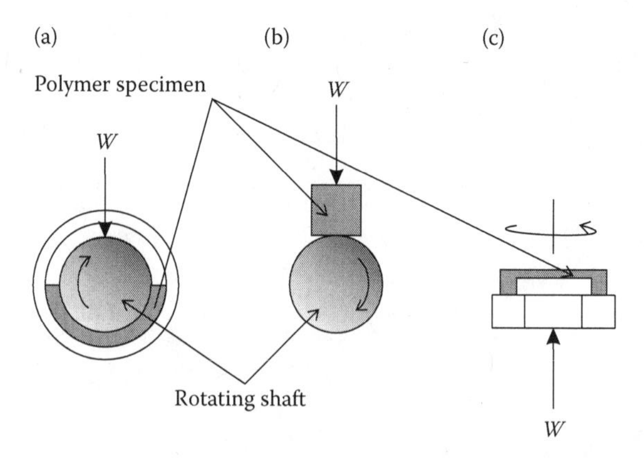

FIGURE 13.5 Test configurations for PV limit determination of polymeric materials. (a) Bushing, (b) block, and (c) thrust washer.

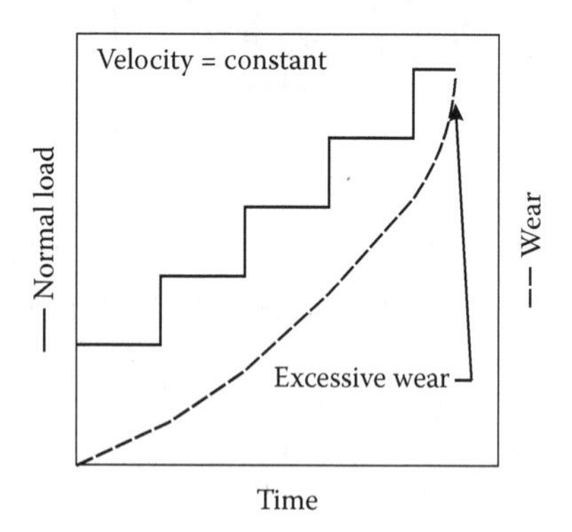

FIGURE 13.6 PV limit test procedure and results.

this point is taken to be the PV limit of the material. Note that in general this product is not a constant value for different ranges of pressure or velocity.

13.7 ROLLING AND SLIDING

Most of us think of friction and wear when there is relative motion between surfaces. Theoretically, when there is perfect rolling there is no relative tangential motion between surfaces. In reality, there is always a small amount of microslip, if not macroscopic slip that occurs. However, even excluding the slip that does occur in rolling, there is still friction and also the possibility of wear due to adhesion. This adhesive wear is referred to as spontaneous adhesion and wear. Friction (or traction) is obviously required for rolling. If the coefficient of friction were zero it would be impossible to accelerate a rolling object. This simple example explains why an automobile tire is made of rubber. Elastomers rolling or sliding against other materials have the highest coefficients of friction and thus offer the best traction of any mating material for roadway surfaces.

The friction in rolling systems generally results from the hysteretic losses in the materials. Thus, the amount of friction or rolling resistance can be designed into a wheel based on geometry and material properties. An example of this is in-line skate wheels, made of PUR, which have a wide variety of durometers for different applications. For fast or racing skates, a hard wheel is desired. Why? In general, the harder the wheel, the less the losses or hysteretic the material, and thus the lower the friction. The result is a fast wheel. The downside to this approach is that harder wheels are much harsher due to their inability to absorb vibration and impacts. From this discussion it can be seen that the optimum wheel for widespread use will balance adequate traction, low forward frictional losses, and comfort.

13.8 MODIFICATION OF POLYMERS FOR FRICTION AND WEAR PERFORMANCE

Polymers in the unfilled state, often need an enhancement of properties to perform adequately in tribological situations. This can be seen if the application is viewed by the classic engineering approach of factor of safety. In the factor of safety approach to design, the factor of safety is defined as the strength over the stress.

$$\eta = \frac{S}{\sigma}, \tag{13.18}$$

where η = factor of safety, S = strength, σ = applied stress.

Obviously, the factor of safety has to be >1 for the component to perform satisfactorily. Therefore, consider a polymer part that is to be used in conjunction with a metallic part. The part as a homopolymer is wearing too fast. What are the options? There are two obvious options if the base homopolymer is kept constant. Reduce the applied stress or increase the strength of the material. It is relatively obvious that if the coefficient of friction is reduced, the contact stresses on the polymer will be reduced. This leads to the conclusion that the homopolymer should be modified to decrease the coefficient of friction, which in turn leads to the topic of internal lubricants. The second alternative is to increase the strength of the polymer and this can be accomplished through the use of reinforcement.

13.8.1 Internal Lubricants

The two most often used fillers in thermoplastics are PTFE and PDMS or silicone. The respective functions of the two materials are quite different. PTFE is almost the "wonder" additive for tribological applications, whether in a solid polymer or liquid lubricant. The mechanical properties of PTFE hold over a range of –260°C to 260°C with a melt temperature of ~327°C. It is chemically inert and hydrophobic. The effect of PTFE as a solid lubricant can be understood from its performance as a homopolymer. As a homopolymer, PTFE has the lowest coefficient of friction of all polymeric materials sliding against metal surfaces. This is due to the lack of bulky side groups on the molecular chain and the formation of a low friction transfer film on the opposing counterface. However, due to the poor mechanical properties, PTFE alone also

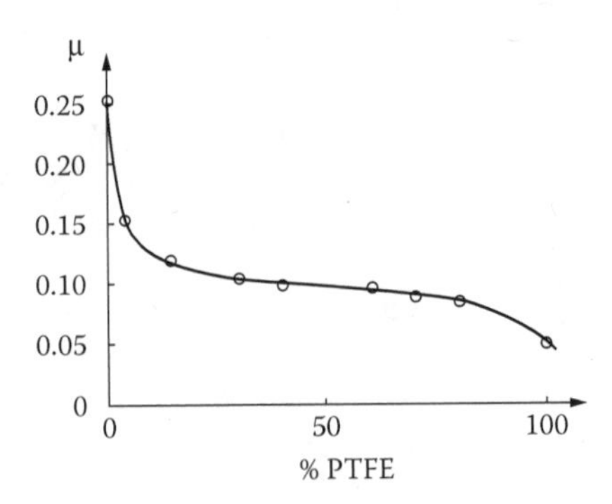

FIGURE 13.7 The friction coefficient of a PTFE/PEEK composite. (Reprinted from *Friction and Wear of Polymer Composites, Composite Materials Series*, Friedrich, K., ed., 1, B. J. Briscoe, p. 51, Elsevier, New York, NY, © 1986, with permission from Elsevier.)

has an extremely high wear rate relative to other polymers. What is expected for PTFE performance based on a composite's approach? Consider PTFE in combination with any other polymer. As the mixture approaches 100% PTFE/0% of another polymer, the friction is low and the wear is high. The reduction of the coefficient of friction of a PTFE-filled PEEK composite is shown in Figure 13.7. However, addition of PTFE reduces both the friction and wear of the composite below the 100% homopolymer level. This optimum is not predicted by a rule of mixtures approach due to the competing effects of wear resistance and reduction in applied stress. Commercially,[2] the optimum level of PTFE in other polymers has been shown to be in the range of 10–15%. This is shown in Figure 13.8.

Silicone also has wide use as an internal lubricant for polymers. While both silicone and PTFE reduce the coefficient of friction and thus the applied combined stresses, silicone functions in a completely different mode than PTFE. Silicone is used in an oil form compared to the dry state of PTFE. The concentration of PTFE throughout the polymer stays constant with time. One of the modes that PTFE functions is that of filling the opposing surface voids and providing a low friction transfer film. Silicone

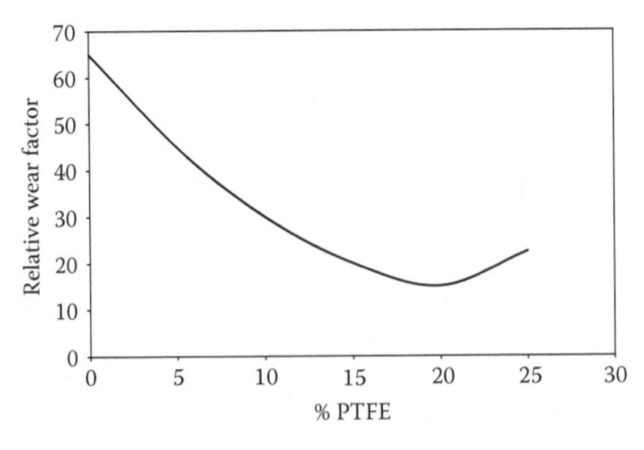

FIGURE 13.8 Effect of PTFE on wear rate. (Adapted from Lubricomp® Internally lubricated reinforced thermoplastics and fluoropolymer composites, *LNP Bulletin*, 254–691.)

functions in the more traditional manner of lubricants by providing a boundary liquid lubricant film between the mating surfaces. Due to the incompatibility of silicone with the base polymer and the concentration gradient, silicone continuously diffuses to the surface. Occasionally, the two lubricants are used in combination due to the beneficial synergy provided. However, there are cases when it is desirable to use only PTFE. This often occurs in polymers used in food handling where the presence or migration of lubricants is a concern. Additionally the presence of silicone can cause problems with other materials (e.g., polymer crazing).

13.8.2 Reinforcements

Carbon, aramid, copper, glass fibers, and glass spheres are all used to reinforce polymers for wear resistance. These materials primarily function by increasing the resistance of the base polymer to the combined normal and tangential applied stresses. Reinforcements will often not lower the coefficient of friction. In fact, the coefficient of friction may increase with the incorporation of rigid fillers. In fact, increased wear resistance may be obtained at the expense of reduced wear resistance of the opposing metallic counterface. In many design situations this can be unacceptable. In certain designs it is desirable to have a sacrificial component that is intended to be replaced periodically. This enables more expensive components to last the lifetime of the entire system with minimal or no modification. An example of this is brakes and clutches (PF binders with polymeric and metallic reinforcements), where the anticipated modification is resurfacing and replacement at reduced intervals when compared with the sacrificial component. The wear factors for various polymers are shown in Table 13.3. Figures 13.9 and 13.10 illustrate the effects of various fillers and additives for Nylon 6,6.

13.9 COMPOSITES

Similar to the rule of mixtures calculation for composite modulus, there is an analogous calculation for the coefficient of friction of a composite (Friedrich). A unidirectional lamina has three principle directions for sliding as shown in Figure 13.11. These directions are sliding transverse to the normal direction, parallel to the fibers, and perpendicular or antiparallel to the fibers.

The following relationship is theoretically obtained for a fiber-reinforced composite:

$$\mu = \frac{F}{W} = \frac{F_f + F_m}{W_f + W_m}, \tag{13.19}$$

where F_f = friction force component of fibers, F_m = friction force component of matrix, W_f = normal load carried by fibers, and W_m = normal load carried by matrix.

The following definitions and assumptions will be necessary for the development of a composite friction force:

$$A_f = v_f A, \tag{13.20}$$

and

$$A_m = v_m A, \tag{13.21}$$

TABLE 13.3
Wear Factors and PV Limits for Various Polymers

Polymer	$K10^{-10}$ $\left(\frac{in.^3 min}{ftlbhr}\right)$	PV Limit (psi fpm)		
		10 fpm	100 fpm	1000 fpm
ABS, 15% PTFE	300	18,000	4000	2000
Acetal	65	4000	3500	<2500
Acetal, 20% PTFE	17	>40,000	12,500	5500
Acetal, 30% glass, 15% PTFE	200	12,500	12,000	8000
Fluorocarbon	2500		1800	
Fluorocarbon filled	1–20		30,000	
Nylon type 6 30% glass, 15% PTFE	17	17,500	20,000	13,000
Nylon type 6/10 30% glass	15	20,000	15,000	12,000
Nylon type 6/6	200	3000	2500	2500
Nylon type 6/6 20% PTFE	12	>40,000	27,500	8000
Nylon type 6/6 30% glass, 15% PTFE	16	17,500	20,000	13,000
Nylon graphite filled	50		4000	
PEEK, 30% carbon	60			
PEEK, 5% carbon, 15% PTFE	60		40,000	
PF	250–2000		5000	
PF, PTFE filled	1020		40,000	
PC	250–2100	750	500	
PC, 30% glass, 15% PTFE	30	27,500	30,000	13,000
Polyester	210			
Polyester, 30% glass, 15% PTFE	20	>40,000	30,000	5500
Polyimide	150		100,000	
Polyimide, graphite filled	15		100,000	
Polyimide, SP211 filled	33		500,000	200,000
PP, 20% glass, 15% PTFE	36	14,000	12,000	7500
Polysulfone	1500	5000	5000	3000
Polysulfone, 30% glass, 15% PTFE	70	2000	35,000	15,000
PUR	240	2000	1500	<1500
PUR, 30% glass, 15% PTFE	35	7500	1000	5000
PPS (polyphenylene sulfide), 30% glass, 15% PTFE	100	27,000	30,000	>30,000
Styrene acrolonitrile, 30% glass, 15% PTFE	65	17,500	10,000	10,000

Source: Adapted from Bayer, R. G., *Mechanical Wear Prediction and Prevention*, Marcel Dekker, Inc., New York, 1994.

Note: 10^{-10} is corrected from reference.

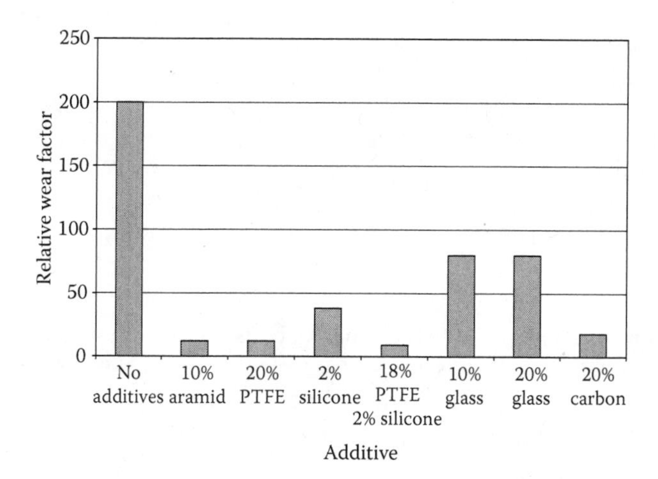

FIGURE 13.9 Effect of various fillers on the wear rate of Nylon 6,6. (Adapted from Lubricomp® Internally lubricated reinforced thermoplastics and fluoropolymer composites, *LNP Bulletin*, 254–691.)

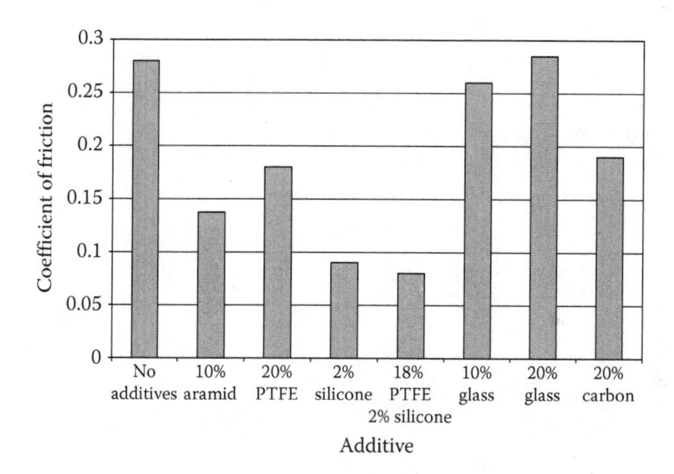

FIGURE 13.10 Effect of various fillers on the coefficient of friction of Nylon 6,6. (Adapted from Lubricomp® Internally lubricated reinforced thermoplastics and fluoropolymer composites, *LNP Bulletin*, 254–691.)

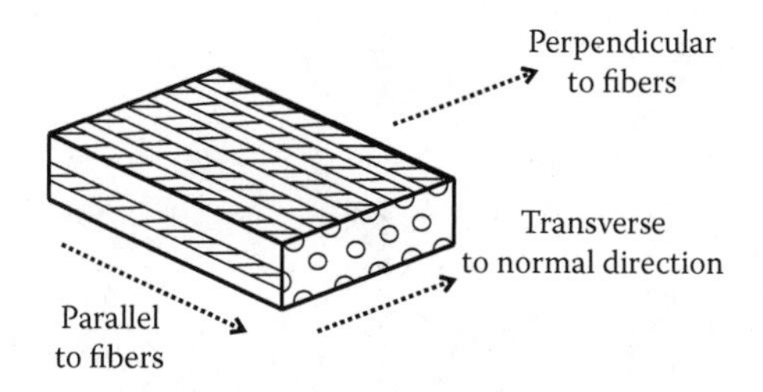

FIGURE 13.11 Sliding directions on a unidirectional composite laminate.

where A_f = area of fiber relative to direction of sliding, A_m = area of matrix relative to direction of matrix, v_f = volume fraction of the fiber, and v_m = volume fraction of the matrix.

The following assumptions are made about the state of stress and strain in the composite material. The first assumption is that the state of shear strain in the fibers is the same as the matrix,

$$\gamma_f = \gamma_m. \tag{13.22}$$

The second assumption is that the shear modulus of the fiber is the same as the matrix. While, conceptually this is a poor assumption, comparison of the final results of the theoretical development with experimental results does not seem to suffer because of this. Thus,

$$G_f = G_m. \tag{13.23}$$

Restating Equation 13.24 in the following manner for the composite properties:

$$\frac{F}{\mu} = W = W_f + W_m, \tag{13.24}$$

where μ = composite coefficient of friction.

Recognizing that

$$F_f = v_f F, \tag{13.25}$$

and

$$F_m = v_m F, \tag{13.26}$$

the following relationship is obtained for a fiber-reinforced composite:

$$\frac{1}{\mu} = \frac{v_f}{\mu_f} + \frac{v_m}{\mu_m}, \tag{13.27}$$

where μ_f = coefficient of friction of the fiber and μ_m = coefficient of friction of the matrix.

Experimental agreement with theory is quite good for polyester resin with E-glass, steel, and carbon fiber as shown in Figure 13.12. However, for other composite combination this is not universally true.

Equation 13.27 can be modified to take into account the presence of more than one fiber in the matrix as shown in Equation 13.28.

$$\frac{1}{\mu} = \frac{v_{f_1}}{\mu_{f_1}} + \frac{v_{f_2}}{\mu_{f_2}} + \cdots + \frac{v_{f_i}}{\mu_{f_i}} + \frac{v_m}{\mu_m}. \tag{13.28}$$

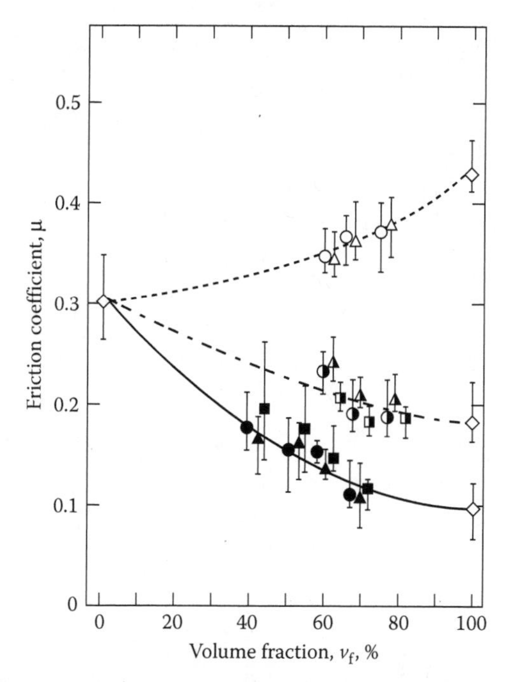

FIGURE 13.12 Composite coefficient of friction for polyester-based matrix material. Symbol legend: Glass fiber-reinforced polymer; ○-parallel, Δ-antiparallel, Steel-reinforced polymer: ◑-parallel, ◭-antiparallel, ◨-normal, carbon fiber-reinforced polymer; ●-parallel, ▲-antiparallel, ■-normal. (Reprinted from *Friction and Wear of Polymer Composites*, *Composite Materials Series*, Friedrich, K., ed., 1, T. Tsukizoe and N. Ohmae, pp. 212–220, Elsevier, New York, © 1986, with permission from Elsevier.)

13.10 WEAR OF COMPOSITES

The wear mode of a unidirectional composite lamina has three stages of wear progression when sliding occurs either parallel or transverse to the fibers:

1. Wear thinning of the fibers
2. Breakdown of fiber continuity
3. Eventual peeling out of the fiber sections

This wear mode is shown in Figure 13.13. Of course, fiber wear occurs with concurrent matrix wear. In general, the fiber reinforcement will be the limiting factor on the wear rate. For this situation the following wear equation has been proposed:

$$Q = k\frac{\mu p}{E}\frac{1}{I_s}WD, \tag{13.29}$$

where k = wear coefficient, $\mu p/E$ = strain of composite, I_s = interlaminar shear strength (MPa), W = normal load, and D = sliding distance.

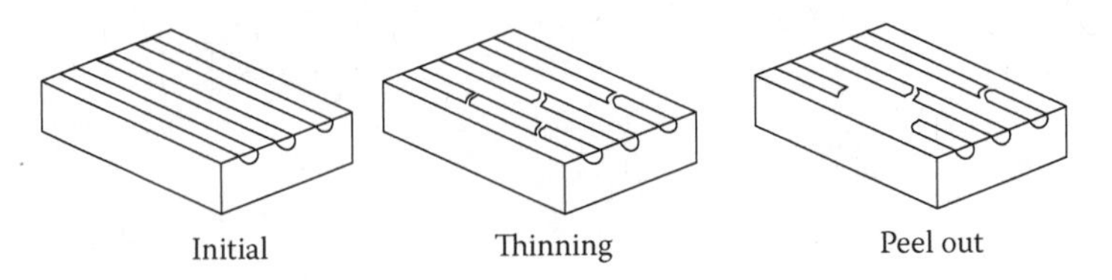

FIGURE 13.13 Types of mating test configurations.

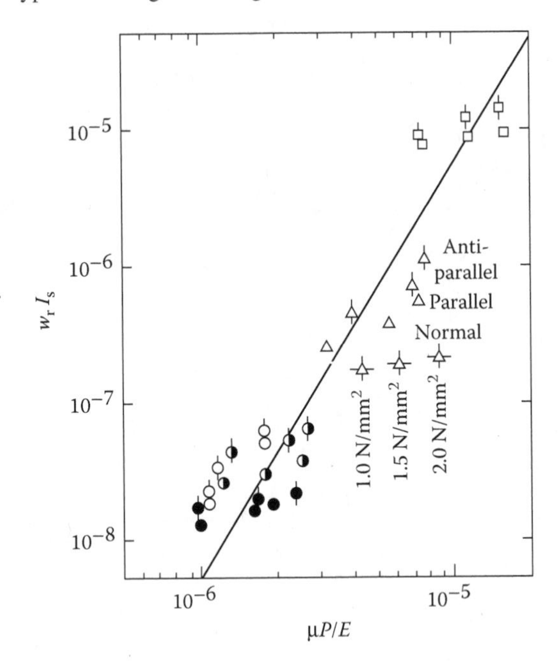

FIGURE 13.14 Specific wear rate of fiber-reinforced polymers. (Reprinted from *Friction and Wear of Polymer Composites, Composite Materials Series*, Friedrich, K., ed., 1, T. Tsukizoe and N. Ohmae, pp. 212–220, Elsevier, New York, © 1986, with permission from Elsevier.)

The specific wear rate is defined as

$$w_r = \frac{Q}{WD}. \tag{13.30}$$

The specific wear rate for various composites is shown in Figure 13.14.

13.11 HEAT GENERATION IN SLIDING POLYMER SYSTEMS

In sliding systems the heat generation results in increase temperatures of the mating materials. The increase in temperature affects the material properties and thus friction and wear. There are two temperatures that are of concern. The first is bulk temperatures that will be obtained from classic thermal calculations, and the second is flash temperatures. Flash temperatures occur due to the fact that contact usually occurs over very small contact areas due to aspirates on the surface. Both surface temperatures can result in melting of the counterface materials. The result of bulk melting is more obvious than that of localized melting due to flash temperatures. Bulk melting often results in catastrophic failure of one or both counterfaces. An example of just this situation occurred where two pieces of nylon, fused in operation, due to high interface temperatures, when they experienced oscillating sliding. When one of the nylon components was replaced with a metal component, melting of the nylon component no longer occurred.

13.11.1 Bulk Surface–Temperature Calculations

As stated earlier, the calculation of the bulk or average surface temperature between two mating surfaces is performed by classic heat transfer methods. In fact the only unique aspect to this problem is the heat generation term Q comes from frictional heat and is assumed to occur at the interface. In order to begin a thermal analysis, we calculate the total heat:

$$Q = \mu WV(\mathrm{J}), \tag{13.31}$$

where μ = coefficient of friction, W = normal load (N), and V = relative sliding velocity (m/s).

Now, the total heat must be partitioned between the two surfaces. In an exact solution, the heat partitioning would be determined by the complete physical problem. The exact partitioning would include the geometry and all the thermal properties and conditions. Why is the calculation of surface temperature important? While a measurement of temperature somewhere else in a test structure or application is indicative of the bulk surface temperature, it is often necessary to more closely estimate the interfacial temperature. As the temperature of operation approaches one of the material transitions (e.g., T_g or T_m) the mode of wear can change as well as the magnitude of friction. In fact, as the melting temperature is approached at the surface catastrophic failure or seizure can ensue. Once an approximate geometry is determined in a design and an estimate of the coefficient of friction is obtained, the interfacial temperatures can be estimated.

13.11.2 Flash Temperature

Occasionally, there will be chemical changes to the mating materials that are not explained by the bulk temperature estimates or other environmental effects. These changes can result in discoloration or reaction by-products that would not be predicted for the mating conditions. One possible reason can be found in flash temperature rise. In general, the mating condition is not perfect but the load is carried by an area less than the calculated real area of contact. This area can approach the dimension scale of the asperities. When this happens, all of the frictional energy is dissipated in a very small volume of material. The instantaneous temperatures can flash up to 50–100°C or more above the bulk interfacial temperature.

13.12 SPECIAL CONSIDERATIONS

13.12.1 Polymer-on-Polymer Sliding

In general, as in metal-on-metal sliding, dry sliding of the identical polymeric materials is not a good idea. Like metals will gall when slid against each other, and like polymeric materials will also gall or seize. This phenomenon results from very similar circumstances. Once the oxidized outer layers have been worn away, the freshly exposed materials are very compatible and will adhere to each other. In general, different polymers should be selected for the opposing counterface surfaces. An example of this would be in a polymeric gear set of three gears. Instead of selecting the same material for all three gears, the idler gear should be selected from another material. However, under lightly loaded situations, an entire gear train may be made of injection-molded nylon.

13.12.2 Coatings

Both thick and thin polymer coatings are applied to surfaces to act as dry bonded lubricants. The coatings are often present to resist both wear and corrosion, as well as reduce the coefficient of friction. The drawback to polymeric coatings is that they will eventually wear through the coating. One of the major difficulties in coatings is to obtain adequate adhesion between the polymer and the substrate. Poor adhesion will result in premature film failure. Unfortunately, the same internal lubricants that yield good tribological properties for the coating also result in poor adhesion to the substrate.

13.12.3 Effect of Surface Topography on Friction and Wear

Surface roughness has a complex effect on both friction and wear. At low surface roughness, both wear and friction and determined by adhesion of the smooth surfaces. As roughness increases, friction and wear will decrease to a minimum dependent on the counterface materials. Further increases in surface roughness will increase both friction and wear due to increased abrasive action of the surfaces. The roughness value at which the minimum occurs will depend on both the materials and methods used to manufacture the surface. In general, the optimum surface roughness to minimize both wear and friction for polymer–metal sliding systems appears to be ~8–16 μ in.

13.12.4 Effect of Environment (Temperature, Humidity, Gases, and Liquids, etc.) on Friction and Wear

Polymeric materials exhibit a great deal of sensitivity to the environmental conditions (temperature, humidity, gases, liquids, etc.). Not only are friction and wear sensitive to the operating environment, they are often more sensitive than bulk properties. Temperature affects friction of polymers in the same way as the complex modulus and other mechanical properties. Studies of both nylon[6] and elastomers[7,8] have shown that the coefficient of friction will transform in the same manner as storage and loss modulus. The coefficient of friction dependence of nylon on temperature is shown in

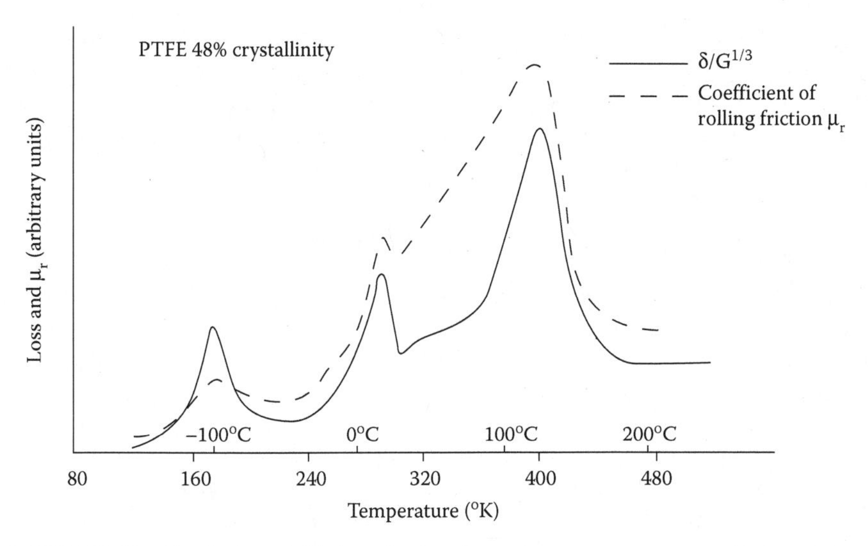

FIGURE 13.15 Coefficient of friction of nylon dependence on temperature. (Adapted from Ludema, K. C. and D. Tabor, *Wear*, 9, 329–348.)

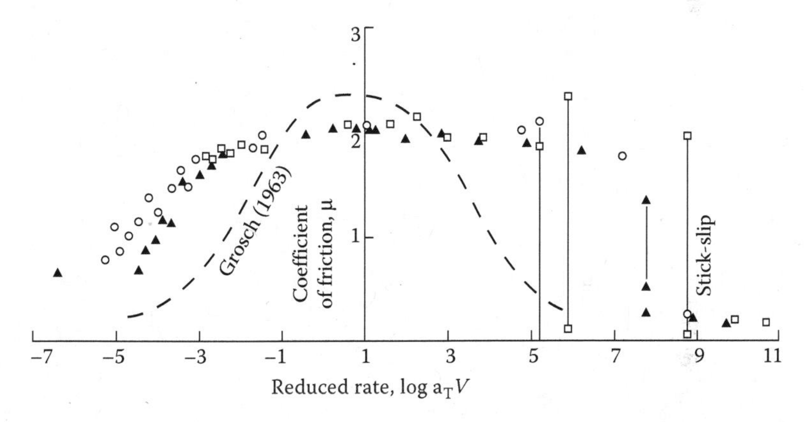

FIGURE 13.16 Comparison of nitrile rubber on wavy glass for Barquins and Roberts with Grosch's results. (Adapted from Barquins, M. and A. D. Roberts, *Journal of Physics D*, 19, 547–563, 1986; Grosch, K. A., *Proceedings of the Royal Society of London A, Mathematical and Physical Sciences*, 274(1356), 1963, 21–39.)

Figure 13.15. Figure 13.16 shows the coefficient of friction dependence of an elastomer on temperature.

13.12.5 Friction-Induced Vibration

Friction-induced vibration, in the most general sense, is the deviation from uniform sliding due to the interaction of friction and system dynamics. This can occur due to surface topography. However, the problem which is of the most concern is self-excited oscillations that occur due to a functional dependence of friction with velocity or position. The oscillations are unwanted noise or vibration, which affect both perceived and actual product performance. Examples occur in machine slideways, bushing, windshield wipers, dashboard components, and so on. The oscillations can occur in both unlubricated and self-lubricated composites. The problem is often exacerbated by the lack of stiffness of the structure, which is quite common in polymeric systems.

13.13 SIMULATIVE LABORATORY TESTING

Often it is desirable to perform screening tests on candidate materials. From the previous discussions of the influence of mating geometry on contact stresses and thermal considerations, it should be obvious that any laboratory test that is not an exact physical duplicate of the application will be a compromise on some aspect. The primary purpose of simulative laboratory tests is to rank candidate materials as they would rank in the application. A secondary purpose is to predict the absolute magnitudes of friction and wear in the application. The only reason that this is secondary and not primary is that is seldom possible to translate laboratory tribology tests directly to absolute performance in the application.

What are the primary issues that should be considered in simulative tribology tests? Applications are usually thought of, first in terms of normal load and relative motion (speed and type). Other considerations are temperature and environment. Often it is possible, and always desirable to perform room-temperature tests that will rank materials. However, while the loading in the application is often referred to in units of force, the concern is really of applied stress. This can be understood by revisiting the factor of safety approach to design. When the applied stress is above the yield or ultimate material properties, wear modes can change or become catastrophic. Therefore, laboratory testing should be in the same nominal pressure range as the application. In considering relative velocity, the strain rate dependence must be taken into account. Typically, this is the one parameter that is matched to the application. Of course, as in the discussion of PV limits, matching the combination of PV is also advantageous to the applicability of the testing. Additionally, the type of motion should be duplicated. For example, there are unidirectional sliding, oscillatory, rolling, and a combination of rolling/slip motions.

While it is typically possible to control the environmental temperature, due to frictional heating it is virtually impossible to control the interfacial temperatures. Slight changes in coefficient of friction can result in significant temperature rises (10–20°C). The environmental temperature is usually set and the surface or subsurface temperatures are monitored throughout the test. Since many of the materials (e.g., nylon) are sensitive to humidity, control of humidity will enhance the quality of the results.

Typically, the tests fall into the following categories of which there are often commercially available testers.

1. Unidirectional or oscillatory slip with thrust washer
2. Unidirectional or oscillatory slip with pin-on-disk
3. Unidirectional slip with rod-in-bearing
4. Pure rolling, rolling/slip with cylinder-on-cylinder

Often, selection of the type of test device is based on the relevance to the application. For example, in designing polymer bushing for shafts selection of a device such as three is the most simulative. However, devices one and two would also suffice.

HOMEWORK PROBLEMS

For the following problems first consider properties listed in this chapter and the polymer database provided with the text. If you still need information and cannot find it within the chapter then search the internet. The purpose of these problems is to have you demonstrate an understanding of the solution method. In an industrial application you should use the material properties provided by the material supplier or those determined from your own testing.

13.1. What is the effect of creep on modulus and subsequently on area of contact? How would this effect the perceived static coefficient of friction?

13.2. Comment on the rolling resistance for an elastomeric wheel as a function of forward velocity with frequency dependent modulus.

13.3. Consider a rod that can both slide and rotate inside of a polymer bushing. Assume that $d_{rod} = 15$ mm, bushing $d_{inner} = 15$ mm, $1 = 25$ mm, the radial load = 15 N. Assume that the coefficient of friction is 0.2 and 0.3:

What are both the force required to slide the rod and the torque required to rotate the rod?

Additionally how long does it take for bushing to generate 0.01 mm^3 of wear debris?

For the conditions stated what is the maximum rotational speed to avoid excessive wear? If the maximum rotational speed is 100 rpm, what is the maximum load that can be accommodated to avoid excessive wear?

Consider only the materials that are selected by your instructor.

1. Nylon 6,6
2. Nylon 6,6 with 20% PTFE
3. PEEK
4. Acetal
5. Acetal with 20% PTFE

13.4. Calculate the difference in real area of contact for cylinder (1 in. diameter) in a bushing of steel ($E_{steel} = 30(10^6)$ psi), brass ($E_{brass} = 1.6(10^6)$ psi), aluminum ($E_{al} = 10(10^6)$), and nylon (see Polymer.xls database) for a load of 25 lb/in.

REFERENCES

1. Greenwood, J. A. and J. B. P. Williamson, Contact of nominally flat surfaces, *Proceeding of the Royal Society London A*, 205, 1966, 300–319.
2. Lubricomp® Internally lubricated reinforced thermoplastics and fluoropolymer composites, *LNP Bulletin*, 254–691.
3. Friedrich, K., ed., *Friction and Wear of Polymer Composites, Composite Materials Series*, 1, B. J. Briscoe, p. 51, Elsevier, New York, NY, 1986.

4. Bayer, R. G., *Mechanical Wear Prediction and Prevention*, Marcel Dekker, Inc., New York, NY, 1994.
5. Friedrich, K., ed., *Friction and Wear of Polymer Composites*, *Composite Materials Series*, 1, T. Tsukizoe and N. Ohmae, pp. 212–220, Elsevier, New York, 1986.
6. Ludema, K. C. and D. Tabor, The friction and visco-elastic properties of polymeric solids, *Wear*, 9, 329–348.
7. Barquins, M. and A. D. Roberts, Rubber friction variation with rate and temperature: Some new observations, *Journal of Physics D*, 19, 547–563, 1986.
8. Grosch, K. A., The relation between the friction and visco-elastic properties of rubber, *Proceedings of the Royal Society of London A, Mathematical and Physical Sciences*, 274(1356), 1963, 21–39.

14 Damping and Isolation with Polymers and Composites

14.1 INTRODUCTION

All materials and hence structures contain damping. For applications where sound and vibration are present, increased damping is desired to attenuate the unwanted vibration. However, in other applications such as rolling wheels and belt drive system high levels of damping are undesired. While the primary focus of this section will be to look at situations where damping is desired, the same analysis and application principles will apply to systems where damping is detrimental to performance. Metallic materials often have very low levels of damping. Typically, the loss modulus will be on the order of 0.1–1.0% of the storage modulus. However, polymeric materials will typically have loss moduli in the range of 1–200% of the storage moduli. Thus, a polymeric or polymeric composite material will damp vibration much more effectively than a metallic material.

For example, consider bicycle frames constructed from steel, aluminum, or carbon fiber/epoxy. Since the modulus of elasticity of aluminum is less than that of both steel and carbon fiber, one might expect that the increased flexibility of the frame would result in less transmission of road inputs. Furthermore, one might also expect based on stiffness that the steel frame would be less harsh than the carbon fiber frame. However, rankings of the harshness by a bicycle rider in order of decreasing harshness would be aluminum, steel, and finally carbon fiber. This order occurs because of decreased damping follows the order of increased perceived harshness. While carbon fiber will impart more transmitted force from a large impact, much of the impression of ride harshness occurs due to the sustained vibration of either an impact event or continual vibration input from the road surface. Additionally, fatigue of cyclists is often partially attributed to sustained vibration of the bicycle.

Typically polymers are used as dampers and isolators to mitigate mechanical vibration, noise, shock, and impact. Applications include

1. Automobile drive train isolators, such as engine and transmission mounts
2. Isolators for transportation of materials and equipment
3. Machine tool isolators
4. Acoustical insulation for rooms and equipment enclosures
5. Equipment motion stops
6. Vibration damping material on space and aerospace structures

7. Replacement for metallic components (structural and nonstructural) for improved vibration absorption
8. Reduction of vibration to enhance the fatigue life of structures

14.2 RELEVANCE OF THE THERMOMECHANICAL SPECTRUM

Damping or hysteretic losses in polymers come from the motion of the polymeric chains. The primary source of damping comes from motion of the backbone of the polymer chain. Additional motions of side groups result in the β and γ transitions as discussed earlier in the text. Most polymers have the maximum damping in the transition zone around the glass transition temperature (α transition) as shown in Figure 14.1. The β and γ transitions are important because of their role in the dissipation of energy during impact events, where the resulting frequencies can be of the order of megahertz (MHz).

In previous sections, time–temperature superposition was discussed in terms of time (s) and temperature (°C). With issues such as creep and stress relaxation the interest is in long times or low frequencies. As an arbitrary breakpoint, 1 Hz (Δt = 1 s) will be considered the transition between creep/stress relaxation and vibration issues. Practically, even though there are large changes in creep/stress relaxation response under 1 s, it is very difficult to apply the loading sufficiently smooth and fast enough to test at time scales <1 s. Thus, the transients associated with load application, typically make data obtained under 1 s subject to debate. Viscoelastic data for use in vibration analysis are obtained in test devices that apply oscillation frequencies in the range of 1–1000 Hz. Most servohydraulic and other testers will be able to cycle up to 100 Hz. There are a few specialized servohydraulic testers that will reach frequencies as high as 1000 Hz. There has been work performed to obtain direct measurements of the viscoelastic properties upto 5–10 kHz.[1] In general, the viscoelastic data are obtained from 1 to 100 Hz and then time–temperature superposition is performed to obtain the response at higher frequencies. The storage and loss moduli and tan δ will transform with respect to frequency in the same way that creep and stress relaxation data transform with respect to time. An example of time–temperature

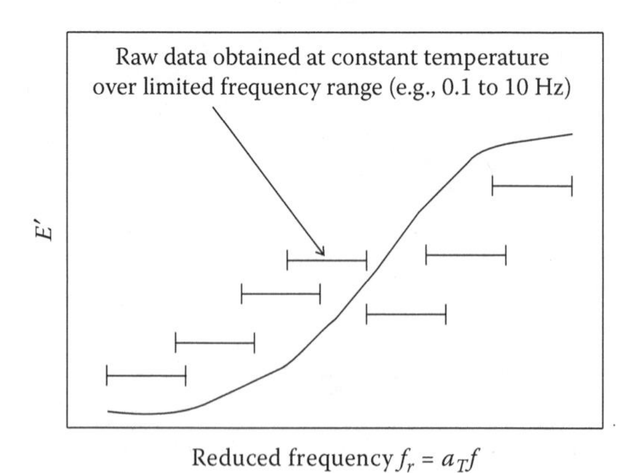

FIGURE 14.1 Time–temperature superposition of storage modulus.

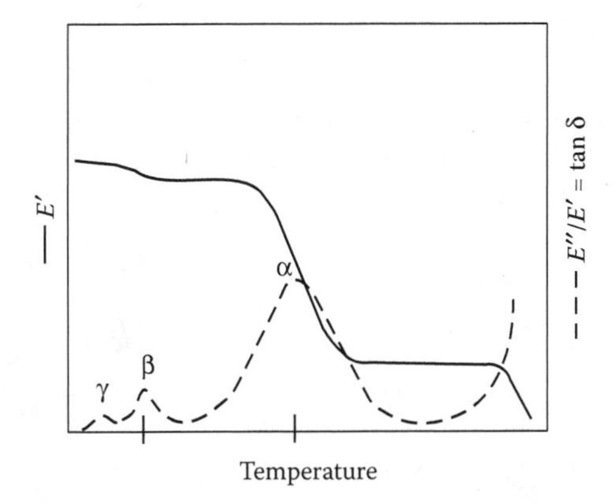

FIGURE 14.2 Damping as demonstrated by the thermomechanical spectrum.

superposition of viscoelastic data as a function of frequency is shown in Figure 14.1. While data are often extrapolated well beyond 20,000 Hz, this value is a realistic upper end to the frequency range of interests. Beyond 20,000 Hz sound waves are inaudible and the energy of vibration results in minimal amplitude of vibration. From Figure 14.2, it can be seen that the maximum damping occurs around the glass transition temperature. The glass transition temperature determined from a 1 Hz tests will superpose to a different temperature at any other frequency of oscillation.

14.3 DAMPING METHODS OF MATERIAL MODIFICATION (CROSS-LINKING, *M*w, STRUCTURE) POLYMERS, AND COMPOSITES USED IN DAMPING AND ISOLATION

Throughout the text, the complex modulus has been expressed as

$$E^* = E' + iE'' \tag{14.1}$$

where E' and E'' are the storage and loss modulus, respectively. However, most mechanics and vibrations work represent Equation 14.1 as

$$E^* = E'(1 + i\eta), \tag{14.2}$$

or

$$E^* = E'(1 + i\delta), \tag{14.3}$$

where, η or δ are referred to as the loss factor. Either of these loss factors is equivalent to

$$\tan\delta = \frac{E''}{E'}. \tag{14.4}$$

Care should be taken to not confuse δ and $\tan\delta$, when working with data from various sources or references.

From theory of rubber elasticity (when $\mu = 0.5$),

$$G = \frac{E}{3}, \tag{14.5}$$

where G = shear modulus.

It will be shown later, that for effective isolation from vibration, the system natural frequency should be less than the driving frequency of vibration. A simple way to reduce the natural frequency of a system is to reduce the stiffness, while holding mass relatively constant. Thus, even without analysis, it is seen that polymers perform better as isolators in shear compared to extension or compression.

14.3.1 Reduced Frequency Nomograph

In vibration and acoustic analysis the viscoelastic spectrum are often represented as a master curve known as the reduced frequency nomograph.[2] Figure 14.3 is an example of a nomograph constructed with simulated data. In addition to the thermomechanical information that is usually presented, both absolute temperature and frequency axes are included in the nomograph. The information presented by the inclusion of the temperature and frequency axes is already contained in the master curve of the viscoelastic data. However, it is convoluted within the master curve. In order to obtain frequency data at other temperatures, a new master curve must be created from the raw data or transformed from the original master curve.

The nomograph is used to quickly obtain data at a given frequency and temperature. For example, if the loss tangent and storage modulus are desired at a temperature of T_{-1} and a frequency, f, the reduced frequency, point x, is found by the intersection of the desired temperature and frequency isoclines. The result is an equivalent reduced frequency. Once the equivalent reduced frequency is known, the viscoelastic properties are then determined. Graphically, this procedure is performed by finding the intersection point of the desired temperature and frequency, and then following the

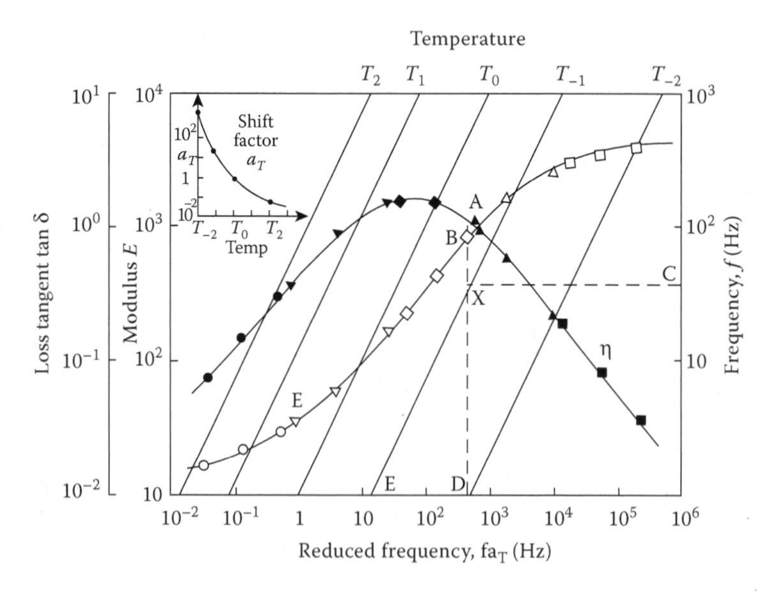

FIGURE 14.3 Reduced frequency nomograph of simulated data. (Adapted from Corsaro, R. D. and L. H. Sperling, eds., *Sound and Vibration Damping with Polymers*, ACS Symposium Series 424, P. T. Weissman and R. P. Chartoff, p. 115, 1990.)

reduced frequency isocline to the intersections with the desired viscoelastic property (e.g., points A and B).

14.4 MATERIALS FOR DAMPING AND ISOLATION

Natural rubber is undoubtedly the first material that comes to mind for use as an isolator. This occurs for many reasons. Historically, natural rubber has been used in isolation applications for over 100 years. Natural rubber has excellent fatigue resistance. However, natural rubber's damping is low and it is not particularly oil or solvent resistant. The low damping is actually beneficial. Low damping means that during use there is minimal internal heat generated. The lack of heat generation contributes to long product lifes due to minimal thermal decomposition over time. The viscoelastic responses for unfilled and 50 parts of carbon black-filled natural rubber are shown in Figures 14.4 and 14.5, respectively. For comparison, butyl, a high-damping rubber is shown in Figure 14.6. Note the difference in both shear modulus and loss factor when butyl is compared to natural rubber.

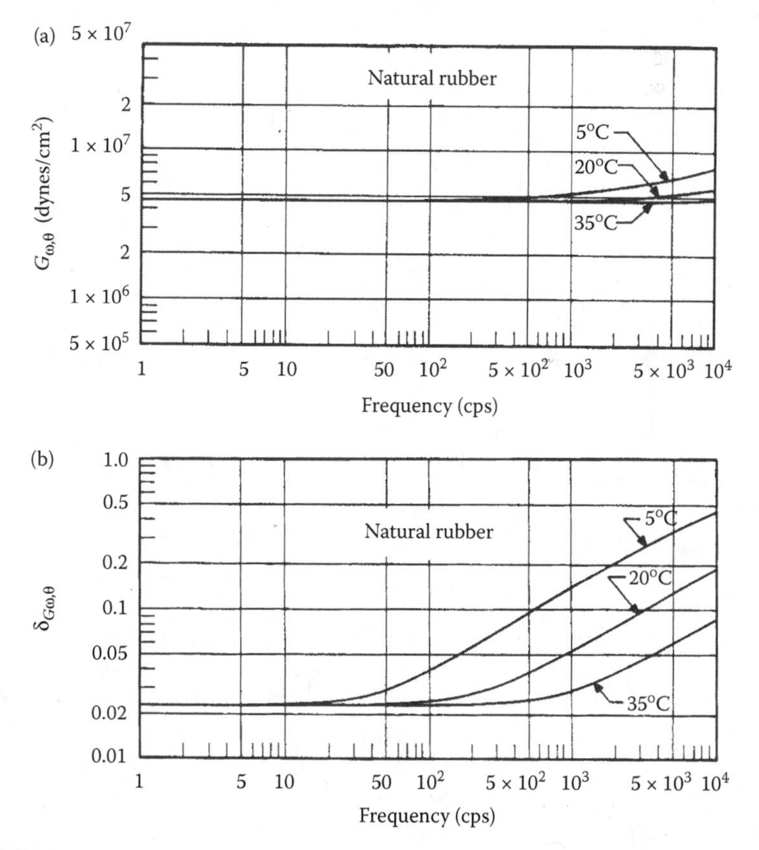

FIGURE 14.4 Viscoelastic spectrum for unfilled natural rubber. (From Snowdon, J. C., *Vibration and Shock in Damped Mechanical Systems*, 1990. Reprinted with permission of John Wiley & Sons, Inc.)

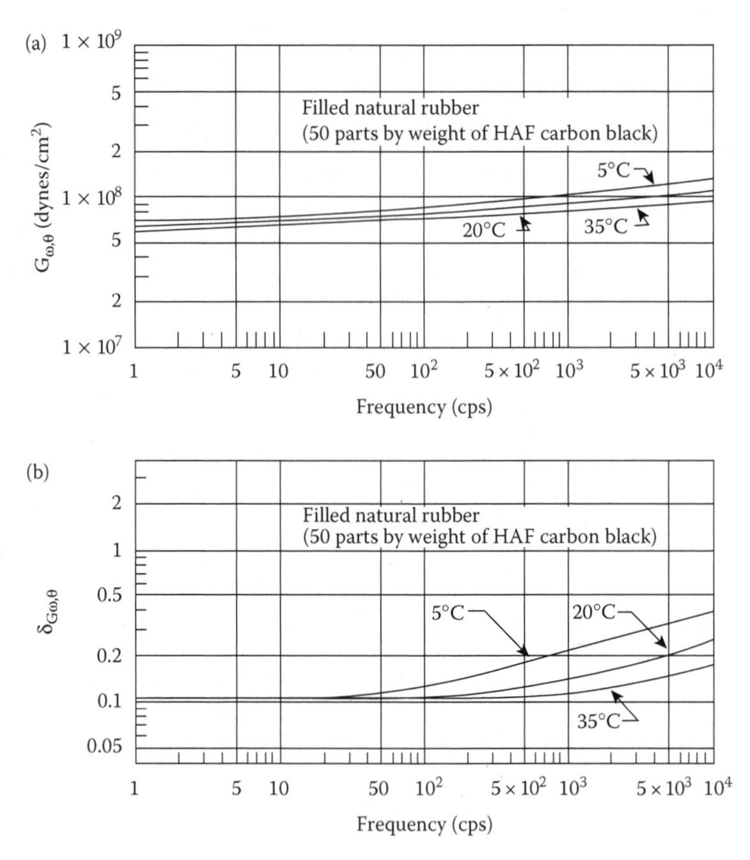

FIGURE 14.5 Viscoelastic spectrum for natural rubber filled with 50 phr carbon black. (From Snowdon, J. C., *Vibration and Shock in Damped Mechanical Systems*, 1990. Reprinted with permission of John Wiley & Sons, Inc.)

Mixtures of two materials can produce blends that have a single narrow glass transition temperature between the two materials, a broad transition encompassing both T_g's and in some instances transitions that lie outside of the original materials T_g's. The location of the transition is dependent on the final morphology of the blend. Materials that separate into distinct phases when blended will yield two transitions. Materials that are miscible will generally lead to a material that has a single transition, typically between the original glass transition temperatures. An example of a material blend which leads to the third behavior, with a single T_g outside the original T_g's is shown in Figure 14.7. The materials blended were polyisoprene (PIP) and polybutadiene (PBD). The blend was 25% PBD and 75% PIP. PIP is synthetic natural rubber.

PURs are often used in mitigating vibration. PURs can have a wide range of material properties with minor processing modifications, while holding the chemistry of the material constant. One factor that leads to the wide use of thermoplastic PURs, is that they can be injection molded, allowing a large number of parts to be made

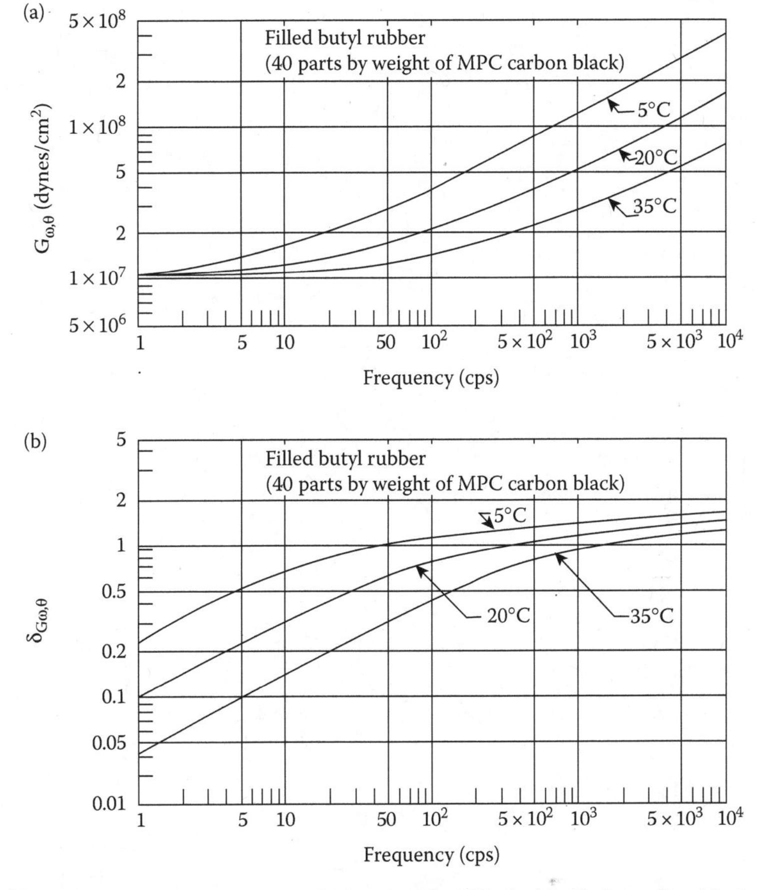

FIGURE 14.6 Viscoelastic spectrum for butyl rubber filled with 40 phr carbon black. (From Snowdon, J. C., *Vibration and Shock in Damped Mechanical Systems*, 1990. Reprinted with permission of John Wiley & Sons, Inc.)

easily. Compression molding of thermosets is labor intensive in comparison. This makes PUR parts ideal for applications requiring large numbers of inexpensive components. Figure 14.8 shows the viscoelastic results for PUR blends.

14.5 FUNDAMENTALS OF VIBRATION DAMPING AND ISOLATION

14.5.1 Dynamics of Vibrating Structures (Continuous and Discrete or Point)

The single-degree-of-freedom oscillator shown in Figure 14.9 is the starting point for consideration of the fundamentals of vibration damping and oscillation. From a

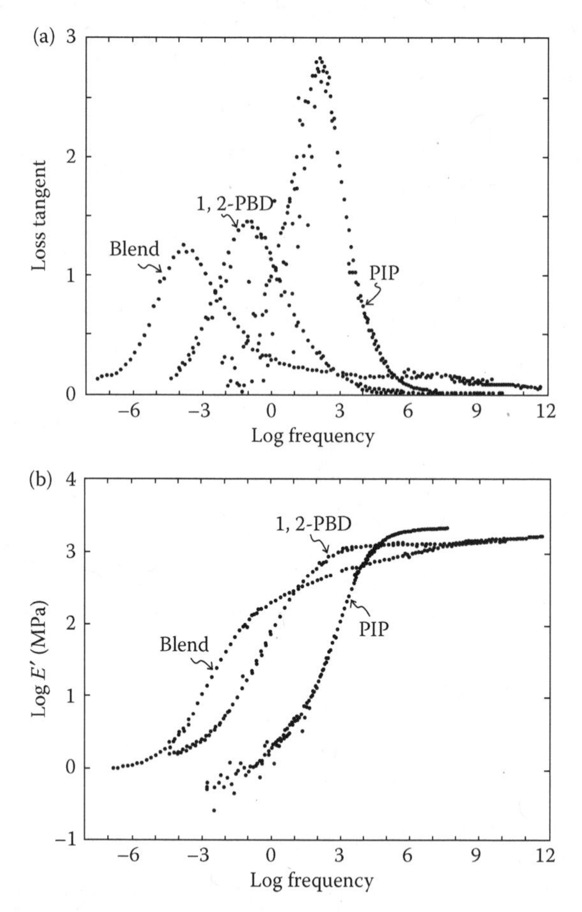

FIGURE 14.7 Viscoelastic spectrum for 75% PIP/25% PBD blends. (Adapted from Corsaro, R. D. and L. H. Sperling, eds., *Sound and Vibration Damping with Polymers*, ACS Symposium Series 424, C. M. Roland and C. A. Trask, Washington, DC, p. 312, 1990; Corsaro, R. D. and L. H. Sperling, eds., *Sound and Vibration Damping with Polymers*, ACS Symposium Series 424, C. M. Roland and C. A. Trask, Washington, DC, p. 310, 1990.)

practical viewpoint it is also a first-order approximation to a lumped device with isolation "feet." The equation of motion for this simple system acted on by a sinusoidally varying force is shown in Equation 14.6;

$$m\ddot{x} + c\dot{x} + kx = \underline{F}, \tag{14.6}$$

where m = lumped mass (kg), c = damping (N s/m), k = stiffness (N/m), $\underline{F} = F_0 \sin \omega t$. Obviously the isolation device material and geometry determine c and k. For an isolator of constant cross-sectional area A, length L, and material described by complex modulus, $E^* = E' + iE''$. In general, most analyses start with the implicit assumption that the material is a Kelvin solid where,

$$k = \frac{AE'}{L}\,(\text{N/m}) \tag{14.7}$$

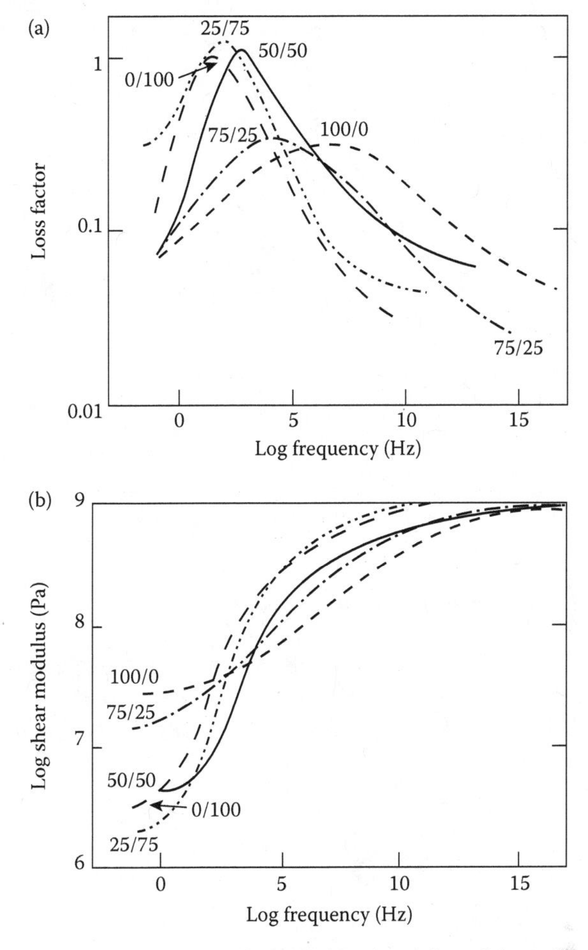

FIGURE 14.8 Viscoelastic spectrum for PUR blends. (Adapted from Corsaro, R. D. and L. H. Sperling, eds., *Sound and Vibration Damping with Polymers*, ACS Symposium Series 424, J. V. Duffy, Washington, DC, p. 292, 115, 1990.)

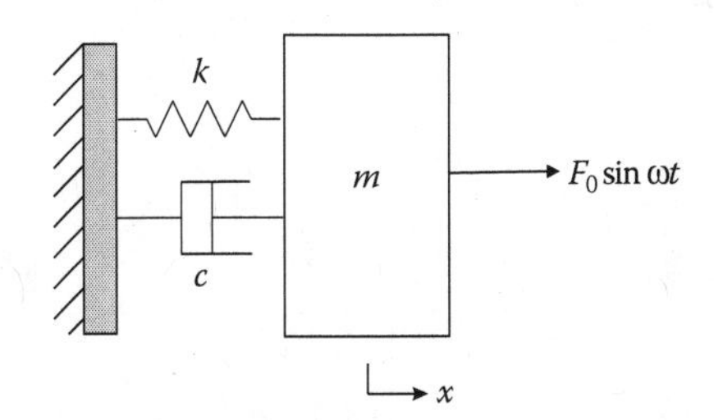

FIGURE 14.9 Single-degree-of-freedom oscillator with damping.

and

$$c = \frac{AE''}{\omega L} \text{ (Ns/m)} \tag{14.8}$$

Rewriting Equation 14.6 in the typical form results in Equation 14.9.

$$\ddot{x} + \frac{c}{m}\dot{x} + \frac{k}{m}x = \frac{F}{m}, \tag{14.9}$$

or,

$$\ddot{x} + 2\xi\omega_n\dot{x} + \omega_n^2 x = \frac{F}{m}, \tag{14.10}$$

where

$$\omega_n = \sqrt{\frac{k}{m}} \quad \text{natural frequency(r/s)}, \tag{14.11}$$

$$\xi = \frac{c}{2m\omega_n} \quad \text{damping ratio (nondimensional).} \tag{14.12}$$

This leads to the classic vibration steady-state solution for forced vibration that is shown in many texts. The results are presented here because they lead to an understanding of the fundamental difference between the concepts of damping and isolation. It is assumed that both the forcing function,

$$F(t) = Fe^{i(\omega t - \phi)} \tag{14.13}$$

and solution are sinusoidal,

$$x(t) = Xe^{i(\omega t - \phi)} \tag{14.14}$$

Solution of Equation 14.10 leads to the following relationships for amplitude, X, and phase angle, φ, of the response:

$$X = \frac{\frac{F}{K}}{\sqrt{\left[1 - \left(\frac{\omega}{\omega_n}\right)^2\right]^2 + \left[2\xi\left(\frac{\omega}{\omega_n}\right)\right]^2}} \text{ (m)} \tag{14.15}$$

or

$$\frac{Xk}{F} = \frac{1}{\sqrt{\left[1 - \left(\frac{\omega}{\omega_n}\right)^2\right]^2 + \left[2\xi - \left(\frac{\omega}{\omega_n}\right)\right]^2}} \tag{14.16}$$

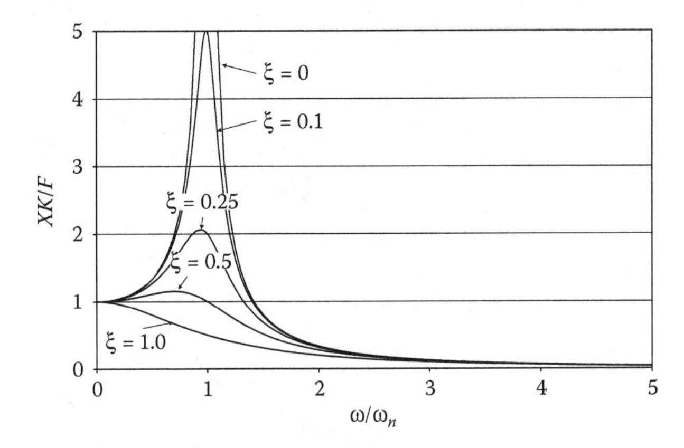

FIGURE 14.10 Frequency–response function for single-degree-of-freedom oscillator.

$$\tan\varphi = \frac{2\xi(\omega/\omega_n)}{1-(\omega/\omega_n)^2}, \tag{14.17}$$

where, ϕ = phase angle between forcing function and displacement response.

The peak amplitude occurs at the damped natural frequency,

$$\omega_\mathrm{d} = \omega\sqrt{1-\xi^2} \tag{14.18}$$

which is always lower than the undamped natural frequency ω_n.

The response of Equation 14.15 as a function of driver frequency is shown in Figure 14.10. Figure 14.10 demonstrates the effect of various levels of damping on the system response. Note that for driving frequencies >2.5 times the natural frequency the differences in damping have negligible effect on the system response. Below that level, especially near the natural frequency the damping has the greatest influence on the system response. The difference between isolation and damping is relatively straightforward. In vibration isolation the system is designed such that the natural frequency is less than $\omega_{\mathrm{driving}}/2.5$. Vibration dampers can be designed as resonant dampers to operate at the system natural frequency. However, it is more common to apply a damper to an existing system. It should be apparent that vibration dampers are most useful when used on systems that are vibrating near the system natural frequency.

Another important lumped mass model is the two-degree-of-freedom oscillator with base motion as shown in Figure 14.11.

The resulting equation of motion is

$$m\ddot{x}_1 = -c(\dot{x}_1 - \dot{x}_2) - k(x_1 - x_2), \tag{14.19}$$

$$m\ddot{x}_1 + c\dot{x}_1 + kx_1 = c\dot{x}_2 + kx_2. \tag{14.20}$$

There are various dynamic situations that could be modeled and solved with this system. For example, sinusoidal vibrational input to either the base, x_2, or the mass element, x_1 could be considered, or impact of this system on either the base or the mass.

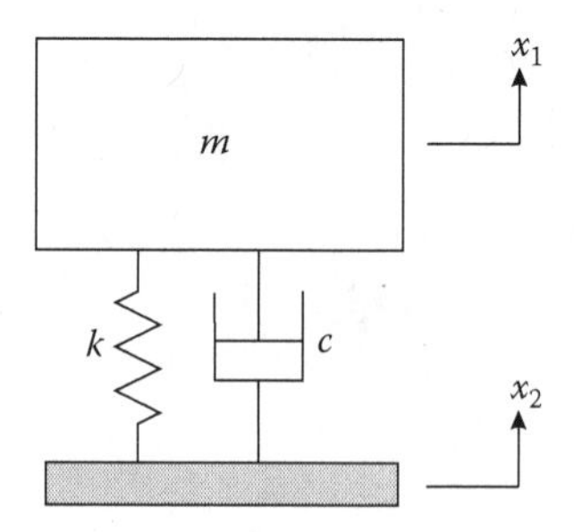

FIGURE 14.11 Two-degree-of-freedom oscillator.

Additionally, the solution path could utilize the correspondence principle discussed previously, where the elastic solution is obtained and the real modulus or stiffness is replaced with the complex modulus or stiffness. However, let us explore the classic method of directly solving the equation of motion. For the first case, consider sinusoidal vibration of the base.

$$x_2 = X_2 e^{i\omega t} \tag{14.21}$$

$$\dot{x}_2 = i\omega\, X_2 e^{i\omega t} \tag{14.22}$$

Assume

$$x_1 = X_1 e^{i(\omega t-\phi)} \tag{14.23}$$

$$\dot{x}_1 = i\omega\, X_1 e^{i(\omega t-\phi)} \tag{14.24}$$

$$\ddot{x}_1 = -\omega^2\, X_1 e^{i(\omega t-\phi)} \tag{14.25}$$

Thus, the equation of motion (Equation 14.18) in steady state becomes

$$-m\omega^2 X_1 e^{i(\omega t-\phi)} + \omega c X_1 e^{i(\omega t-\phi)} + kX_1 e^{i(\omega t-\phi)} = i\omega c X_2 e^{i\omega t} + kX_2 e^{i\omega t}, \tag{14.26}$$

or rearranging terms,

$$(-m\omega^2 + i\omega c + k)X_1\, e^{i\omega t} e^{-i\phi} = (i\omega c + k)X_2 e^{i\omega t}. \tag{14.27}$$

Rearranging Equation 14.27 results in the transmissibility relationship

$$T = \left|\frac{X_1}{X_2}\right|, \tag{14.28}$$

where

$$\frac{X_1}{X_2} = \left(\frac{-i\omega c + k}{m\omega^2 - i\omega c + k}\right) e^{-i\varphi}. \tag{14.29}$$

Multiplying by the complex conjugate of the denominator, Equation 14.29 can be expressed as a magnitude and phase angle as shown in Equations 14.30 and 14.31.

$$\left|\frac{X_1}{X_2}\right| = \sqrt{\frac{k^2 + (\omega c)^2}{(k - m\omega^2)^2 + (\omega c)^2}}. \tag{14.30}$$

$$\phi = \tan^{-1}\left(\frac{mc\omega^3}{k(k - m\omega^2) + (\omega c)^2}\right) \tag{14.31}$$

Note that due to the presence of damping represented by c, there is a phase shift between the input and output. However, both the base and the mass are vibrating at the same frequency. This solution is only valid for linear systems. While many isolation and damping systems are linear, there are nonlinear systems, which can result in more complex motions. Nonlinear systems can possess motions that evidence period doubling or chaos. However, such systems are beyond the scope of this text.

14.6 ROLE OF DAMPERS

The role of dampers, or damping layers, is to attenuate the amplitude and energy of a vibrating structure by dissipation of energy through heat. Figure 14.12 shows that the more the damping, the faster the system will ring down in free vibration. Thus, a transient input is washed out by the system damping. For vibrations, material or composite damping can be viewed to act in parallel with stiffness as in the Kelvin representation. If two structures that have the same stiffness (storage modulus) but different damping (loss modulus), for the same energy input at one end of a structure, less energy will be transmitted at the other end for the one with greater damping. The magnitude of displacement of a structure decreases as damping is increased. This decrease was shown by the classic result of the frequency response function of a single-degree-of-freedom oscillator shown in Figure 14.10. As a footnote, damping will also shift the natural frequency to the damped natural frequency.

When designing metallic structures, the incorporation of damping is often through the use of damping layers or discrete dampers (such as automobile shock

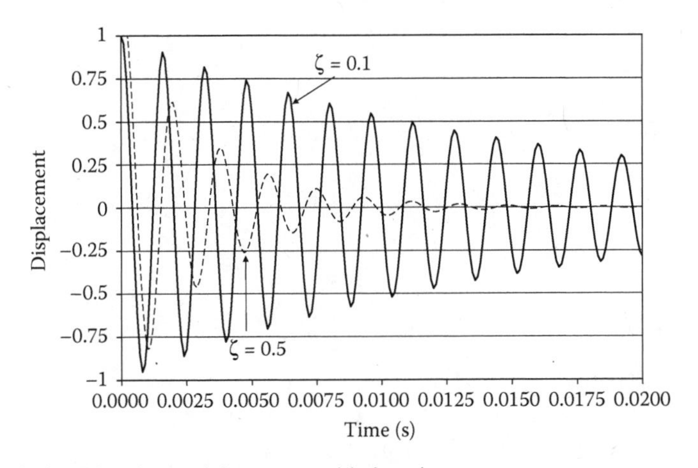

FIGURE 14.12 Ring down of the system with damping.

absorbers). However, when using a composite, useful structural damping can be obtained by simply changing polymer components or modifying the damping behavior of the selected components.

Even though elastomers are used as both dampers and isolators, the function of an isolator is not to provide any significant damping, but to isolate the transmission of dynamic loads. Isolators can function simply as a recipient and translator of kinetic to potential energy or to change the system natural frequency and transfer function.

14.7 DAMPING LAYERS

In many cases it is desired to damp out vibration or noise radiation of metallic structures that have already been designed or constructed. In this situation damping layers can be applied to the existing structure to damp out vibrations. The damping layers can take the form of spray coatings or adhesive pads of viscoelastic materials. How does a damping layer perform? The primary purpose is to absorb energy and for the material to do this it must dissipate the vibrational energy as heat.

14.7.1 Application of Dampers and Isolators: Discrete Design of Dampers and Isolators for Equipment

Application of damping layers to structures has been utilized for more than 50 years. Essentially, there are two methods of applying damping layers as shown in Figure 14.13. The most obvious and simplest method is to bond a damping layer to the outside skin of the material or structure to be damped. This is referred to as a free layer. Due to environmental, abrasion, heat resistance, chemical resistance and other reasons, it is often desired to have the damping layer internal to the outer layer of the structure. This is referred to as constrained layer (shown in Figure 14.13) damping.

Free layer damping primarily adds damping. The increase in stiffness and mass loading will often be negligible for the composite system. Since the damping layers are on the outer surface they experience the largest extension and also the highest rate of extension of any position through the structure. This is important since damping is often dependent on both extension and extension rate. Free damping layers are often added to metallic structures where the incorporation of constrained layers is either infeasible or cost prohibitive. Additionally, free damping layers will be added

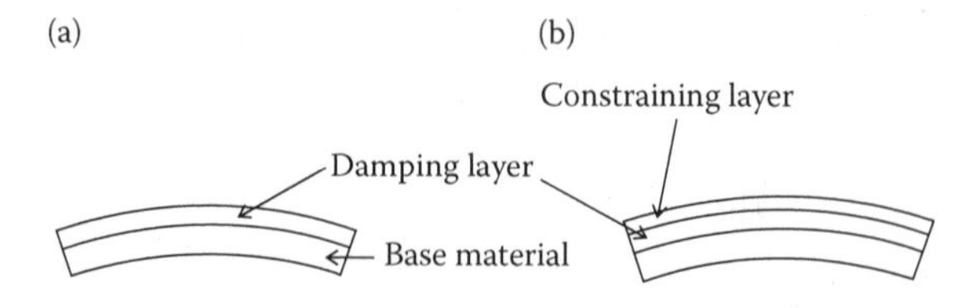

FIGURE 14.13 Free and constrained viscoelastic-layer damping treatments. (a) Unconstrained and (b) constrained. (Adapted from Corsaro, R. D. and L. H. Sperling, eds., *Sound and Vibration Damping with Polymers*, ACS Symposium Series 424, P. T. Weissman and R. P. Chartoff, Washington, DC, p. 115, 1990.)

as corrective action to a component where vibration was not initially thought to be a design consideration. The damping layer is under a fluctuating state of extension and compression due to flexure of the beam or plate to which it is applied. The composite effect of adding a free damping layer is shown in Equation 14.32,

$$\tan\delta_{\text{composite}} = K\,\frac{E_2}{E_1}\left(\frac{H_2}{H_1}\right)^2 \tan\delta_2, \tag{14.32}$$

where

E_1 = storage modulus of the base,
E_2 = storage modulus of the viscoelastic layer,
H_1 = height of the base,
H_2 = height of the viscoelastic layer,
$\tan\delta_2$ = $\tan\delta$ of the viscoelastic layer,
K = constant, often the contact area is used as this value.

Figure 14.14 demonstrates the effect of damping layer thickness, modulus, and loss factor on the composite loss factor.

An interesting example, albeit low-technology example, is the common stainless-steel kitchen sink. Historically, stainless-steel kitchen sinks were sprayed with a damping material that covered the entire bottom of the sink. Recently, sinks have been produced that have pads of damping material adhered to the bottom of the pans. The pads are placed in locations of high vibration, where they will be most effective.

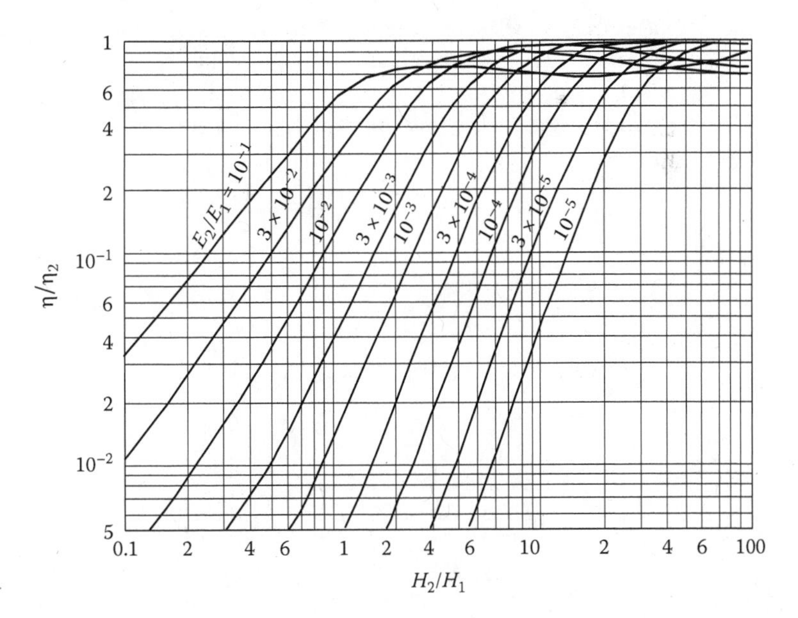

FIGURE 14.14 Effect of free damping layer thickness on loss factor. (Adapted from Oberst, H., *Acustica*, 2, p. 186, 1952.)

The effectiveness of the damping pads can be demonstrated by running the garbage disposal with the pads on and then running the disposal with one or both pads off. There is a dramatic difference in the acoustical response. This is a rather interesting design solution for a sink possessing a relatively thin gauge of stainless steel. By the way, this experiment should be performed at someone else's house.

The kitchen sink example of free damping layers is a one-dimensional vibration (normal to pan) application. For this example the amplitude of vibration will depend spatially on the two directions perpendicular to the normal vibration. Often damping layers are applied to long slender beams or rods where the transverse vibration can also be described as one dimensional (normal to the long axis). In this case, the amplitude of vibration typically is dependent only on the distance along the long axis.

An example of constrained layer damping, or more appropriately isolation, is earthquake bearings for buildings. These bearings are a classic example of a laminate composite. The bearings are designed to absorb surface waves which cause the greatest damage to structures in earthquakes. The maximum damage is caused not by transverse vibration of the ground, but by extension/compression and shear of the earth's surface layers. Thus, the bearings must be flexible in shear, but stiff in compression to support the static loads of the structure. The bearings are made by alternating layers of elastomer and steel as shown in Figure 14.15. The resulting stiffness can be calculated from the methods presented in Chapter 4. The composite effect of different stiffness in orthogonal directions is demonstrated by earthquake bearings. This concept has also been used in bushings to increase the radial stiffness, while not significantly increasing the torsional stiffness.

Another example of constrained layer damping is oil pans for automobile engines, where a layer of polymer is formed between two layers of steel. Constrained layer damping places the viscoelastic material primarily in a shear state of stress during the transverse deflection of the substrate. Thus, the relationship for constrained layer damping is not identical to that for free layer damping. The approximate relationship for constrained layer damping is shown in Equation 14.33:

$$\tan \delta_{\text{composite}} = K \frac{E_3}{E_1} \left(\frac{H_3}{H_1} \right)^2 \tan \delta_2, \qquad (14.33)$$

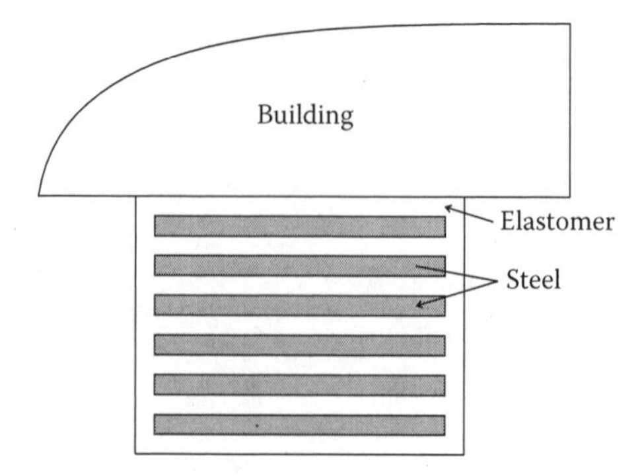

FIGURE 14.15 Earthquake bearings.

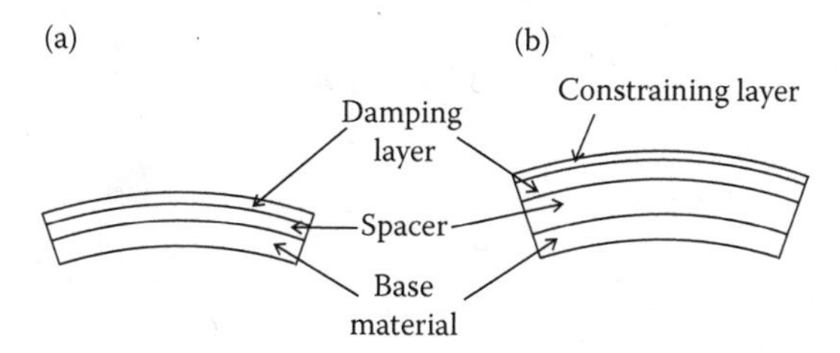

FIGURE 14.16 Use of spacers in free and constrained layer damping. (a) Unconstrained with spacer and (b) constrained with spacer.

where

E_1 = storage modulus of the constraining layer,
E_3 = storage modulus of the base,
H_1 = height of the constraining layer,
H_3 = height of the base,
$\tan \delta_2 = \tan \delta$ of the viscoelastic layer,
K = constant, often the contact area is used as this value.

A point made by many researchers for both free and constrained layer damping is that the damping material must experience deformation to be effective. For this reason, spacers are sometimes used to increase the energy input into the viscoelastic material as shown in Figure 14.16.

HOMEWORK PROBLEMS

14.1. A sinusoidal load of amplitude 1 N at 1000 Hz is centered on a cabinet, which weighs 15 N. The cabinet is supported by four feet. The diameter of the feet is 14.5 mm and the height is 14.5 mm. What is xk/F for this condition for the following polymers natural rubber 0, 50 phr carbon black, and butyl rubber 40 phr carbon black (use the figures presented in the text for material properties).

14.2. Plot the transmissibility for the 2 dof system with base excitation for $\zeta = 0.5$, and $0.0 < \omega/\omega_n < 4.0$.

REFERENCES

1. Agee, B. L. and L. D. Mitchell, Frequency dependent viscoelastic property measurement via modal analysis techniques, *9th International Conference on Experimental Mechanics*, Copenhagen, Denmark, pp. 1978–1988, 1990.
2. Jones, D. I. G., *Handbook of Viscoelastic Damping*, John Wiley & Sons, New York, NY, 2001.
3. Corsaro, R. D. and L. H. Sperling, eds., *Sound and Vibration Damping with Polymers*, ACS Symposium Series 424, P. T. Weissman and R. P. Chartoff, Washington, DC, p. 115, 1990.
4. Corsaro, R. D. and L. H. Sperling, eds., *Sound and Vibration Damping with Polymers*, ACS Symposium Series 424, C. M. Roland and C. A. Trask, Washington, DC, p. 312, 1990.

5. Corsaro, R. D. and L. H. Sperling, eds., *Sound and Vibration Damping with Polymers*, ACS Symposium Series 424, C. M. Roland and C. A. Trask, Washington, DC, p. 310, 1990.
6. Corsaro, R. D. and L. H. Sperling, eds., *Sound and Vibration Damping with Polymers*, ACS Symposium Series 424, J. V. Duffy, Washington, DC, p. 292, 115, 1990.
7. Oberst, H., Uber die dämpfung der biegeschwingungen dunner bleche durch fest haftende bläge, *Acustica*, 2, p. 186, 1952.

15 Rapid Prototyping with Polymers

15.1 INTRODUCTION

Rapid prototyping (RP) is a modern method of producing physical prototypes using digital data from CAD programs downloaded to digital manufacturing devices using computer-aided manufacturing software. The 3D physical models can be used to communicate style, shape, as well as to test the function of a model. They may be a simple approximation of the product focusing on only one or two features as a proof of concept, or they may include more complex details of features. Several prototypes may be produced for comparison in the early design stage. Prototypes may be made of tooling as well as of the product. Sometimes "soft tooling" may be used for pilot production to test tooling concepts before "hard" tooling is produced for mass production.

These RP applications are called rapid because of the fast transfer of electronic design data. However, the production of a polymer part may require hours of production time once the input data are loaded into the RP machine.

In earlier times physical prototypes may have been hand carved out of wood or molded from clay. In modern times, polymers are the materials used most widely in RP applications. These materials include ABS, nylon, and UV-curable photopolymer resins such as epoxy, acrylates, and vinyl-ethers. This chapter describes the various techniques and devices used to produce RP parts and details of the polymer materials that are used. Advantages and disadvantages of different techniques such as accuracy and strength of the product are considered.

This is not an exhaustive treatment. For more details Jacobs[1] and Chua et al.[2] are recommended as texts devoted to RP. Some of the information presented here is taken from these references.

15.2 RAPID PRODUCT DEVELOPMENT, TOOLING, AND MANUFACTURE

Product development can be divided into three phases: Phase 1, conceptual design and prototyping; Phase 2, production design and tooling production; and Phase 3, production. The cost of changes is much less in the conceptual and prototype phase than if required later. Urlich and Eppinger[3] show how a prototype can increase the probability of success from 70% to 95% as shown in Figure 15.1. Taking time to build and test a prototype, they say, may help a development team detect a problem early rather than detecting the problem later after building a costly injection mold.

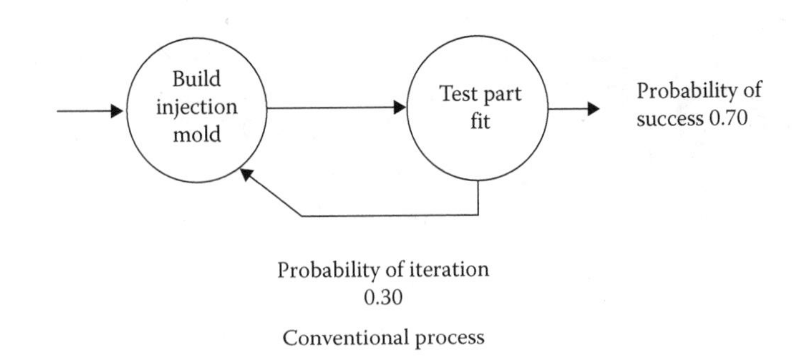

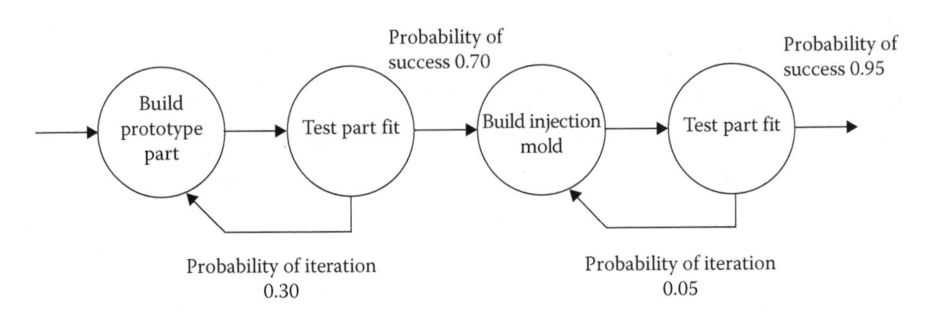

FIGURE 15.1 A prototype may reduce the risk of costly iteration. (Adapted from Urlich K. T. and S. D. Eppinger, *Product Design and Development*, 4th ed., McGraw-Hill, Exhibit 12-9, 2008.)

RP has progressed beyond the simple show and tell of a physical prototype to the ability to create functional prototypes for testing, customized products, and short production runs of standardized products. In addition, low-cost tooling can be built for fixtures, assembly tooling, and molds.

An example of an individually customized product is the transparent braces made by more than one company. A mold is made of your teeth and then laser scanned. A series of brace pairs are then constructed via RP out of materials such as polyurethane.

Functional prototypes can be constructed via RP that either perform static functions, such as the snowmobile rack shown in Figure 15.2 or dynamic functions such as the surgical ratchet made from polycarbonate shown in Figure 15.3.

15.3 RP TECHNIQUES

The basic prototype modeling techniques can be classified as either subtractive or additive depending on whether material is removed or added during the manufacturing process. For example, an early version of the subtractive type is a manual milling machine where material is successively removed by cutting. In modern times, this is the method that is the basis for automated manufacturing using Computer Numerical

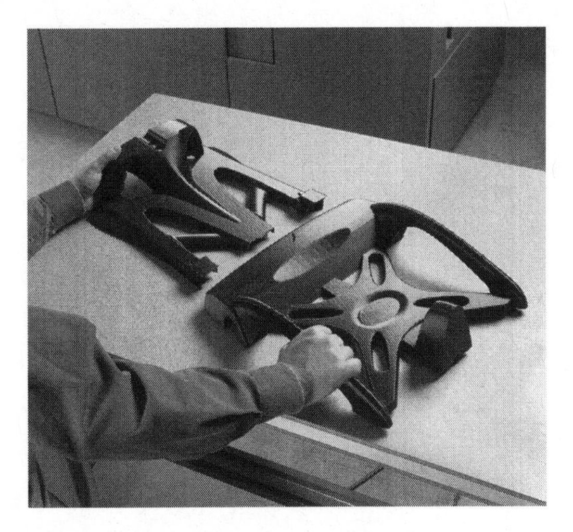

FIGURE 15.2 Polaris snowmobile rack rapid prototyped by Fortus FDM. (Courtesy of Fortus.)

Control (CNC) machines. An additional technique is the formative method where material is reshaped rather than added or subtracted. However, material forming is more a manufacturing production technique rather than a prototyping method. For example, sheet forming and blow molding considered in Chapter 10 are forming methods.

The two techniques, subtractive or additive, are illustrated in Figure 15.4. The shape shown is a stepped staircase shape. In (a), material is removed from a solid block. In the additive method (b), one can consider gluing separate layers of wood of different widths to build up to the final shape. Method (b) is the common method of

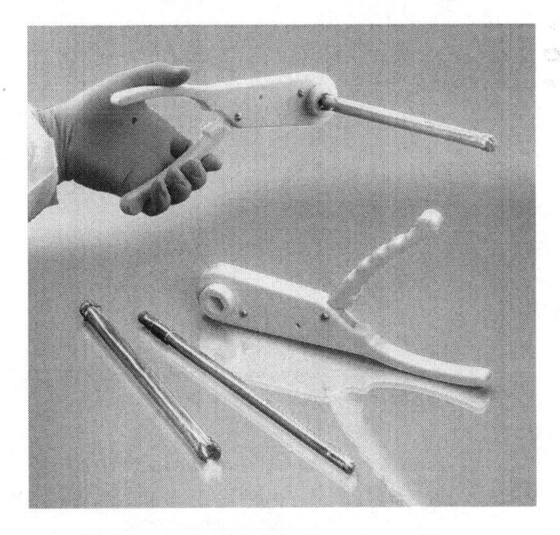

FIGURE 15.3 Medtronic polycarbonate surgical ratchet rapid prototyped with Fortus FDM. (Courtesy of Fortus.)

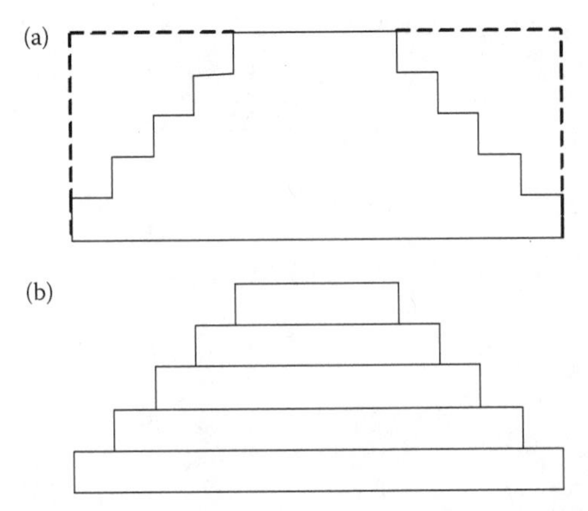

FIGURE 15.4 (a) Subtractive and (b) additive methods of physical model development.

most RP devices used today. Considering (b) in Figure 15.4, let the layers be very small. Then one can imagine making any contour shape as shown in Figure 15.5. The accuracy depends on the thinness of each layer.

Each of the 2D layers in Figure 15.5 is called a slice. Slice software is an added software in most solid modeling software programs. The standard file format adopted by the RP industry is "file.STL" (STereoLithgography file format). STL files are faceted models of the surface geometry and are the output from 3D CAD solid modeling software programs. The triangular facets are described by a set of x, y, and z coordinates of the three nodes and also the components of the unit normal vector to the face of the facet. The process of creating the STL file is called tessellation.[1] The STL file is a very large file, much larger than the CAD data file. The size depends on the accuracy desired in the RP model. The STL files are then sliced into the 2D planar slices called SLI (SLIce) files. The SLI file is the input to the RP machine. Any 3D object can be divided into a large number of 2D slices.

The STL files are not always perfect depending upon the complexity of the part. There can be gaps, degenerate facets, and/or overlapping facets. The STL file should be checked and repaired if necessary before producing the RP model. The SLI software programs should have the capability of displaying the STL geometry for examination.

The RP techniques can be further subdivided into whether the starting material is in liquid, solid, or powder form. Chua et al.[2] lists 15 different liquid-based, 9

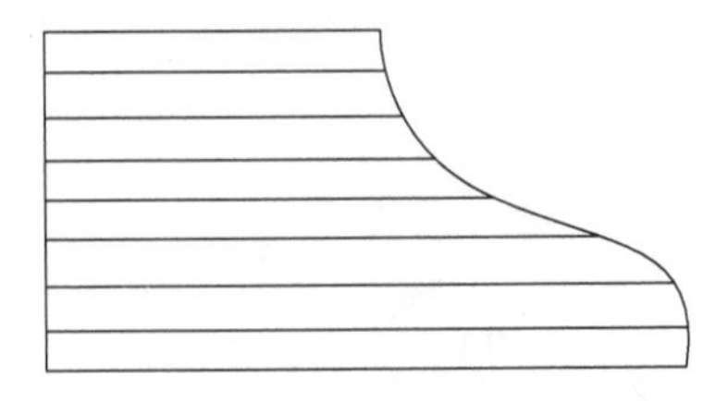

FIGURE 15.5 Matching any shape with thin layers using the additive method.

solid-based, and 16 powder-based systems that have been used or are in current use. In this text, only a few of the systems that are most popular are considered.

Fused deposition modeling (FDM) is an additive solid-based process as shown in Figure 15.6. A solid thermoplastic polymer is heated to the melting temperature and deposited through a nozzle in narrow strips in the y-direction. The nozzle is moved in the x-direction as subsequent strips are laid down (or the platform can be moved in the $-x$ direction). The thickness t of each layer can be as small as 0.004 in. (0.10 mm) and the width as small as 0.01in. (0.25 mm). The x, y motion of the process for each layer is similar to that of an inkjet printer and thus the extension of this to RP is also known as 3D printing.

Stratasys is the largest manufacturer of FDM equipment. They supply the polymer material in large spools of filament. A large number of different thermoplastics are available, including ABS, PC, PC-ABS, and polyphenylsulfone (PPSF).

The strengths of FDM include

1. Fabrication of functional parts
2. Minimal waste of material
3. Ease of support removal
4. Ease of material change
5. Large build volume up to $2772 \times 1683 \times 2281$ mm^3

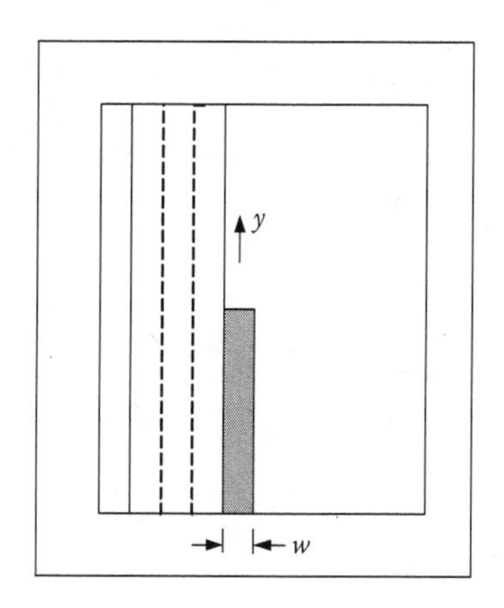

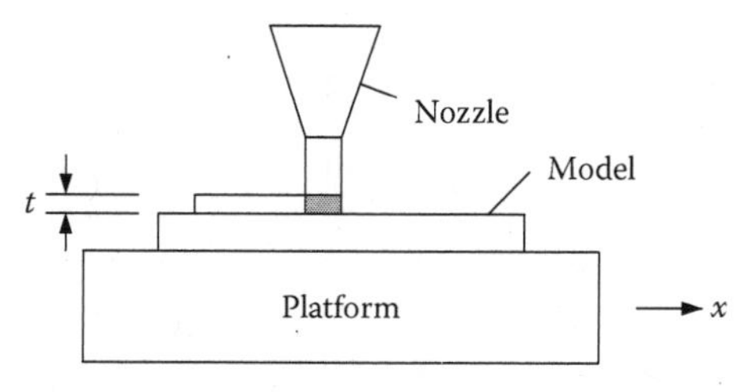

FIGURE 15.6 Fused deposition modeling.

The weaknesses of FDM include

1. Restricted accuracy
2. Slow process
3. Unpredictable shrinkage

Stereolithography (SLA) is an additive method and uses a liquid-based system. The liquid polymer is photocured layer by layer. The liquid polymers are (UV) ultraviolet-curable photopolymer (thermoset) resins such as epoxy, acrylates, and vinyl ethers. The system is illustrated in Figure 15.7. A tub (vat) is filled with the liquid resin. A platform supports the model as it is built. The platform is first raised to just below the liquid surface at a depth of the first layer to be cured. Then a computer-controlled laser is focused so that it solidifies the first layer to the required depth as it is moved in the *x*, *y* directions within the 2D area of the first slice. Subsequently, the platform is lowered a distance equal to the second layer thickness to be cured and the process is continued. Layer thicknesses as small as 0.025 mm (0.001 in.) up to 0.5 mm (0.020 in.) can be cured in this manner. 3D Systems is the company that supplies its patented SLA equipment to a worldwide market.

The strengths of the SLA method include

1. Continuous round-the-clock operation
2. A range of build volumes from $250 \times 250 \times 250$ mm^3 to $737 \times 635 \times 533$ mm^3
3. Good accuracy (root mean square (RMS) error of user parts of +/−45 μm (0.0018 in.)[1])
4. The best surface finish
5. A wide range of materials

The weaknesses of SLA include:

1. Requires support structures
2. Postprocessing is time consuming to remove support structures
3. Requires postcuring

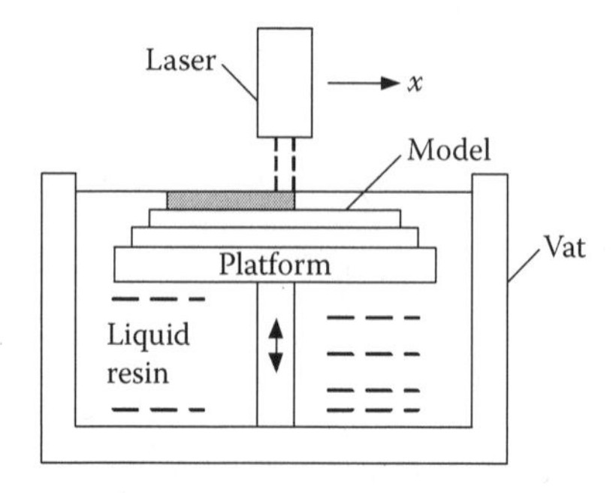

FIGURE 15.7 SLA process.

Laminated object manufacturing (LOM) is a subtractive solid-based method where the starting material is in the form of sheet material. Potentially any sheet material can be used. However, Kraft paper with a polyethylene-type adhesive is the most popular and cost effective. Sheet paper is applied to the LOM machine continuously from a large roll. It is advanced one model width at a time for each layer. A heated roller then laminates the layer onto the previous layer. Then the periphery (outline) of the cross section of the new layer is cut by a CO_2 laser as shown in Figure 15.8 and the process is continued. The result is a solid rectangular block of the overall model width. Finally, a postprocessing step is needed. The shaded material in Figure 15.7 is separated from the model. Layer thicknesses of 0.08–0.25 mm (0.003–0.008 in.) can be accommodated with this process.

The strengths of LOM include

1. A wide range of materials
2. Fast build time
3. High precision
4. No need of additional support structure
5. No postcuring required

The weaknesses of LOM include

1. Precise power adjustment needed of the laser
2. Fabrication of thin walls is difficult
3. Integrity of functional prototypes is low
4. Removal of support structure is labor intensive

Selective laser sintering (SLS) is a powder-based system. This system was first developed by the DTM Corporation in 1987, but later was acquired by 3D Systems in 2001. This process can build models out of a variety of materials in addition to polymers including metals, ceramics, and composites. The process is similar to the SLA process except that a powder is used instead of a liquid. The powder is sintered layer by layer with a CO_2 laser. An advantage of sintering over melting is that particles are only fused together at their mating surfaces without going through the liquid phase resulting in less distortion. The polymer materials include PC, nylon (PA),

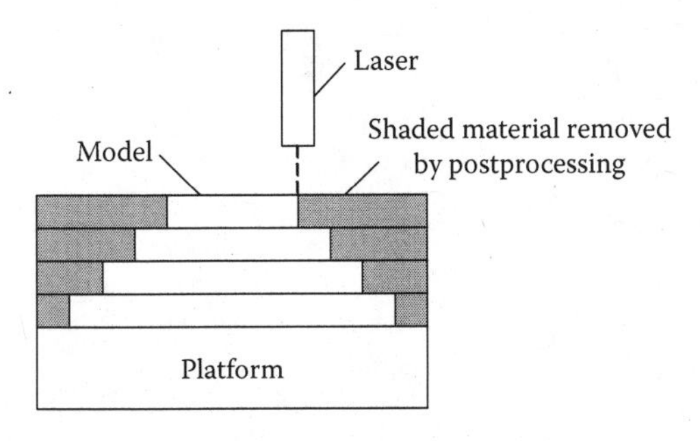

FIGURE 15.8 LOM process.

glass-filled nylon, impact-resistant ABS and PP, thermoplastic elastomers, and a PS casting material used for molds for casting of metals.

The strengths of SLS include

1. Good part stability
2. Large range of materials
3. No supports
4. Little postprocessing
5. No postcuring

The weaknesses of SLS include

1. Large physical size of the RP unit
2. High power consumption
3. Poor surface finish

15.4 RP MATERIALS

15.4.1 Materials Used in FDM by Stratasys

Mechanical properties and thermal properties are listed in Tables 15.1 and 15.2. ABS-M30 is 25–70% stronger than conventional ABS and is used for functional parts, tooling, and end-use parts. It has greater tensile strength, impact, and flexural strength. Specimens were 0.3175 cm (0.125 in.) thick. The properties listed for FDM Material PC show that the PC is stronger than the ABS material. Another FDM material is a PC-ABS copolymer that offers the superior strength of PC and the

TABLE 15.1
Mechanical Properties of FDM Materials

Mechanical Property	Test Rate, in./min (cm/min)	FDM, ABS–M30	FDM, PC	FDM, PC-ABS	FDM, PPSF
Tensile strength	0.2 (0.508)	5200 psi (36 MPa)	9800 psi (68 MPa)	5900 psi (41 MPa)	8000 psi (55 MPa)
Tensile modulus	0.2 (0.508)	350 ksi (2.4 GPa)	330 ksi (2.28 GPa)	278 ksi (1.92 GPa)	300 ksi (2.07 GPa)
Tensile elongation	0.2 (0.508)	4%	4.8%	6.0%	3.0%
Flexure strength	0.5 (1.27)	8800 psi (61 MPa)	15,100 psi (104 MPa)	9800 psi (68 MPa)	15900 psi (110 MPa)
Flexure modulus	0.5 (1.27)	336 ksi (2.3 GPa)	324 ksi (2.23 GPa)	280 ksi (1.93 GPa)	320 ksi (2.21 GPa)
Izod impact, notched	23°C	2.6 ft-lb/in. (139 J/m)	1.0 ft-lb/in. (53 J/m)	3.7 ft-lb/in. (196 J/m)	1.1 ft-lb/in. (58.7 J/m)
Izod impact, unnotched	23°C	5.3 ft-lb/in. (283 J/m)	6.0 ft-lb/in. (320 J/m)	9.0 ft-lb/in. (481 J/m)	3.1 ft-lb/in. (166 J/m)

TABLE 15.2
Thermal and Other Properties of FDM Materials

Thermal Property	Test Condition	FDM, ABS–M30	FDM , PC	FDM, PC-ABS	FDM, PPSF
Heat deflection temperature (HDT)	66 psi, unannealed	204°F (96°C)	280°F, 138°C	230°F, 110°C	NA
HDT	264 psi, unannealed	180°F(82°C)	261°F, 127°C	205°F, 96°C	372°F, 189°C
Coefficient of thermal expansion	Flow direction	4.9(10^{-5}) in./in./F (8.82/°C)	3.8(10^{-5}) in./in./°F, 6.84/°C	4.1(10^{-5}) in./in./°F, 7.38/°C	3.1(10^{-5}) in./in./°F, 5.5/°C
Glass transition (T_g)		226°F, 108°C	322°F, 161°C	257°F, 125°C	446°F, 230°C
Specific gravity		1.04	1.2	1.2	1.28
Rockwell hardness		109.5	R115	R110.5	M86
Dielectric strength		28.0 kV/mm	15.0 kV/mm	35.0 kV/mm	14.6 kV/mm

flexibility of ABS. An FDM material that is useful for medical devices in addition to industrial applications is PPSF that has the highest heat and chemical resistance. For example, it is resistant to antifreeze, gasoline (unleaded), motor oil, power steering fluid, transmission fluid, and windshield washer fluid.

There are many more materials available from Stratasys. See the website: www.Fortus.com.

15.4.2 Materials Used in SLA

Accura 10 is the basic material used by 3D Systems, (www.3dsystems.com). Its appearance is clear amber and it has a liquid specific gravity at 25°C (77°F) of 1.16. The laser penetration depth of the layers is 6.3–6.9 mils. Mechanical properties are listed in Table 15.3 and thermal properties are listed in Table 15.4 for postcured material (at 90-min UV Curing). Accura 25 simulates the flexibility of polypropylene and ABS, its appearance is white, and it has a liquid specific gravity at 25°C (77°F) of 1.13. The laser penetration depth of the layers is 4.2 mils.

Accura 40 simulates the properties of nylon 6,6. Its appearance is clear amber and it has a liquid specific gravity at 25°C (77°F) of 1.16. The laser penetration depth of the layers is 6.6–6.8 mils. Accura 60 simulates the properties and appearance of polycarbonate. Its appearance is clear and it has a liquid specific gravity at 25°C (77°F) of 1.13. The laser penetration depth of the layers is 6.3 mils. There are many other SLA materials available from 3D Systems. See the web site: www.3dsystems.com.

15.4.3 Materials Used in LOM

The main supplier of LOM equipment was Helisys Inc. until 2000 when Cubic Technologies succeeded it. For reference and comparison, the properties[1] of the sheet paper product are shown in Tables 15.5 and 15.6.

15.4.4 Materials Used in SLS

DuraForm® GF from 3D Systems simulates the properties of glass-filled polyamide (nylon). Mechanical properties are listed in Table 15.7 and thermal properties are listed in Table 15.8.

15.5 APPLICATIONS

Using the Internet and checking the websites cited earlier or carrying out additional searches, one can find many examples of the types and kinds of parts made by the different RP techniques. There are hundreds of RP service companies that use the systems described earlier that will produce a prototype for a customer. A small company may want to use such a service. Siegert[4] gives advice on selecting one. Large companies may want to purchase their own RP system. Using the data in this chapter will help compare the various techniques. RP systems can be expensive and require training in their use. Therefore, a deliberate and careful approach should be taken before deciding on a purchase.

TABLE 15.3
Mechanical Properties of SLA Materials

Mechanical Property	SLA Accura 10	SLA Accura 25	SLA Accura 40	SLA Accura 60
Tensile strength	9010–10,940 psi, 62–76 MPa	5540–5570 psi, 38 MPa	8270–8920 psi, 57–61 MPa	8410–9860 psi, 58–68 MPa
Tensile modulus	440–510 ksi, 3.05–3.53 GPa	230–240 ksi, 1.59–1.66 GPa	380–480 ksi, 2.63–3.32 GPa	390–450 ksi, 2.69–3.10 GPa
Tensile elongation	3.1–5.6%	13–20%	4.8–5.1%	5–13%
Flexure strength	12,900–16,600 psi, 89–115 MPa	7960–8410 psi, 55–58 MPa	13,400–14,000 psi, 93–97 MPa	12,620–14,650 psi, 87–101 MPa
Flexure modulus	410–460 ksi, 2.83–3.19 GPa	200–240 ksi, 1.38–1.66 GPa	380–440 ksi, 2.62–3.04 GPa	392–435 ksi, 2.70–3.00 GPa
Izod impact, notched	0.28–0.52 ft-lb/in., 14.9–27.7 J/m	0.4 ft-lb/in., 19–24 J/m	0.42–0.56 ft-lb/in., 22–30 J/m	0.3–0.5 ft-lb/in., 15–25 J/m

TABLE 15.4
Thermal and Other Properties of SLA Materials

Thermal Property	Test Condition	SLA Accura 10	SLA Accura 25	SLA Accura 40	SLA Accura 60
HDT	66 psi, unannealed	136°F, 58°C	136–145°F, 58–63°C	124–129°F, 51–54°C	127–131°F, 53–55°C
HDT	264 psi, unannealed	122°F, 50°C	124–131°F, 51–55°C	109–120°F, 43–49°C	118–122°F, 48–50°C
Coefficient of thermal expansion	Thermal mechanical analysis (TMA) ($T < T_g$, 0–20°C)	3.6 (10^{-5}) in./in./°F, 6.4/°C	5.0 (10^{-5}) in./in./°F, 10.7/°C	4.83 (10^{-5}) in./in./°F, 8.7/°C	5.5–7.28 (10^{-5}) in./in./°F, 7.1–13.1/°C
	TMA ($T > T_g$, 90–150°C)	9.4 (10^{-5}) in./in./°F, 17/°C	8.39 (10^{-5}) in./in./°F, 15.1/°C	10.4 (10^{-5}) in./in./°F, 18.7/°C	8.50(10^{-5}) in./in./°F,15.3/°C
Glass transition (T_g)		143°F, 62°C	140°F, 60°C	144–150°F, 62–65.6°C	136°F, 58°C
Specific gravity		1.21	1.19	1.19	1.21
Shore hardness		D86	D80	D84	D86

TABLE 15.5
Mechanical Properties of Models Produced by LOM

Mechanical Property	Value
Tensile strength, in plane	9.5 ksi, 66 MPa
Tensile modulus, in plane	971 ksi, 6.7 GPa
Tensile elongation, in plane	2%
Compressive strength, in plane	3.8 ksi, 26 MPa
Compressive strength, transverse	0.57 ksi, 3.9 MPa
Compressive modulus, in plane	1350 ksi, 9.3 GPa
Compressive modulus, transverse	118 ksi, 814 MPa
Compressive strain at failure, in plane	1%
Compressive strain at failure, transverse	12.9%

Source: Adapted from Jacobs, P. F., *Stereolithography and Other RP&M Technologies*, Society of Manufacturing Engineers, Dearborn, 1996.

TABLE 15.6
Thermal and Other Properties of Models Produced by LOM

Thermal Property	Test Condition	Value
Thermal conductivity, in plane	RT	0.002256 W/cm K
Thermal conductivity, transverse	RT	0.000703 W/cm K
Coefficient of thermal expansion, in plane transverse	RT-140°F (60°C); RT-151°F (66°C)	0.61(10^{-5}) in./in./°F, 1.1/°C; 8.39(10^{-5}) in./in./°F, 15.1/°C
Glass transition (T_g)		194°F, 90°C
Specific gravity		1.449

Source: Adapted from Jacobs, P. F., *Stereolithography and Other RP&M Technologies*, Society of Manufacturing Engineers, Dearborn, 1996.

TABLE 15.7
Mechanical Properties of SLS Material DuraForm GF

Mechanical Property	Value
Tensile strength, ultimate	3771 psi, 26 MPa
Tensile modulus	590 ksi, 4.07 GPa
Tensile elongation	1.4%
Flexure strength	5366 psi, 37 MPa
Flexure modulus	450 ksi, 3.11 GPa
Izod Impact, notched	0.8 ft-lb/in., 41 J/m

TABLE 15.8
Thermal and Other Properties of SLS Material DuraForm GF

Thermal Property	Test Condition	Value
HDT	66 psi	354°F, 179°C
HDT	264 psi	273°F, 134°C
Coefficient of thermal expansion	TMA (0–50°C) TMA (85–145°C)	3.46 (10^{-5}) in./in./°F, 6.23/°C 6.9 (10^{-5}) in./in./°F, 12.4/°C
Thermal conductivity		3.26 BTU-in./h ft^2°F 0.47 W/m K
Specific gravity		1.49
Shore hardness		D77

Alternatively, a simpler approach and a less costly way may be beneficial depending on the type of prototype needed. One may consider CNC machining of a metal, wood, or solid plastic model. Machining plastic must be done with very small cuts to avoid residual stresses and distortion however. If the material to be used in the final product is a polymer another approach is to make a wood or metal pattern. Then make a silicone rubber mold by pouring liquid rubber over the pattern in a box.[1,4] Then make prototype castings with a resin-like polyurethane. A silicone mold can make up to 20 parts before it begins to deteriorate.

A case study of how the design process can work with concept development using functional prototyping and iterations are presented in the reference by de Beer and Campbell.[5] The product was a self-tensioning device for tensioning banners of size 1 × 2 m. The device was an assembly of several parts. The design team used a concurrent engineering approach. The first prototype was developed using laser cutting, CNC punching and bending, and manual assembly. This proved too costly, it was labor intensive, the unit was unstable, and the results did not meet expectations. Additional prototypes (three) were produced using laser sintering of some of the parts. A total of 10 functional prototypes of the final design were produced for marketing purposes resulting in an order of 45,000 units by the industrial client.

RP has found use as well in the bioengineering field. For example, at the University of Colorado at Denver, the CAT scan data of a human skull with a tumor was converted into STL format and a polymer model of the skull was made to help guide surgeons before an operation was made (see Figure 15.9).

Reverse Engineering is another application of RP in producing after-market parts when production has ceased. The geometry of an existing physical part can be copied using a coordinate measurement machine (CMM). The x, y, z geometry data then can be input into CAD software and an STL file can be output to an RP machine. Of course, the CMM can be used to measure the RP part as well to check for accuracy.

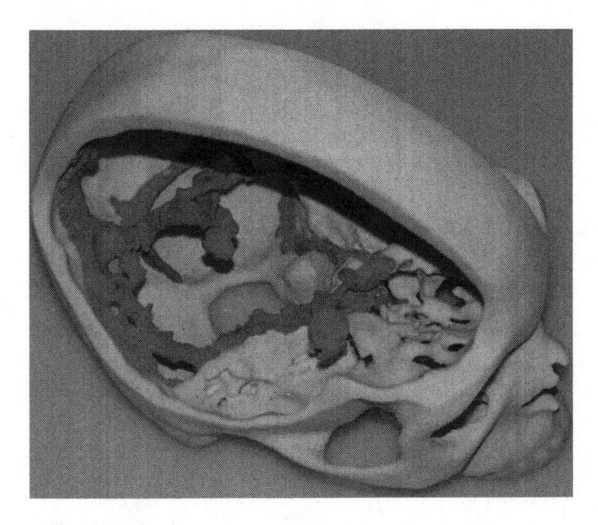

FIGURE 15.9 FDM model of human skull with tumor.

HOMEWORK PROBLEMS

15.1. Assume that your university has a CAD lab as well as an RP lab. Develop a 3D solid model of a selected part. (Possibly selected by your instructor.) Produce an STL file and download to an RP machine. Construct a prototype of the part. Report on the following: Design specifications, time required to produce the part, measurements of the final dimensions of the part, and comparison with the specifications.

15.2. In Chapter 5, the combined properties of impact toughness and flexure modulus were considered. Using the data in Tables 15.1 through 15.8 above for RP materials, construct a figure like Figure 5.5. (Omit the LOM material.)

15.3. A vacuum cleaner has a fan impeller blade made of PC that frequently is broken when the vacuum picked up hard objects like stones. The manufacturer wants to replace the material of the fan blade. Based on the results from Problem 15.2 above, what material(s) do you suggest as a prototype for replacement? Are there any other factors besides impact toughness and flexure modulus that should be considered in the choice?

15.4. Assume that your university has a CAD lab, RP lab, and a materials testing lab. Develop a 3D solid model of a standard tensile specimen. Choose one of the materials from Tables 15.1 through 15.8 Produce an STL file and download to an RP machine. Construct several duplicate prototypes of the specimen. Conduct tensile tests and compare average results with those reported by the manufacturer.

15.5. Assume that your university has a CAD lab, RP lab, and a materials testing lab. Develop a 3D solid model of a standard Izod impact specimen.

Choose one of the materials from Tables 15.1 through 15.8. Produce an STL file and download to an RP machine. Construct several duplicate prototypes of the specimen. Conduct impact tests and compare average results with those reported by the manufacturer.

REFERENCES

1. Jacobs, P. F., *Stereolithography and Other RP&M Technologies*, Society of Manufacturing Engineers, Dearborn, 1996.
2. Chua, C. K., K. F. Leong, and C. S. Lim, *Rapid Prototyping, Principles and Applications*, 3rd ed., World Scientific Publishing Co., Ltd., Singapore, 2010.
3. Urlich K. T. and S. D. Eppinger, *Product Design and Development*, 4th ed., McGraw-Hill, New York, Exhibit 12-9, 2008.
4. Siegert, A., Selecting a rapid-prototyping bureau, *Machine Design.com*, 63–68, 2009.
5. de Beer, D. J. and R. I. Campbell, Concept development through functional RP, *Rapid Prototyping*; Fourth Quarter 2005; 11, 4; www.sme.org/rtam.

16 Piezoelectric Polymers

16.1 INTRODUCTION

Piezoelectric polymers are one kind of piezoelectric material, and piezoelectric materials are one of the subsets of ferroelectric materials. Ferroelectric materials are materials that become polarized when subjected to an applied electric field. The dipoles consisting of plus and minus electric charges within the material become aligned with the electric field. The prefix "ferro-" originally means iron and because iron is known to become magnetized (polarized) under a magnetic field, "ferro-" has been used as a synonym for the ability to become polarized.

Piezoelectric materials are pressure-sensitive materials. When subject to pressure (stress) they become polarized and produce an electric field. The prefix "piezo-" comes from the Greek word pressure. The reverse is true when these materials are subject to an electric field; the result is a mechanical strain and displacement. These materials have many industrial applications as transducers, accelerometers, and sensors, and have found use in robotic and biomedical applications. Composites of these materials are also known as smart materials and are used in adaptive structures.

There are also pyroelectric materials that exhibit an electric field when subject to heating or cooling. Some materials are both piezoelectric and pyroelectric. (The prefix "pyro-" comes from the Greek word for fire.)

Muscles, bone, and tendons in the human body are examples of natural organic materials that exhibit piezoelectric effects. For example, one of the authors (Gerdeen) has a medical defect called super ventricular tacardia (SVT), wherein an electric short between the left ventricular chamber and the sinus node in the heart causes an irregular heartbeat. He takes medicine to block this short. Some people have the opposite problem. They have artificial pacemakers installed with electrodes contacting the heart muscles to regulate the heart rate. Normally the human body regulates the heart beats with its own pacemaker when it is in a healthy condition.

Fukada[1] gives a thorough review of the history of piezoelectric polymer research in Japan. Shear-type piezoelectric effects were detected and measured in a mixture of carnauba palm wax and resin, in cellulose and in collagen protein tissues, and in wood in the 1950s; bending-type piezoelectric effects in bone in 1953; shear in bone and tendons in 1953; and combined effects in copolymers of polyvinylidene fluoride (PVDF)/Trek) in 1995. Fukada[1] also presents appreciable data.

The most common types of piezoelectric materials are piezoelectric ceramics and the most common of these ceramics is lead zirconate titanate (PZT). PVDF is the most popular piezoelectric polymer material. It is polymerized from the vinylidene fluoride monomer [−CH2 − CF2−]. A linear chain of PVDF is shown in Figure 16.1. Figure 16.1 shows electric dipoles in a polarized thin film of this material. If a voltmeter is connected to the top and bottom of this material a voltage, V_0 would be detected.

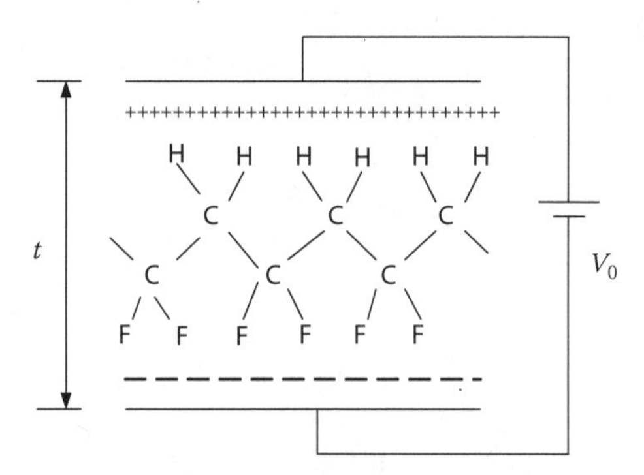

FIGURE 16.1 Electric dipoles in linear chain of PVDF.

16.2 PIEZOELECTRIC STRAIN BEHAVIOR

First consider one-dimensional strain behavior in a polymer material. Let ε represent the total strain where

$$\varepsilon = \sigma J(t) + \Lambda, \tag{16.1}$$

where $\sigma J(t)$ is the mechanical viscoelastic strain due to external stress. Let Υ represent the nonmechanical strain. The nonmechanical strain due to an external applied voltage V is

$$\Lambda = \mathrm{d}V/t, \tag{16.2}$$

where d is the piezoelectric coefficient with units length/voltage and t = thickness of the sample being considered. There are analogous nonmechanical strains. For example, the strain due to a temperature change is $\Lambda = \alpha \Delta T$. Nonmechanical strains can occur due to a magnetic field, humidity, nuclear swelling, and phase changes in a material from a temperature transition. The latter occurs in some polymers and in shape-memory metal alloys. This chapter focuses on piezoelectric effects.

Figure 16.2 shows the mechanical strains in an unrestrained piezoelectric material subject to an external electric field $E3 = V3/t$ in the vertical polling direction. A positive thickness change Δt occurs and a negative width change Δw occurs. The opposite happens when the electric field is reversed. The strains due to $E3$ are

$$\Lambda_3 = \Delta t/t, \quad \Lambda_2 = \Delta w/w, \quad \text{and} \quad \Lambda_1 = \Delta L/L \tag{16.3a–c}$$

If an external electric field $E2$ is applied, then a shear strain also occurs between axes 2 and 3:

$$\Lambda_4 = \gamma_{23} = u/t, \tag{16.4}$$

where u is the shear displacement in the 2 direction. A similar shear strain $\Lambda_5 = \gamma_{13}$ occurs between axes 1 and 3 if an external electric field $E1$ is applied as shown in Figure 16.3. Because of the shear strains, the material can be considered anisotropic

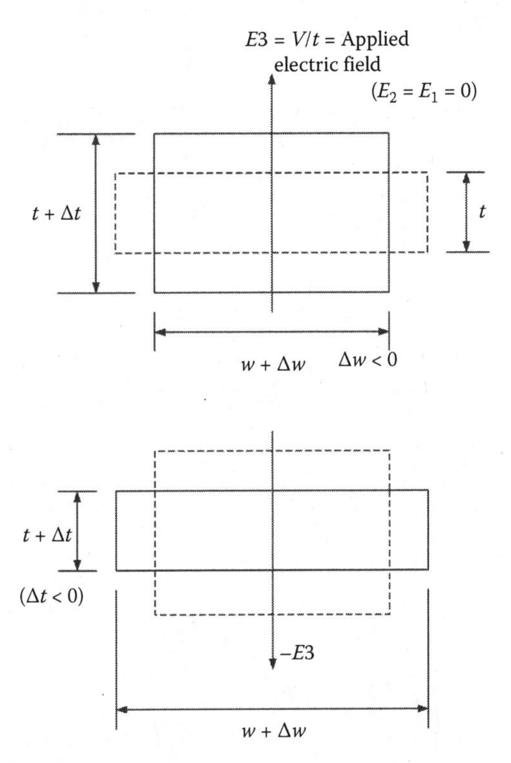

FIGURE 16.2 Axial and lateral strains in a mechanically unconstrained piezoelectric material.

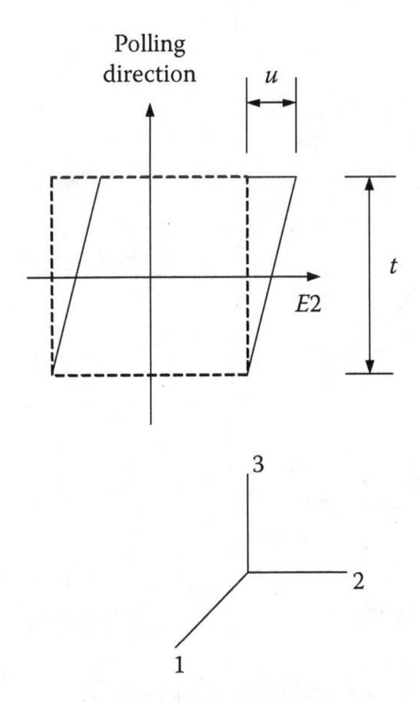

FIGURE 16.3 Shear strain in piezoelectric material.

and is similar to the shear strain effect exhibited in a laminate material (see Figure 8.16b).

For 3D analysis, the stress–strain relationships are expressed in matrix form similar to Equation 8.39 for anisotropic materials with the addition of the nonmechanical strain as follows:

$$\{\varepsilon\} = [S]\{\sigma\} + \{\Lambda\}, \tag{16.5}$$

where

$$\{\Lambda\} = \begin{Bmatrix} \Lambda_1 \\ \Lambda_2 \\ \Lambda_3 \\ \Lambda_4 \\ \Lambda_5 \\ \Lambda_6 \end{Bmatrix} = [d]^{\mathrm{T}}\{E\} \tag{16.6}$$

and

$$[d]^{\mathrm{T}} = \begin{bmatrix} 0 & 0 & d_{31} \\ 0 & 0 & d_{32} \\ 0 & 0 & d_{33} \\ 0 & d_{24} & 0 \\ d_{15} & 0 & 0 \\ 0 & 0 & 0 \end{bmatrix}, \quad \{E\} = \begin{Bmatrix} E1 \\ E2 \\ E3 \end{Bmatrix} = \begin{Bmatrix} V_1/L_1 \\ V_2/L_2 \\ V_3/L_3 \end{Bmatrix}, \tag{16.7a,b}$$

where d_{ij} are the piezoelectric coefficients, in this case written for an orthotropic polarized material, and where $\{D\} = [e]\{E\} + [d]\{\sigma\}$, $E1$, $E2$, and $E3$ are the external electric fields. (To avoid confusion in notation, the symbol $E3$ is used for voltage and E_3 is used for modulus of elasticity in this chapter.)

When stress is applied, an electric current is developed. The inverse relation is expressed as

$$\{D\} = [e]\{E\} + [d]\{\sigma\}(\mathrm{C/m^2}). \tag{16.8}$$

For references on the above equations, see Marin-Franch et al.[2] and also Nye.[3] D is called electric displacement in electrical texts, since d is a displacement change in the dipoles, but the resulting unit of D is charge per area and that is what can be measured. Since the units of d, the piezoelectric coefficient, are displacement/volt/thickness, these units must be kept clear in later calculations. Thus, d = displacement/V.

Here the symbol e is used for the dielectric constant, that is, for the permittivity ε. The symbol ε is used in electrical texts but to avoid confusion of ε with strain, e is used here. Also in electrical texts S is used for strain instead of ε. The permittivity e is an electrical term first used for capacitors. It is defined as

$$e = \frac{C}{V}\left(\frac{t}{A}\right)(\text{F/m}), \tag{16.9}$$

where the units of a farad (F) = Coulomb/volt. Thus, the terms $[e]\{E\}$ have units of (Coulomb/volt) × (thickness/area) (volt/thickness) = Coulomb/area = C/m^2.

The second terms $[d]\{\sigma\}$ have units of (displacement/volt) × (force/area). Since work is force times displacement (N m) is equivalent to one Joule of energy, and one joule is also volt-Coulomb, then (m/V) × (N/m^2) = Joule/(V m^2) = Coulomb/area = Coulomb/m^2.

16.3 PIEZOELECTRIC MATERIAL PROPERTIES

Some properties of the piezoelectric ceramic PZT and the piezoelectric polymer PVDF are listed in Table 16.1 for comparison. These data are for poling in the z-direction normal to the thin film sheet. PVDF has less electrical transducing power, but is more sensitive to mechanical pressure. However, cellular PP has higher values for d_{33} around 200 pC/N but d_{31} and d_{32} are very small of the order of 1 pC/N. (Cellular piezoelectric polymers have dipoles formed in oval-shaped cells. The cells also have voids and are softer than the polymer matrix surrounding the cells.)[2]

However, for uniaxial stretching of thin-film PVDF, Bauer and Bauer[4] give the following data: $d_{31} = 16$ pm/V, $d_{33} = -20 - 23$ pm/V, and $d_{32} = 3$ pm/V. For biaxial stretching they have $d_{31} = 5$ pm/V, $d_{33} = -20 - 25$ pm/V, and $d_{32} = 5$ pm/V. Copolymers of vinylidene fluoride and trifluoroethylene P(VDF–TrFE) depict slightly

TABLE 16.1
Material Properties of Ceramic PZT and Polymer PVDF for Field $E3$

	PZT Ceramic	PVDF Polymer Film
d_{33} (pm/V)	360	33
d_{31} (pm/V)	−166	−23
d_{15} (pm/V)	585 pm/V	−27
d_{24} (pm/V)		−23
E_3 (MPa)	4.9(10^4)	
E_1 (MPa)	6.3(10^4)	2.0(10^3)
S_{uc} (MPa)	500	60
S_{ut} (MPa)	70	160–300
Temperature range (°C)	300	−40–80
Maximum voltage	8 kV/cm	750 V/μm

Note: pm, picometer = 10^{-12} m; E, modulus of elasticity; S_{uc}, ultimate compressive strength; S_{ut}, ultimate tensile strength.

TABLE 16.2
Piezoelectric Shear Properties of Some Biomaterials

Class	Material	Shear Piezoelectric Coefficient, $-d_{14}$, pC/N
Cellulose	Wood	0.1
	Ramie	0.2
Chitin	Crab Shell	0.2
Amylose	Starch	2.0
Collagen Proteins	Bone	0.2
	Tendon	2.0
	Skin	0.2
Keratin	Wool	0.1
	Horn	1.8

Source: Fukada, E., *IEEE Transactions on Ultrasonics, Ferroelectrics, and Frequency Control*, 47(6), 1277–1290, 2000.

larger values of $d_{33} = -24\text{–}30$ pm/V and $d_{31} = d_{32}$ (pm/V) and an extended temperature range of use up to 110°C.

Thus, it is found that the piezoelectric properties depend on the stress state and the polling direction. The properties also depend on the amount of crystallites and the amount of amorphous regions in the semicrystalline polymer.

Permittivity data are less reported in the literature. For PVDF, Bauer and Bauer[4] give data for the voltage coefficient $g_{33} = d_{33}/e_{33} = 0.2$ V m/N with $d_{33} = 20$ pm/V. This gives a result for the permittivity $e_{33} = 100$ pC/V m $= 0.1(10^{-9})$ C/V m. Fukada[1] also shows how the properties vary with frequency and with temperature.

Shear properties (d_{14}) of some biomaterials are listed in Table 16.2. In some fibrous biomaterials when the polymers form oriented crystallites with antiparallel fibers, the longitudinal strains cancel out and only some shear strains result represented by the property d_{14}, where $-d_{14} = d_{25}$.[1]

16.4 HYSTERESIS

Polymers exhibit creep and hysteresis as previously indicated in Chapter 4, Section 4.3. The hysteresis results in energy loss and the polymers are frequency sensitive under cyclic loading. There is also a corresponding hysteresis piezoelectric effect under cyclic electric AC current or the reverse by the current resulting from cyclic stress. Figure 16.4 shows a hysteresis loop for PVDF at room temperature (20°C) with a cyclic voltage E at a frequency of 1 MHz. Takase et al.[4] also shows plots of hysteresis at lower temperatures down to −100°C well below the glass transition temperature of −50°C. The residual polarization remained nearly constant at all temperatures, about 50–60 mC/m^2.

In order to produce a stable poling direction in the first place, the cyclic process that is most widely used is the Bauer[5] process. This room-temperature method, called the corona poling method, involves applying a large electric field up to 8 kV.

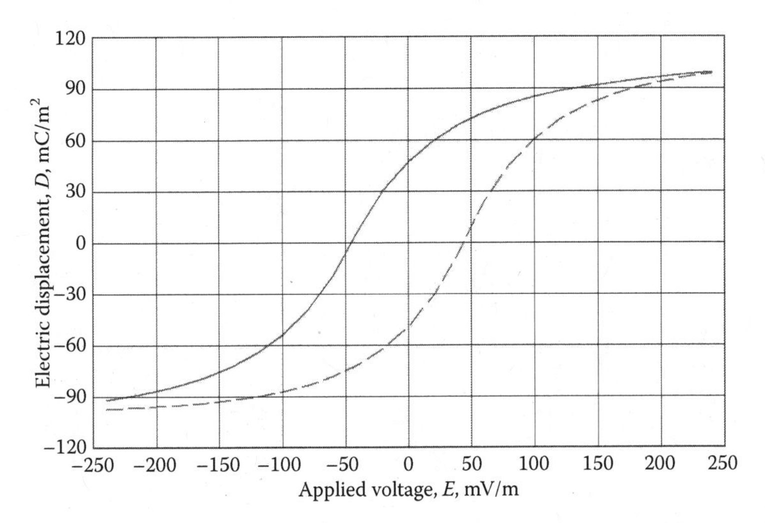

FIGURE 16.4 Hysteresis loop for PDVF at room temperature (20°C) at 1 MHz. (Experimental data from Takase, Y., J. W. Lee, J. I. Scheinbeim, and B. A. Newman, *Macromolecules*, 24(25), 6644–6652, 1991.)

Piezoelectric properties between samples can be reproduced within 2% with this process. Bauer and Bauer[5] describe the process in detail.

In order to analyze the hysteresis, it is useful to use complex algebra. Define a complex piezoelectric coefficient as

$$d^* = d' - id'' = d_0 e^{-i\delta t}, \tag{16.10}$$

where, $d' = d_{33}$, for example. Equation 16.10 is similar to Equation 2.38 for the complex creep compliance. Let the applied electric field be represented by

$$E = E_0 \sin(\omega t). \tag{16.11}$$

Then by analogy with Equation 2.35, the electric displacement D will lag the electric field and is given by

$$D = D_0 \sin(\omega t - \delta). \tag{16.12}$$

And where

$$\tan\delta = \frac{d''}{d'}. \tag{16.13}$$

Again by analogy, the energy dissipation per volume per cycle is

$$\delta H = d_0 D_0^2 \sin\delta. \tag{16.14}$$

Figure 16.5 shows a plot of one cycle of the linear relation $D = D_0 \sin(\omega t - \delta)$ and $E = E_0 \sin(\omega t)$, where $E_0 = 200$ V/μm, $D_0 = 100\ \mathrm{mC/m^2}$, and $\delta = 0.175$ rad for $\omega t = 0$

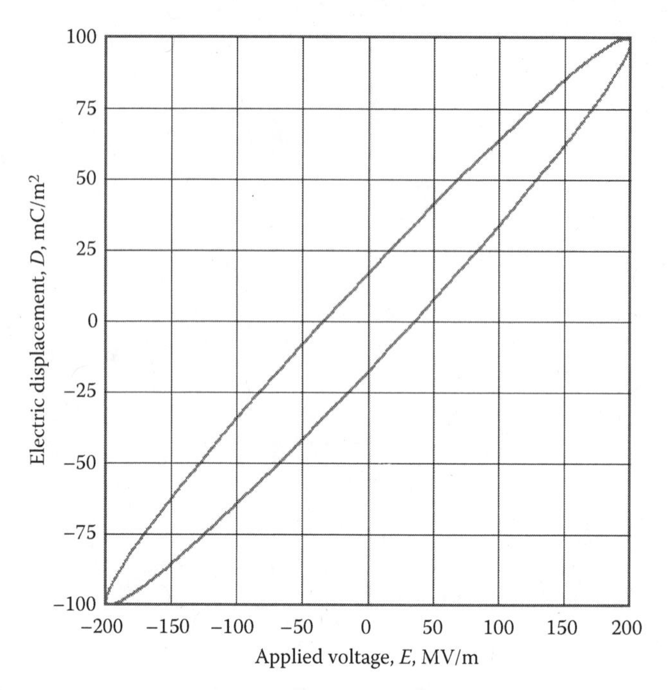

FIGURE 16.5 Linear model of hysteresis loop with $D = D_0 \sin(\omega t - \delta)$ and $E = E_0 \sin(\omega t)$ with $E_0 = 200\,\text{V}/\mu\text{m}$, $D_0 = 100\,\text{mC/m}^2$, and $\delta = 0.175$ rad.

to 2π. This does not look like the experimental data in Figure 16.4. The reason is that the actual behavior is nonlinear. In an attempt to model the nonlinearity, assume

$$D/D_0 = (\sin(\omega t - \delta))^n$$

or

$$E = E_0 \sin(\omega t)$$

and

$$D = D_0 \sin(\omega t - \delta)\left(\frac{D}{D_0}\right)(|\sin(\omega t - \delta)|)^n \tag{16.15}$$

with $n = 1/5$ and $\delta = 0.175$ rad. Figure 16.6 shows the result that more resembles the data in Figure 16.4. It is not an exact fit, and more analysis is required, but this does verify that there is a nonlinear effect. For small D and E, the linear relations can be used. For large D and E, a secant relation D/E can be obtained with a DC voltage applied. For application, a particular device should be calibrated.

16.5 COMPOSITES

Most applications will require a composite structure, because to be most effective the electrodes must cover maximum surface area to collect or transmit the electric charge where total charge $q = D\,A$, where A = area resulting in units of Coulombs.

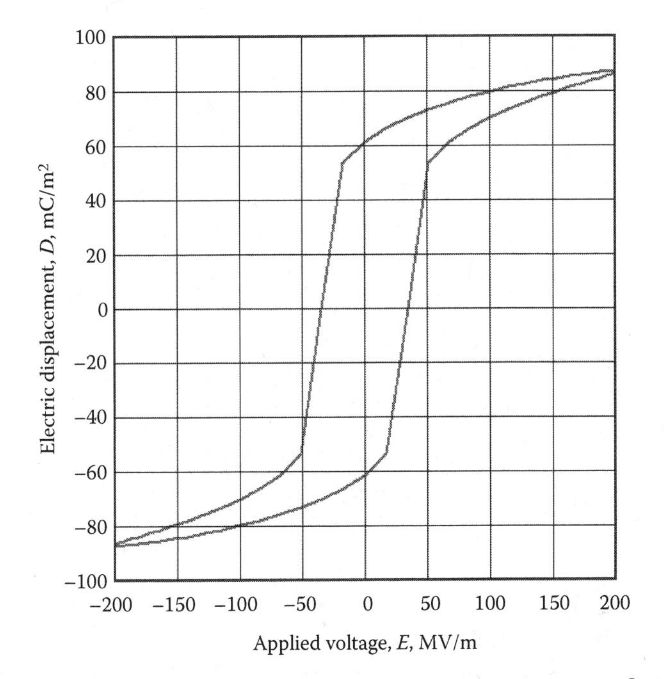

FIGURE 16.6 Nonlinear model of hysteresis loop with $D = D0*\text{sign}(\sin(\omega t - \delta)) \times \text{power}(0.5\ \text{abs}(\sin(\omega t - \delta)), n)$ and $E = E_0 \sin(\omega t)$ with $E_0 = 200\,\text{V}/\mu\text{m}$, $D_0 = 100\,\text{mC/m}^2$, $\delta = 0.175$ rad, and $n = 0.5$.

The electrodes bonded to the top and bottom surfaces of the polymer film become part of the structure as shown in Figure 16.7. For example, the electrodes are commonly thin strips of gold (for good conductivity) that are much stiffer than the polymer material. Figure 16.7a and b shows the structure of a piezoelectric stress gauge and a bending actuator.

The stress gauge can be used for compression in the vertical z-direction normal to the thin strips or for measuring tension or compression in the x- and y-directions parallel to the thin strips when glued to the surface of another structure to be analyzed. The thicknesses of the layers are very thin. A typical gauge from one manufacture has thicknesses of 25 μm for both the gold electrodes and the piezoelectric PDVF element.

The piezoelectric material itself may be a composite. For example, combinations of piezoelectric polymers and piezoelectric ceramics have been made. Sporn and Schoenecker[6] discuss ceramic fibers in a polymer matrix. First, PZT fibers with diameters <30 mm oriented uniaxially in a planar fiber architecture along with interdigital electrodes. Then the fiber/electrode architectures are embedded within glass fiber-reinforced polymers and the fibers are poled and become piezoelectric.

Sakamoto et al.[7] used a ceramic powder imbedded in the polymer polyurethane and Marin-Franch et al.[2] used thin films of ceramic and polymer composites of calcium-modified lead titanate (PTCa) and a polar copolymer, P(VDF-TrFE).

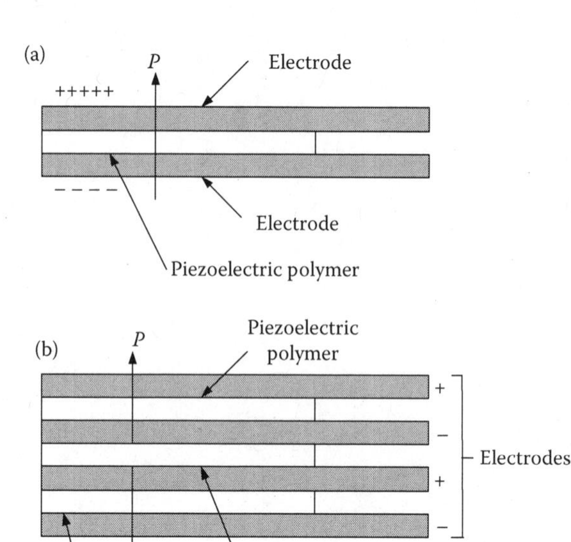

FIGURE 16.7 Structure of a piezoelectric stress gauge. (a) Composite structure for stress gauge and (b) structure for bending actuator.

EXAMPLES

Compressive Stress Gauge

Consider the structure shown in Figure 16.7a, with the thicknesses of the same value at $t_1 = t_2 = 25\ \mu\text{m}$, where t_1 is the thickness of each electrode and t_2 is the thickness of the piezoelectric polymer PVDF. The surface area of the gauge element is 1 cm × 1 cm. Assume the gauge is subject to a compressive stress σ_3 normal to the gauge in the poling direction. Assume a voltage drop, a potential difference of $V = -100$ V is detected across the gauge. (No external voltage is applied.) Determine the magnitude of the stress in the gauge.

SOLUTION

Use data from Table 16.1, $d_{33} = 33(\text{pm/V or pC/N})$. Take $e_{33} = 100$ pC/Vm from Marin-Franch et al.[2]

Calculate

$$\Delta E3 = \Delta V/t_2 = -100/25 = -4\ (\text{V}/\mu\text{m}).$$

Next, use Equation 16.8. The external voltage applied is zero, $E3 = 0$. The gauge will act like a capacitor with the polymer being the dielectric with an emissivity change. The capacitor will exhibit no change in charge. From Equation 16.8,

$$e_{33}\Delta E3 + d_{33}\sigma_3 = 0 = (0.1\times10^{-9}\ (C/V)\times(t/A))\times(4V/(t,\mu\text{m}) + (33\ \text{pC/N})(\sigma_3, N/A).$$

Solve for the stress

$$\sigma_3 = \left(\frac{-400}{33}\right),$$

$$\sigma_3 = \left(\frac{-400}{33}\right)((\mu C/A)/(pC/N)) = -12.1\,MPa,$$

where one Joule = volt Coulomb = N m.

Discussion

The prior solution is based on the assumption that the polymer deforms without restraint. However, the electrode layers are much stiffer than the polymer layer and if they are perfectly bonded to the polymer then they will prevent the polymer from expanding in the lateral x- and y-directions. A biaxial compressive stress develops in the polymer. This in turn causes a positive Poisson tensile strain in the z-direction and reduces the compressive stress strain that will be detected by the gauge. Therefore, to obtain maximum sensitivity, a compliant conductive adhesive should be used between the layers. Also, a prestress can be applied to improve contact.

In order to obtain more accurate experimental data, each gauge should be calibrated. However, the above analysis gives an estimate of the sensitivity of such a gauge.

Design Consideration

In addition to adhesive properties discussed above, an important design parameter is considered for the piezoelectric material itself. From the above solution note that the voltage output is proportional to the stress as follows:

$$\Delta E3 = \frac{d_{33}}{e_{33}}\sigma_3 = g_{33}\,\sigma_3.$$

The material property g_{33} is known as the voltage coefficient and can be called a design figure-of-merit (FOM) in this case, where

$$\begin{aligned} \text{FOM} &= g_{33} \\ &= \frac{d_{33}}{e_{33}} \\ &= \frac{E3}{\sigma_3}. \end{aligned} \tag{16.17}$$

$$\text{FOM} = g_{33} = (\text{displacement/V})/\big((C/V)\times(t/A)\big) = (\Delta t/t)/(C/A)$$
$$= \Lambda_3/(C/A) = (V/N)\times(A/t) = E3/\sigma_3.$$

Cellular PP has a higher FOM, higher than PVDF in this case, 30 (V/N) (A/t) compared to 0.2 (V/N) (A/t) for PVDF. Substituting cellular PP for PVDF in the above example gives a greater sensitivity of output voltage by a factor of 150 times greater for the same thickness. This means that a larger thickness of the polymer PP can be used to achieve the same voltage gradient $\Delta E3$. This is not the same for tension as indicated below because the d_{31} and d_{32} values and the g values are very small in cellular polymers. Because of its large g_{33} factor, cellular PP has found much use in flexible key pads.

Tensile Stress Gauge

Assume the same configuration as Example 16.1, but now assume that the bottom electrode layer is perfectly bonded to a structure under tension. Also, assume in this case that there is perfect bonding between the PVDF and electrode layers in order to transfer the tensile strain in the structure to the gauge. Assume that a potential difference of $V = 12.5$ V is detected across the gauge in the normal direction. What is the magnitude of strain detected in the structure?

SOLUTION

The piezoelectric parameters needed now are $d_{31} = 16\,\text{pm/V}$, $d_{33} = -20$ to 23 pm/V, and $d_{32} = 3$ pm/V taken from Marin-Franch et al.[2] for uniaxial tension. Take $e_{33} = 100$ pC/V m from Marin-Franch et al.[2] The elastic modulus from Table 16.1 is $E_1 = 2.0(10^3)\,\text{MPa}$.

Calculate

$$\Delta E3 = \Delta V/t_2 = 12.5\text{V}/25\,\mu\text{m} = 0.5\text{V}/t\,\mu\text{m}$$

and $d_{31}\sigma_1 = e_{33}\Delta E3$. Thus $\sigma_1 = 0.1(10^{-9})\,(C/V)(t/A) \times (0.5\text{ V}/\text{t}, \mu\text{m})/(16\text{ pC/N}) =$ 3.1 MPa.

The strain due to stress in the polymer is

$$\varepsilon_1 = \frac{\sigma_1}{E_1} = \frac{3.1}{2.0(10^3)} = 1550.$$

In contrast, for comparison, calculate the piezoelectric strain Λ_1 that would result if an unbonded PVDF film was subjected to an external voltage of $E3 =$ 0.5 V/t, μm. $\Lambda_1 = d_{31}E3 = 16\,(\text{t,pm/V}) \times (0.5\text{ V/t}, \mu\text{m}) = 8.0\,\mu$.

Conclusion: This configuration is not practical as an actuator unless the voltage is very high. The voltage is limited by the dielectric strength of 750 V/μm. See the following example for an actuator. A tensile-type gauge can also be used to measure pressure in a pipe or hose.

Pressure in a Cylinder

If a thin piezoelectric film is formed into the shape of a cylinder and is used as a layer in a composite cylindrical pipe or hose, and if the pipe or hose is subjected to internal pressure, then the film will experience tension and can be used as a gauge to measure the pressure.

The pressure tensile hoop stress in a cylinder is given by Equation 7.31 as $\sigma = PD/2t$, where P = pressure, D = diameter, t = thickness, and $D \gg t$. Thus, for a thin wall pipe, $\sigma \gg P$. For the pipe, an FOM, which is related to design pressure, can be found.

First assume an ideal case where the piezoelectric film comprises the entire pipe.

The hoop stress

$$\sigma = \frac{PD}{2t} = \sigma_1 = \frac{e_{33}}{d_{31}}\left(\frac{V_3}{t}\right).$$

In this case the thickness t drops out when solving for pressure, that is,

$$P = \frac{2}{D}\left(\frac{e_{33}}{d_{31}}\right)V_3$$

and inverting and solving for voltage gives

$$V_3 = \frac{D}{2}\left(\frac{d_{31}}{e_{33}}\right)P,$$

where the

$$\text{FOM} = \frac{d_{31}}{e_{33}}.$$

But electrodes still are needed to detect the voltage. They will form a composite, and will carry most of the stress and reduce the stress in the polymer and thus reduce the sensitivity of the gauge. For a composite wall assume that σ is the average stress and t is the total thickness. The stresses in each layer will be different and related by Equations 8.3 through 8.6 in Chapter 8. Assume that the configuration through the wall is similar to the stress gauges above but now in a cylindrical shape. See Figure 16.8.

Let subscript 2 denote the outer and inner electrode layers, and subscript 1 denote the intermediate polymer layer. Then it can be shown that

$$\sigma = \sigma_1\left(\frac{2t_2E_2}{tE_1} + \frac{t_1}{t}\right).$$

From this equation, σ can be determined from σ_1, which in turn can be related to pressure. For the stresses in a composite pipe, see Figure 16.8.

Problem

Determine the pressure in a pipe of diameter $d = 1\,\text{cm}$, and with thicknesses $t_1 = t_2 = 25\,\mu\text{m}$. Assume the same voltage change and the same piezoelectric properties as in the previous example. Thus, $\sigma_1 = 3.1$ MPa. Assume copper electrodes with $E_2 = 80$ GPa. Then

$$\sigma = \sigma_1\left(\frac{2t_2E_2}{tE_1} + \frac{t_1}{t}\right) = 3.1\left(\frac{2(1)80}{3(2)} + \frac{1}{3}\right) = 83.7\text{ MPa},$$

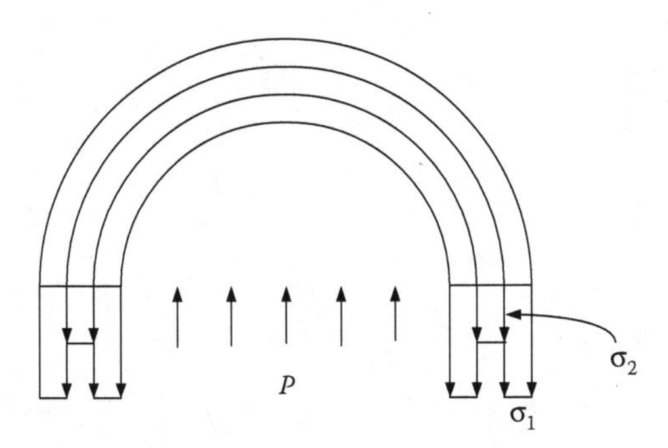

FIGURE 16.8 Stress distribution in piezoelectric wrapped around a cylinder.

from which the pressure is

$$P = \sigma \frac{2t}{D} = 83.7 \frac{2(75)}{1} = 1.25\,\text{MPa}.$$

For design, a softer polymer material, which will carry little stress, can be used as a liner and jacket to protect and insulate the cylindrical gauge.

Human-Powered Electronics

Piezoelectric polymer film can be used for supplying low power to submicron CMOS electronic devices. Klimiec et al.[8] show that mechanical energy can be converted from walking, by using a copolymer polyethylene–polypropylene (PE–PP) shoe insole. They claim that the amount of electric energy obtained from a PE–PP foil of a thickness of 11 μm for a single step of a duration of 1 s amounts to 340 nJ.

Let us check this out to see if it is possible. The energy of 340 nJ ($340(10^{-9})$ N m). Assume that each foot applies a force of 100 lb (445 N) to the insole. For linear behavior the work done is

$$W = \frac{1}{2} F\delta = \frac{1}{2}(445)\delta = 340(10^{-9}).$$

Rearranging and solving for the deflection gives

$$\delta = 2\frac{W}{F} = 2\frac{340(10^{-9})}{445} = 1.53(10^{-9})\text{ m}.$$

Assume $d_{33} = 200$ (pC/N) or (pm/V) for cellular polypropylene. Since

$$d = \frac{\delta}{V},$$

this gives

$$V = \frac{\delta}{d} = \frac{1.53(10^{-9})}{200} = 7.65\text{V}.$$

Alternatively, we can solve for the charge $C = d_{33} F = (200 \text{ pC/N}) \times 445 = 8.9(10^{-8})$.

Next, use $e_{33} = 100$ pC/V m = (C/V) × (t/A) = $(8.9\ (10^{-8})/7.65) \times (11(10^{-6}))/A$.

Solving for the area gives $A = 12.8(10^{-4})\ \text{m}^2 = 12.8\ \text{cm}^2$. This is the predicted area of the insole in the shoe. It is equivalent to a rectangular area 2 cm × 6.4 cm which is reasonable. Therefore, it is concluded that the result of Klimiec et al.[8] is realistic.

Bending Actuator (A Bimorph)

If a composite beam as shown in Figure 16.7b is used with metal electrodes the deflection will be very low because the metal with higher stiffness will greatly restrict the piezoelectric strain in the polymer. Therefore, a conductive polymer is recommended to be used as the electrodes. Researchers at Montana State University have used the conducting polymer poly(ethylene dioxythiophene) (PEDOT) with success.[9] They have also constructed a composite beam with two electrode layers (see Figure 16.9) instead of four as shown in Figure 16.7.

(a)

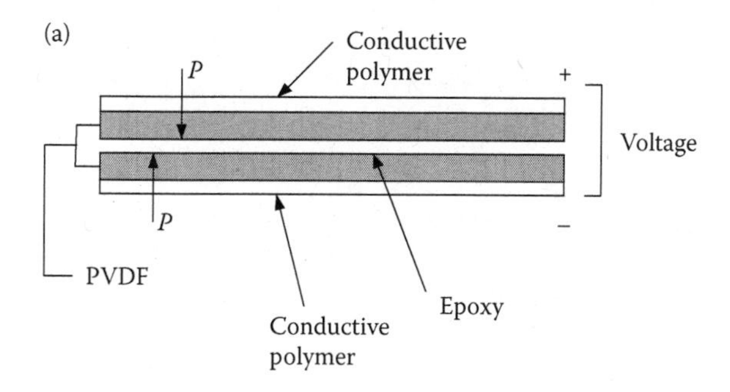

(b)

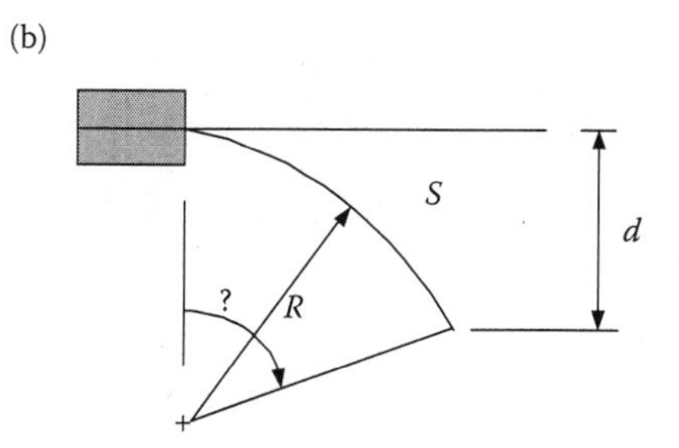

FIGURE 16.9 Beam actuator with conductive polymer electrodes. (a) Structure for bending actuator and (b) deflection of beam.

Assume the configuration in Figure 16.9 with two PVDF layers of thickness 32 μm each.

Assume that the total thickness of the beam is 80 μm.

Problem

Find the maximum strain in the beam and find the deflection of the beam mounted as a cantilever if the beam is originally 5 cm long as is bent to a 5 cm radius. Also, find the external voltage V required to achieve this result. The piezoelectric polymer is PVDF.

SOLUTION

The maximum longitudinal strain at the surface of the beam is

$$\varepsilon = \frac{h}{2R} = \Lambda_1 = 80(10^{-6}) / (2 \times 5 \times 10^{-2}) = 800\ \mu.$$

From Figure 16.9b, the angle of bending can be calculated as

$$\theta = \frac{S}{R} = \frac{5}{5} = 1.0\,\text{rad}.$$

The deflection

$$\delta = R(1 - \cos\theta) = 5(1 - \cos(1)) = 2.3\,\text{cm}.$$

The electric field

$$E3 = \frac{\Lambda_1}{d_{31}} = \frac{800\ \mu}{30(10^{-12})} = 26.7(\text{MV}/t),$$

from which

$$V = (E3)t = (26.7)32 = 853.$$

This result is in the range found in Schmidt et al.[9]

HOMEWORK PROBLEMS

16.1. Confirm the discussion on Example Problem 16.1 by calculation using the following assumptions and data. Assume that the polymer is fully constrained in the x- and y-directions. Then the maximum strain in the z-direction is limited by the compressibility of the polymer determined by its bulk modulus. Assume that the polymer is isotropic with a modulus of elasticity of 2 GPa and a Poisson ratio of 0.45.

16.2. Zhang et al.[12] gives complex values of d for (PVDF-TrFE) 75/25 film as $d^* = d' - id''$, as shown

Parameter	Real	Imaginary
d_{31} (pC/N)	10.7	0.18
d_{32} (pC/N)	10.1	0.19
d_{33} (pC/N)	–33.5	–0.65
d_{15} (pC/N)	–36.3	–0.32
e_{33}/e_0	7.9	0.09

1. Using these data and assuming voltage applied in the normal direction, plot a hysteresis loop like Figure 16.5 for $D3$ as a function of $E3$. (The reference permittivity in a vacuum is $e_0 = 8.85(10^{-12})$ C/V m.) What is the conclusion?
2. Using these data and assuming a pressure applied in the normal direction, plot a hysteresis loop for the strain Λ_1 as a function of $E3$.

16.3. Design a microphone using a circular diaphragm of thin film PVDF. Assume the diameter of the diaphragm is 10 mm. Determine the thickness required to avoid resonance and to give an output between 10 and 100 μV over the human audio range. Assume that the density of PVDF is 1.75 g/cm^3. Use other data for PVDF in the text above. (Note to instructor: this is a graduate-level problem. You may want to give the equations to use if you have undergraduates in your class. See solution manual.)

16.4. The polymer PVDF is to be used in a beam-type bimorph actuator as shown in Figure 16.10. The bottom layer is a permalloy, a nonpiezoelectric soft magnetic material ($Ni_{80}Fe_{20}$).[12] The metal alloy adds some stiffness to the bimorph, and being a magnetic material it can be used to add additional force to the actuator if needed. (The electrode layers are a

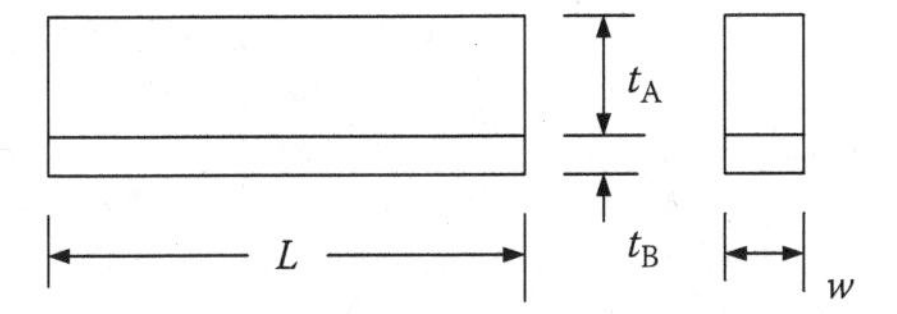

FIGURE 16.10 Dimensions of a bimorph actuator.

nickel–copper alloy of 40 nm thickness that are electroplated on the top and bottom of the PVDF film. Neglect the stiffness of the electrodes in the analysis.)

The dimensions are $t_A = 28$ μm, $t_B = 5$ μm, $L = 5$ mm, and $W = 1$ mm. A voltage of $V = 100$ V is applied through the thickness of the PVDF. For the moduli, use $E_A = 2.8$ GPa and $E_B = 150$ GPa. Find the deflection at the tip of the beam if mounted as a cantilever.

FURTHER READING

For more information on applications of piezoelectric materials, including ceramics, references Swartz[10] and Choi and Han[11] are recommended in addition to the references cited in the text.

REFERENCES

1. Fukada, E., History and recent progress in piezoelectric polymers, *IEEE Transactions on Ultrasonics, Ferroelectrics, and Frequency Control*, 47(6), 1277–1290, 2000.
2. Marin-Franch, P., D. L. Tunnicliffe, and D. K. Das-Gupta, Dielectric properties and spatial distribution of polarization of ceramic+polymer composite sensors, *Materials Research Innovations*, 4, 334–339, 2001.
3. Nye, J. F., *Physical Properties of Crystals*, Clarendon Press, Oxford (1987), *IEEE Standard on Piezoelectricity*, 1988.
4. Takase, Y., J. W. Lee, J. I. Scheinbeim, and B. A. Newman, High-temperature characteristics of nylon-11 and nylon-7 piezoelectrics, *Macromolecules*, 24(25), 6644–6652, 1991.
5. Bauer, S. and F. Bauer, Piezoelectric polymers and their applications, piezoelectricity, Chap. 6, *Springer Series in Materials Science*, Vol. 114, Springer, Berlin, 2008.
6. Sporn, D. and A. Schoenecker, Composites with piezoelectric thin fibers—First evidence of piezoelectric behavior, *Materials Research Innovations*, Vol. 2, pp. 303–308, Springer-Verlag, Berlin Heidelberg, 1999.
7. Sakamoto, W. K., D. H. F. Kanda, and D. K. Das-Gupta, Dielectric and pyroelectric properties of a composite of ferroelectric ceramic and polyurethane, *Materials Research Innovations*, 5, 257–260, 2002.
8. Klimiec, E. K., W. Zaraska, K. Zaraska, K. P. Gąsiorski, T. Sadowski, and M. Pajda, Piezoelectric polymer films as power converters for human powered electronics, *Microelectronics Reliability*, 48(6), 897–901, 2008.
9. Schmidt, V. H., J. Polasik, L. Lediaev, and J. Hallenberg, *12th International Symposium on Electrets*, pp. 378–381, Department of Physics, Montana State University, Penn State Univ., Bozeman, MT, USA, 2005.

10. Swartz, M., ed., *Smart Materials*, CRC, Taylor & Francis Group, Boca Raton, FL, 2009.
11. Choi, S.-B. and Y.-M. Han, *Piezoelectric Actuators, Control Applications of Smart Materials*, CRC Press, Taylor Francis Group, Boca Raton, FL, 2010.
12. Zhang, Q., Huang, C., Xia, F. and Ji Su, In Bar-Cohen, Y., ed., *Electroactive Polymer (EAP) Actuators as Artificial Muscles: Reality, Potential, and Challenges*, 2nd ed., SPIE, PM136, 2004.

Appendix A: Conversion Factors

Since in most cases it is preferable to use SI units, especially for dynamics problems, the conversion factors are listed to convert English units by multiplying by the listed conversion factor to SI units. Obviously, to convert from SI units to English, one just has to divide the SI value by the conversion factor. The English to SI conversion factors for units of interest are listed in Table A.1. In addition, there are some conversions that are of interest due to their historical use in polymers and mechanics that are listed in Table A.2.

TABLE A.1
English to SI Conversion Factors

English Unit (X)	SI Unit ($Y = CX$)	Conversion Factor (C)
in.	mm	25.4
ft	m	0.305
yd	m	0.914
in.4	cm4	41.6
ft/min	m/s	0.0051
lbf	N	4.45
short ton (2000 lbm)	kg	907
psi	kPa	6.89
Msi	GPa	6.89
lbf s/in.2	kg	0.454
ft lbf	N m	1.36
in. lbf	N m or J	0.113
lbf/in.	N/m	175
gal	L	3.785

TABLE A.2
Other Conversions

X	$Y = CX$	Conversion Factor (C)
dynes	Pa	10
centipoise	Pa s	0.001

Appendix B: Area Moments of Inertia

Rectangular Coordinate System (x, y)

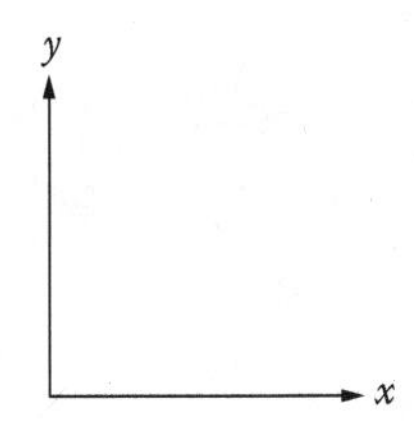

Solid Rectangular Section

$$A = bh, \quad I_x = \frac{bh^3}{12}, \quad I_y = \frac{b^3h}{12}$$

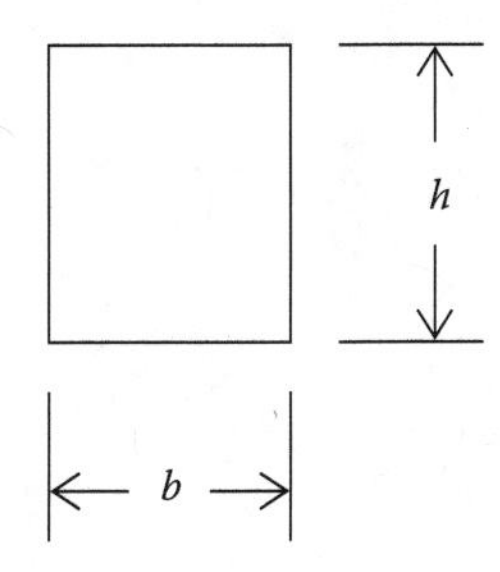

Hollow Rectangular Section

$$A = b_o h_o - b_i h_i, \quad I_x = \frac{b_o h_o^3 - b_i h_i^3}{12}, \quad I_y = \frac{b_o^3 h_o - b_i^3 h_i}{12}$$

b_i

h_i h_o

b_o

Circular Section

$$A = \frac{\pi d^4}{4}, \quad I_x = \frac{\pi d^4}{64}, \quad I_y = \frac{\pi d^4}{64}, \quad J = \frac{\pi d^4}{32}$$

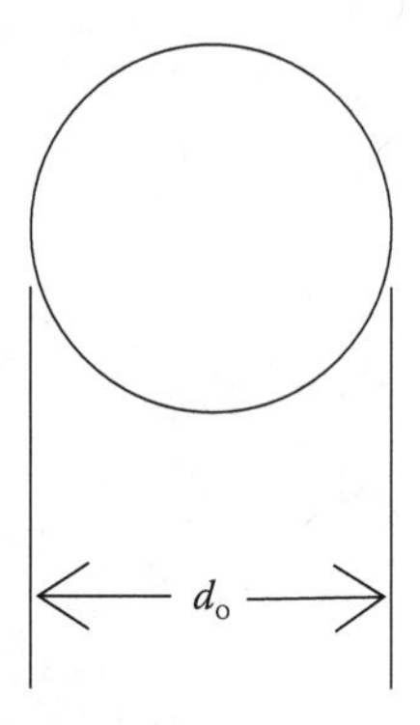

Tubular Section (thin and thick walled)

$$A = \frac{\pi(d_\mathrm{o}^2 - d_\mathrm{i}^2)}{4}, \quad I_x = \frac{\pi(d_\mathrm{o}^4 - d_\mathrm{i}^4)}{64}, \quad I_y = \frac{\pi(d_\mathrm{o}^4 - d_\mathrm{i}^4)}{64}, \quad J = \frac{\pi(d_\mathrm{o}^4 - d_\mathrm{i}^4)}{32}$$

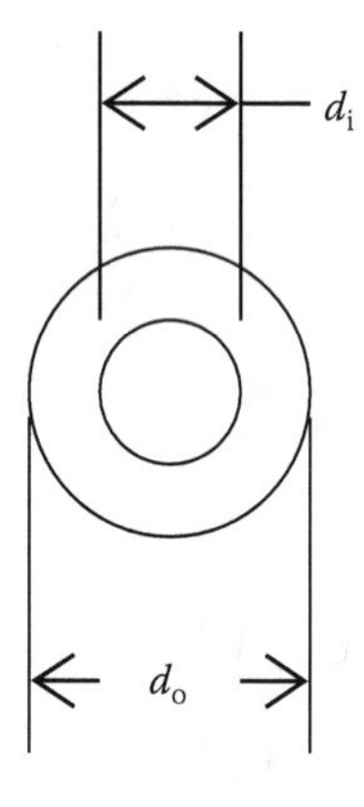

Appendix C: Beam Reactions and Displacements

This appendix lists the reaction loads and displacements for common simple beam configurations. Typically when these solutions are provided it is assumed that the beams are long slender beams (i.e., l/d or l/h), such that the shear contribution to displacement is negligible. However, polymers are often used in applications where the shear contribution to deflection, not only must be considered in some cases, it dominates.

The coordinate system for all of the beams has the axes aligned with the left-hand side of the beam. The y axis is pointing down.

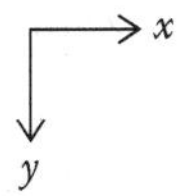

Cantilever beam with transverse end load

$$y(x) = \frac{F(3x^2 l - x^3)}{6EI}$$

$$M(x) = F(x - l)$$

$$R_L = F, \quad M_L = Fl$$

Cantilever beam with transverse intermediate span load

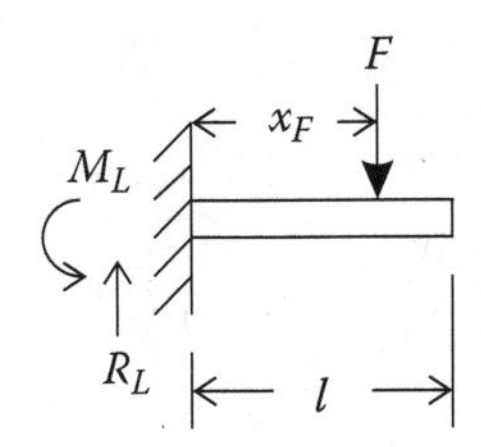

for $x < x_F$

$$y(x) = \frac{F}{6EI}(3x^2 x_F - x_F^3)$$

for $x > x_F$

$$y(x) = \frac{F}{6EI}(3x x_F^2 - x_F^3)$$

$$M(x) = 0$$

$$R_L = F, \quad M_L = F x_F$$

Cantilever beam with distributed load

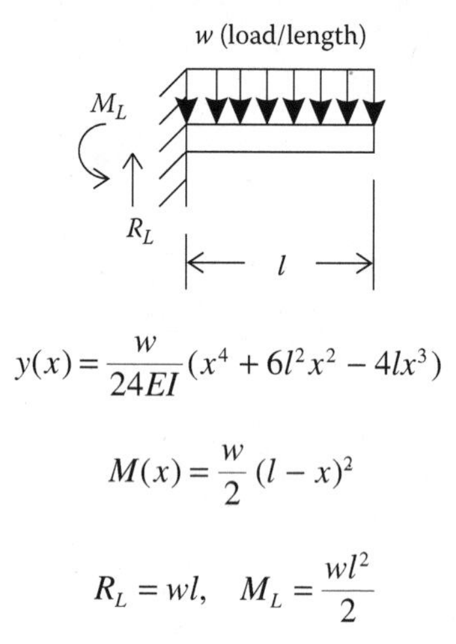

$$y(x) = \frac{w}{24EI}(x^4 + 6l^2x^2 - 4lx^3)$$

$$M(x) = \frac{w}{2}(l - x)^2$$

$$R_L = wl, \quad M_L = \frac{wl^2}{2}$$

Cantilever beam with concentrated moment

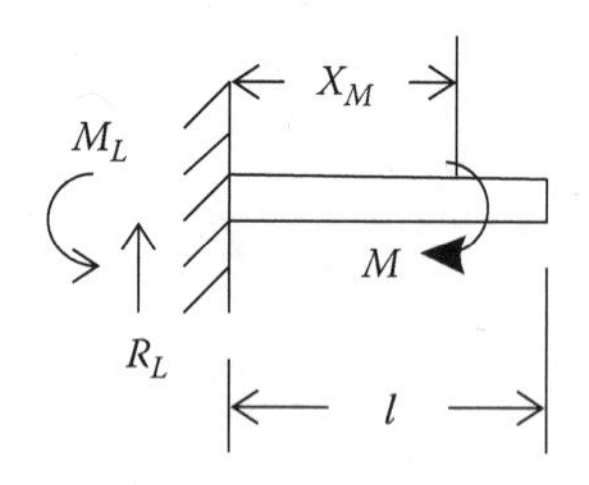

for $x \le x_\mathrm{m}$

$$y(x) = \frac{Mx^2}{2EI}$$

$$M(x) = M$$

$$R_L = 0, \quad M_L = M$$

Simply supported beam with transverse mid-span load

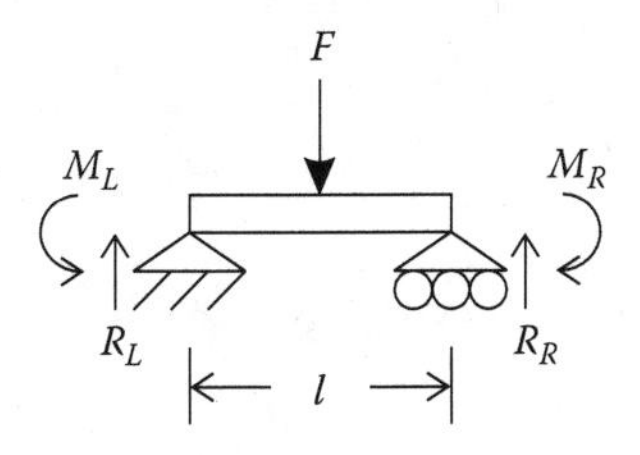

for $x < 1/2$

$$y(x) = \frac{F}{48EI}(3l^2x - 4x^3), \quad M(x) = \frac{Fx}{2}$$

$$R_L = \frac{F}{2}, \quad R_R = \frac{F}{2}, \quad M_L = 0, \quad M_R = 0$$

Simply supported beam with transverse intermediate span load

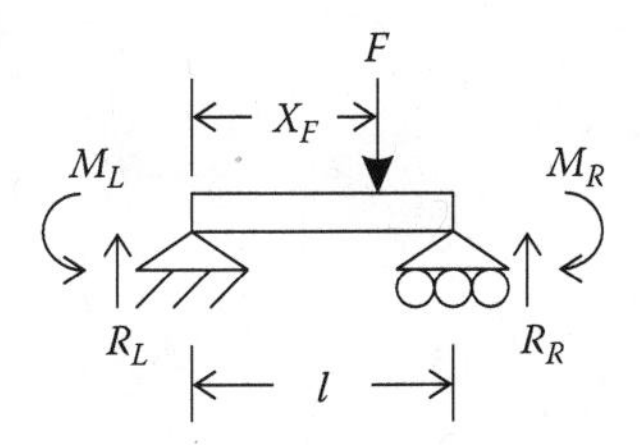

for $x < x_F$

$$y(x) = \frac{F(l - x_F)}{6EIl}((2x_Fl - x_F^2)\,x - x^3),$$

$$M(x) = \frac{Fx}{2}$$

for $x < x_F$

$$y(x) = \frac{Fx_F(x - l)}{6EIl}(x^2 + x_F^2 - 2lx)$$

$$R_L = \frac{F}{2}, \quad R_R = \frac{F}{2}, \quad M_L = 0, \quad M_R = 0$$

Simply supported beam with distributed load

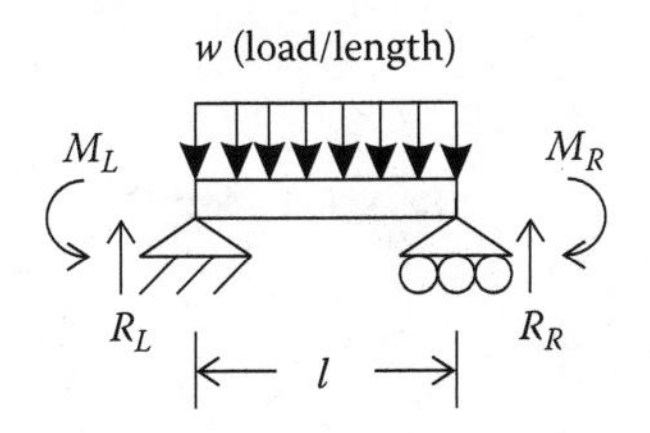

$$y(x) = \frac{xw}{24EI}(x^3 + l^3 - 2lx^2), \quad M(x) = \frac{wx}{2}(l - x)$$

$$R_L = \frac{wl}{2}, \quad R_R = \frac{wl}{2}, \quad M_L = 0, \quad M_R = 0$$

Fixed–fixed beam with transverse intermediate load

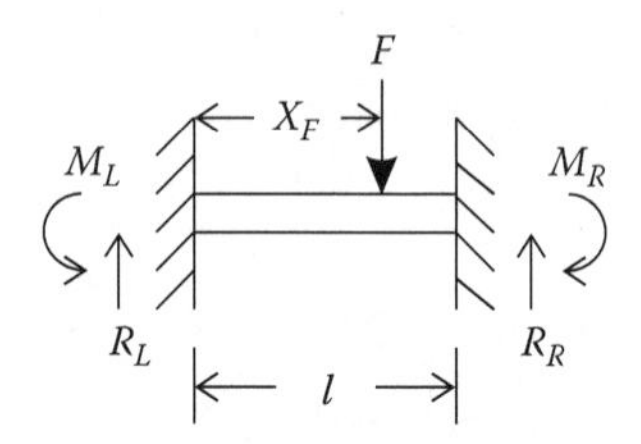

for $x < x_F$

$$y(x) = \frac{F(l - x_F)^2 x^2}{6EIl^3}(3x_F l - x(2x_F + l)),$$

$$M(x) = \frac{F(l - x_F)^2}{l^3}(2xx_f + xl - x_f l)$$

for $x > x_F$

$$y(x) = \frac{Fx_F^2(l - x)^2}{6EIl^3}(3(l^2 - x_F l) + (x - l)(3l - 2x_F)),$$

$$M(x) = F\left(\frac{(l - x_F)^2}{l^3}(2xx_f + xl - x_f l) - (x - x_F\right)$$

$$R_L = \frac{F(l - x_F)^2}{l^3}(2x_F + l), \quad R_R = \frac{Fx_F^2}{l^3}(3l - 2x_F),$$

$$M_L = \frac{Fx_F}{l^2}(l - x_F)^2, \quad M_R = \frac{Fx_F^2}{l^2}(l - x_F)$$

Fixed–fixed beam with distributed load

w (load/length)

$$y(x) = \frac{wx^2}{24EI}(l - x)^2_; \quad M(x) = \frac{w}{12}(6lx - 6x^2 - l^2)$$

$$R_L = \frac{wl}{2}, \quad R_R = \frac{wl}{2}, \quad M_L = \frac{wl^2}{12}, \quad M_R = \frac{wl^2}{12}$$

Appendix D: Laminate MATLAB® or Octave Code

```
clear all;
%
%   Laminate Analysis Program to Accompany
%   "Engineering Design with Polymers and Composites"
%   by J. C. Gerdeen and R. A. L. Rorrer
%
%   published by Taylor and Francis
%
%   copyright 2011 Ronald A. L. Rorrer
%   University of Colorado Denver
%
%   By the way you use this at your own risk!
%   You must verify that both the programs are correct and
additionally it is applicable to your problem.
%
%   Additionally, it may not be included in any other material
without permission.
%
%   This is the "Old School" version of Laminate Analysis!
%
%   It is an easy matter nowadays to invert a 6x6 matrix.
%   However this programs is set up to use the partitioning of
the 6x6 matrix into four 3x3 matrices.
%
%   This was done for 2 reasons:
%
%     The first was nostalgia (It is the way I learned it)!
%     The second is to make it consistent with both the text
%     and the historical way this has been presented in the
literature.
%
%   The output is space deliminted for ease in importing into
a spreadsheet.
%
%   Initialize variables
%
h=0;
A=zeros(3,3);
B=zeros(3,3);
D=zeros(3,3);
N=zeros(3,1);
```

```
M=zeros(3,1);
NT=zeros(3,1);
MT=zeros(3,1);
NE=zeros(3,1);
ME=zeros(3,1);
%
epsilonm=zeros(3,1);
kappa=zeros(3,1);
%
%   Input file with Laminate Properties
%   n-number of layers of laminate
%   material properties and rotations of each layer
%
%   Layer, E1, E2, Nu12, G12, Theta, t, XL, XL', XT, XT',
load compdata.txt;
m=size(compdata);
%   n=number of layers of the laminate
n=m(1);
%
for k=1: n;
   E1(k)=compdata(k,2);
   E2(k)=compdata(k,3);
   nu12(k)=compdata(k,4);
   G12(k)=compdata(k,5);
   theta(k)=(pi/180)*compdata(k,6);
   t(k)=compdata(k,7);
   XL(k)=compdata(k,8);
   XLP(k)=compdata(k,9);
   XT(k)=compdata(k,10);
   XTP(k)=compdata(k,11);
   S(k)=compdata(k,12);
   alpha1(k)=compdata(k,13);
   alpha2(k)=compdata(k,14);
   DT=compdata(k,15);
end
%
%Calculate Reduced Stiffness and Compliance Matrices for each
Lamina
%
for k=1:n;
   %Check the following reciprocity relationship
   nu21(k)=nu12(k)*E2(k)/E1(k);
   %
   S11(k)=1/E1(k);
   S12(k)=-nu12(k)/E1(k);
   S22(k)=1/E2(k);
   S66(k)=1/G12(k);
   %

SR(1,1,k)=S11(k)*(cos(theta(k)))^4+S22(k)*(sin(theta(k)))^4+
(2*S12(k)+S66(k))*(cos(theta(k)))^2*(sin(theta(k)))^2;
```

```
   SR(1,2,k)=(S11(k)+S22(k)-S66(k))*(cos(theta(k)))^2*(sin
(theta(k)))^2+S12(k)*((cos(theta(k)))^4+(sin(theta(k)))^4);
   SR(2,1,k)=SR(1,2,k);
SR(2,2,k)=S11(k)*(sin(theta(k)))^4+S22(k)*(cos(theta(k)))^4+(2
*S12(k)+S66(k))*(cos(theta(k)))^2*(sin(theta(k)))^2;
   %  Note SR(3,3) is SR (6,6) from text
   SR(3,3,k)=4*(S11(k)+S22(k)-2*S12(k))*(cos(theta(k)))^2*(sin
(theta(k)))^2+S66(k)*((cos(theta(k)))^4-(sin(theta(k)))^4);
   %  Note SR(1,3) is SR (1,6) from text
   SR(1,3,k)=(2*S11(k)-S66(k)-2*S12(k))*(cos(theta(k)))^3*(sin
(theta(k)))-(2*S22(k)-2*S12(k)-S66(k))*(cos(theta(k)))
*(sin(theta(k)))^3;
   SR(3,1,k)=SR(1,3,k);
   %  Note SR(2,3) is SR (2,6) from text
   SR(2,3,k)=(2*S11(k)-S66(k)-2*S12(k))*(cos(theta(k)))*(sin
(theta(k)))^3-(2*S22(k)-2*S12(k)-S66(k))*(cos(theta(k)))^3*(sin
(theta(k)));
   SR(3,2,k)=SR(2,3,k);
   %
   Q11(k)=E1(k)/(1-nu12(k)*nu21(k));
   Q12(k)=(nu12(k)*E2(k))/(1-nu12(k)*nu21(k));
   Q22(k)=E2(k)/(1-nu12(k)*nu21(k));
   Q66(k)=G12(k);
   %
QR(1,1,k)=Q11(k)*(cos(theta(k)))^4+Q22(k)*(sin(theta(k)))^4+(2
*Q12(k)+4*Q66(k))*(cos(theta(k)))^2*(sin(theta(k)))^2;
   QR(1,2,k)=(Q11(k)+Q22(k)-4*Q66(k))*((cos(theta(k)))^2)*(sin
(theta(k)))^2+Q12(k)*((cos(theta(k)))^4+(sin(theta(k)))^4);
   QR(2,1,k)=QR(1,2,k);
QR(2,2,k)=Q11(k)*(sin(theta(k)))^4+Q22(k)*(cos(theta(k)))^4+(2
*Q12(k)+4*Q66(k))*(cos(theta(k)))^2*(sin(theta(k)))^2;
   %  Note QR(3,3) is QR (6,6) from text
   QR(3,3,k)=(Q11(k)+Q22(k)-2*Q12(k)-2*Q66(k))*(cos(theta(k)))^
2*(sin(theta(k)))^2+Q66(k)*((cos(theta(k)))^4+(sin(theta
(k)))^4);
   %  Note QR(1,3) is QR (1,6) from text
   QR(1,3,k)=(Q11(k)-2*Q66(k)-Q12(k))*(cos(theta(k)))^3*(sin(
theta(k)))-(Q22(k)-Q12(k)-2*Q66(k))*(cos(theta(k)))*(sin(thet
a(k)))^3;
   QR(3,1,k)=QR(1,3,k);
   %  Note QR(2,3) is QR (2,6) from text
   QR(2,3,k)=(Q11(k)-2*Q66(k)-Q12(k))*(cos(theta(k)))*(sin
(theta(k)))^3-(Q22(k)-Q12(k)-2*Q66(k))*(cos(theta(k)))^3*(sin
(theta(k)));
   QR(3,2,k)=QR(2,3,k);
end
%
%  Calculate Global ABD matrices
%
%Calculate z's for distance
```

```
for k=1:n;
   h=h+t(k);
end
z(1)=-h/2;
for k=1:n;
   z(k+1)=z(k)+t(k);
end
%
for i=1:3;
   for j=1:3;
      for k=1:n;
         A(i,j)=QR(i,j,k)*(z(k+1)-z(k))+A(i,j);
         B(i,j)=(1/2)*QR(i,j,k)*(z(k+1)^2-z(k)^2)+B(i,j);
         D(i,j)=(1/3)*(QR(i,j,k)*(z(k+1)^3-z(k)^3))+D(i,j);
      end
   end
end
   %
   %Calculation of Thermal Loads
   %
for k=1:n;
   alphax(1,k)=alpha1(k)*cos(theta(k))^2+alpha2(k)*sin(theta
(k))^2;
   alphax(2,k)=alpha1(k)*sin(theta(k))^2+alpha2(k)*cos(theta
(k))^2;
   alphax(3,k)=2*(alpha1(k)-alpha2(k))*sin(theta(k))*cos(theta
(k));
end
   %
for k=1:n;
   for i=1:3;
      for j=1:3;
   NT(i)=(QR(i,j,k)*alphax(j,k))*(z(k+1)-z(k))*DT+NT(i);
   MT(i)=(QR(i,j,k)*alphax(j,k))*(z(k+1)^2-z(k)^2)*DT/2+MT(i);
      end
   end
end
%
% Calculation of Global neutral axis strain, epsilon and
curvature, Kappa
%
%
loadstrain=input('Enter 1 if you want to input global loads or
2 if you want to input strains');
if loadstrain==1;
   N(1)=input('Enter Nx');
   N(2)=input('Enter Ny');
   N(3)=input('Enter Nxy');
   M(1)=input('Enter Mx');
   M(2)=input('Enter My');
   M(3)=input('Enter Mxy');
```

```
   %
   AI=inv(A);
   BI=inv(B);
      %
   NE=N+NT
   ME=M+MT
   epsilonm=(AI+AI*B*inv(D-B*AI*B)*B*AI)*NE-(AI*B*inv(D-
B*AI*B))*ME;
%
   kappa=-(AI*B*inv(D-B*AI*B))*NE+inv(D-B*AI*B)*ME;
else
   %
   epsilonm(1)=input('Enter epsilon x')
   epsilonm(2)=input('Enter epsilon y')
   epsilonm(3)=input('Enter epsilon xy')
   kappa(1)=input('Enter Kappa x')
   kappa(2)=input('Enter Kappa x')
   kappa(3)=input('Enter Kappa xy')
   %
end
%
%  Calculation of bottom and top ply strains
%
for i=1:3;
   for k=1:n;
      eb(i,k)=epsilonm(i)+z(k)*kappa(i);
      et(i,k)=epsilonm(i)+z(k+1)*kappa(i);
   end
end
%
%Calculation of bottom and top ply stresses
%
%  zero sigma matrices
%
for i=1:3;
   for k=1:n;
      sigmab(i,k)=0;
      sigmat(i,k)=0;
      sigmapb(i,k)=0;
      sigmapt(i,k)=0;
   end
end
%
for k=1:n;
   for i=1:3;
      for j=1:3;
      sigmab(i,k)=QR(i,j,k)*(eb(j,k)-alphax(j,k)*DT)+sigmab(i,k);
      sigmat(i,k)=QR(i,j,k)*(et(j,k)-alphax(j,k)*DT)+sigmat(i,k);
      end
   end
end
```

```
%
%
%Failure Criteria
%
% Transform x and y to 1 and 2
%
for k=1:n;
   T(1,1,k)=(cos(theta(k)))^2;
   T(1,2,k)=(sin(theta(k)))^2;
   T(2,1,k)=T(1,2,k);
   T(1,3,k)=2*cos(theta(k))*sin(theta(k));
   T(2,2,k)=(cos(theta(k)))^2;
   T(2,3,k)=-2*cos(theta(k))*sin(theta(k));
   T(3,1,k)=-cos(theta(k))*sin(theta(k));
   T(3,2,k)=cos(theta(k))*sin(theta(k));
   T(3,3,k)=(cos(theta(k)))^2-(sin(theta(k)))^2;
end
%
%  Calculation of stress in principle directions
for k=1:n;
   for i=1:3;
      for j=1:3;
         sigmapb(i,k)=T(i,j,k)*sigmab(j,k)+sigmapb(i,k);
         sigmapt(i,k)=T(i,j,k)*sigmat(j,k)+sigmapt(i,k);
      end
   end
end
%
%Calculation of Tsai-Hill Failure Criteria
%
for k=1:n;
   %
   if sigmapb(1,k)>0;
      X1(k)=XL(k);
   else
      X1(k)=XLP(k);
   end
   %
   if sigmapb(2,k)>0;
      X2(k)=XT(k);
   else
      X2(k)=XTP(k);
   end
   THb(k)=(sigmapb(1,k)/X1(k))^2-(sigmapb(1,k)*sigmapb(2,k))/
X1(k)^2+(sigmapb(2,k)/X2(k))^2+(sigmapb(3,k)/S(k))^2;
   %
   if sigmapt(1,k)>0;
      X1(k)=XL(k);
   else
      X1(k)=XLP(k);
   end
```

```
   %
   if sigmapt(2,k)>0;
      X2(k)=XT(k);
   else
      X2(k)=XTP(k);
   end
%
   THt(k)=(sigmapt(1,k)/X1(k))^2-(sigmapt(1,k)*sigmapt(2,k))/
X1(k)^2+(sigmapt(2,k)/X2(k))^2+(sigmapt(3,k)/S(k))^2;
end
%
%Calculation of Tsai-Wu Failure Criteria
%
for k=1:n;
   f1(k)=1/XL(k)-1/XLP(k);
   f11(k)=(1/(XLP(k)*XL(k)));
   f2(k)=1/XT(k)-1/XTP(k);
   f22(k)=1/(XTP(k)*XT(k));
   f66(k)=1/S(k)^2;
   f12(k)=-0.5*(f11(k)*f22(k))^(0.5);
   %
Fb(k)=f1(k)*sigmapb(1,k)+f2(k)*sigmapb(2,k)+f11(k)*sigmapb
(1,k)^2+f22(k)*sigmapb(2,k)^2+f66(k)*sigmapb(3,k)^2+2*f12(k)*
sigmapb(1,k)*sigmapb(2,k);
Ft(k)=f1(k)*sigmapt(1,k)+f2(k)*sigmapt(2,k)+f11(k)*sigmapt(1,k)
^2+f22(k)*sigmapt(2,k)^2+f66(k)*sigmapt(3,k)^2+2*f12(k)*sigmapt
(1,k)*sigmapt(2,k);
   %
end
%
%  Output to data file
%
   fid=fopen('compout.txt','w');
   fprintf(fid, 'Composite_Laminate_Output \r\n');
   fprintf(fid,'\r\n');
%
   fprintf(fid, 'Laminate_Layup \r\n');
   fprintf(fid,'\r\n');
%
   fprintf(fid, 'Layer E1 E2 NU12 G12 theta t XL XLP XT XTP S
alpha1 alpha2 \r\n');
for k=1:n;
   fprintf(fid, '%d %d %d %d %d %d %d %d %d %d %d %d %d
%d\r\n', k,E1(k),E2(k),nu12(k),G12(k),theta(k)*180/
pi,t(k),XL(k),XLP(k), XT(k),XTP(k),S(k),alpha1(k),alpha2(k));
end
%
   fprintf(fid,'\r\n');
   fprintf(fid, 'Externally_Applied_Loads \r\n');
   fprintf(fid,'\r\n');
%
```

```
   fprintf(fid, ' Nx Ny Nz Mx My Mz \r\n');
   fprintf(fid, ' %d %d %d %d %d %d \r\n',
N(1),N(2),N(3),M(1),M(2),M(3));
   fprintf(fid,'\r\n');
%
   fprintf(fid, 'DT=%d Degrees_C \r\n \r\n', DT);
   fprintf(fid, '\r\n');
   fprintf(fid, 'Effective_Thermal_Loads \r\n');
   fprintf(fid, ' Nx  Ny  Nxy  Mx  My  Mz  \r\n');
   fprintf(fid, ' %d %d %d %d %d %d \r\n', NT(1),NT(2),NT(3),M
T(1),MT(2),MT(3));
   fprintf(fid, '\r\n');
   fprintf(fid, 'Total_Effective_Loads \r\n');
   fprintf(fid, '\r\n');
   fprintf(fid, ' Nx Ny Nxy Mx My Mz \r\n');
   fprintf(fid, ' %d %d %d %d %d %d \r\n', NE(1),NE(2),NE(3),M
E(1),ME(2),ME(3));
   fprintf(fid, '\r\n');
   fprintf(fid, 'Global_Strains \r\n');
   fprintf(fid, '\r\n');
   fprintf(fid, 'epsilonx epsilony epsilonxy Kappax Kappay
Kappaxy \r\n');
   fprintf(fid, '%d %d %d %d %d %d \r\n', epsilonm(1),epsilonm
(2),epsilonm(3) ,kappa(1),kappa(2),kappa(3));
   fprintf(fid, '\r\n');
%
%
   fprintf(fid,'\r\n');
   fprintf(fid, 'Failure Criteria \r\n');
   fprintf(fid,'\r\n');
%
   fprintf(fid, 'Layer Tsai-Hill Tsai-Hill Tsai-Wu Tsai-Wu
\r\n');
   fprintf(fid, '\t bottom top bottom top \r\n');
%
for k=1:n;
   fprintf(fid, '%d %d %d %d %d %d %d \r\n', k,THb(k),THt(k),
Fb(k),Ft(k));
   fprintf(fid,'\r\n');
end
%
   fprintf(fid,'\r\n');
   fprintf(fid, 'Ply_Stresses \r\n');
   fprintf(fid,'\r\n');
%
for k=1:n;
   fprintf(fid, 'Layer=%d sigmax sigmay tauxy \r\n', k);
%  fprintf(fid, 'Top_of_Ply_Stresses\r\n');
   fprintf(fid, 'Top %d %d %d \r\n', sigmat(1,k), sigmat(2,k),
sigmat(3,k));
%   fprintf(fid,'\r\n');
```

```
%   fprintf(fid, 'Bottom_of_Ply_Stresses \r\n');
   fprintf(fid, 'Bottom %d %d %d \r\n', sigmab(1,k),
sigmab(2,k), sigmab(3,k));
   fprintf(fid,'\r\n');
end
fclose(fid);
%
%
%
```

Appendix E: Sample Input/ Output for Laminate Program

The following example illustrates the use of the composite laminate program. The program is either utilized in MATLAB® or Octave from the program laminate.m.

Assume that there are four lamina that comprise the laminate. The composite material is carbon fiber in an epoxy matrix with the (0/30/45/90) layup. Notice that this layup is neither symmetric nor antisymmetric.

The input data file for the programs is initially named compdata.txt. Table E.1 shows the format of the data file. Thus, specifically the compdata.txt file is shown in Table E.2.

The load input is prompted from the screen either as loads and moments or strains and curvatures.

When the laminate program is run, the following text in italics is sent to the screen by the program and the text in bold is the input from the user:

Enter 1 if you want to input global loads or 2 if you want to input strains **1**
Enter Nx **100**
Enter Ny **0**
Enter Nxy **0**
Enter Mx **0**
Enter My **0**
Enter Mxy **0**

The output data file is compout.txt and is opened in Excel by importing as delimited with spaces as the delimiter. The output of the program is shown in Table E.3.

TABLE E.1
Input Data for compdata.txt

Layer	E_1	E_2	μ_{12}	G_{12}	θ	t	X_L	X'_L	X_T	X'_T	S	α_1	α_2	ΔT
1	2.10E+07	1.40E+06	0.25	8.50E+05	0	0.005	3.15E+05	2.50E+05	7.80E+03	2.34E+04	1.26E+04	0.88E–6	31E–6	80
2	2.10E+07	1.40E+06	0.25	8.50E+05	30	0.005	3.15E+05	2.50E+05	7.80E+03	2.34E+04	1.26E+04	0.88E–6	31E–6	80
3	2.10E+07	1.40E+06	0.25	8.50E+05	45	0.005	3.15E+05	2.50E+05	7.80E+03	2.34E+04	1.26E+04	0.88E–6	31E–6	80
4	2.10E+07	1.40E+06	0.25	8.50E+05	90	0.005	3.15E+05	2.50E+05	7.80E+03	2.34E+04	1.26E+04	0.88E–6	31E–6	80

TABLE E.2
Example compdata.txt File

1	2.10E+07	1.40E+06	0.25	8.50E+05	0	0.005	3.15E+05	2.50E+05	7.80E+03	2.34E+04	1.26E+04	0.88E–6	31E–680	80
2	2.10E+07	1.40E+06	0.25	8.50E+05	30	0.005	3.15E+05	2.50E+05	7.80E+03	2.34E+04	1.26E+04	0.88E–6	31E–680	80
3	2.10E+07	1.40E+06	0.25	8.50E+05	45	0.005	3.15E+05	2.50E+05	7.80E+03	2.34E+04	1.26E+04	0.88E–6	31E–680	80
4	2.10E+07	1.40E+06	0.25	8.50E+05	90	0.005	3.15E+05	2.50E+05	7.80E+03	2.34E+04	1.26E+04	0.88E–6	31E–680	80

TABLE E.3
Example compout.txt File Imported Into Excel

Composite_Laminate_Output													
Laminate_Layup													
Layer	E_1	E_2	NU_{12}	G_{12}	θ	t	X_L	X_{LP}	X_T	X_{TP}	S	α_1	α_2
1	21,000,000	1,400,000	2.50E–01	850,000	0	5.00E–03	315,000	250,000	7800	23,400	12,600	8.80E–07	3.10E–05
2	21,000,000	1,400,000	2.50E–01	850,000	3.00E+01	5.00E–03	315,000	250,000	7800	23,400	12,600	8.80E–07	3.10E–05
3	21,000,000	1,400,000	2.50E–01	850,000	45	5.00E–03	315,000	250,000	7800	23,400	12,600	8.80E–07	3.10E–05
4	21,000,000	1,400,000	2.50E–01	850,000	90	5.00E–03	315,000	250,000	7800	23,400	12,600	8.80E–07	3.10E–05
Externally_applied_loads													
	Nx	Ny	Nz	Mx	My	Mz							
	100	0	0	0	0	0							
DT =	80	Degrees_C											
Effective_thermal_loads													
	Nx	Ny	Nxy	Mx	My	Mz							
	5.72E+01	6.01E+01	–5.39E+00	4.69E–02	–4.69E–02	–9.67E–04							
Total_effective_loads													
	Nx	Ny	Nxy	Mx	My	Mz							
	1.57E+02	6.01E+01	–5.39E+00	4.69E–02	–4.69E–02	–9.67E–04							
Global_strains													
	epsilonx	epsilony	epsilonxy	Kappax	Kappay	Kappaxy							
	2.35E–03	8.48E–04	–2.77E–03	2.92E–01	–1.82E–01	–1.63E–01							
Failure	Criteria												
Layer	Tsai-Hill	Tsai-Hill	Tsai-Wu	Tsai-Wu									
	bottom	top	bottom	top									
1	8.79E–03	2.14E–02	2.23E–02	–4.91E–02									

2	4.15E–04	3.55E–02	–8.26E–03	3.73E–02
3	1.99E–02	1.14E–01	7.53E–02	2.26E–01
4	1.13E–01	3.05E–01	2.36E–01	5.04E–01

Ply_stresses

Layer = 1	sigmax	sigmay	tauxy
Top	1.71E+04	–7.29E+02	–1.66E+03
Bottom	–1.34E+04	3.43E+01	–9.69E+02
Layer = 2	sigmax	sigmay	tauxy
Top	1.32E+04	1.88E+03	5.24E+03
Bottom	3.18E+03	7.91E+02	1.68E+03
Layer = 3	sigmax	sigmay	tauxy
Top	4.50E+03	–2.08E+03	–4.67E+02
Bottom	3.29E+03	7.34E+02	1.25E+03
Layer = 4	sigmax	sigmay	tauxy
Top	3.56E+03	–2.09E+04	–3.74E+03
Bottom	1.83E+03	–2.27E+03	–3.04E+03

Appendix F: Composite Materials Properties

The mechanical properties of a variety of composite materials are shown in Table F.1. However, it should be noted that due to both advances and variability of mechanical properties the designer should use material properties either provided by the supplier or determined by testing.

The thermal properties of various composite materials are shown in Table F.2.

TABLE F.1
Composite Material Properties

Material	E_1 (Msi)	E_2 (Msi)	G_{12} (Msi)	ν_{12}	X_t (ksi)	X_c (ksi)	Y_t (ksi)	Y_c (ksi)	S (ksi)
E-Glass Epoxy[1]	6.5	1.8	0.8	0.28	150.15	90.02	7.0	20.0	10.0
E-Glass Epoxy[2]	7.8	2.6	1.25	0.25	150	150	4	20.0	6
*S*2-Glass/Epoxy[1]	8.10	2.60	0.9	0.27	259.5	103.8	9	22.49	13.50
Kevlar49-Epoxy[1]	11.	0.8	0.3	0.34	199.98	40.0	4.30	20.0	9.0
Standard carbon/Epoxy[1]	21.0	1.4	0.85	0.25	315.0	249.9	7.80	23.38	12.58
Intermediate carbon/Epoxy[1]	31.49	1.0	0.57	0.3	330.02	155	7.0	24.41	8.50
High carbon/Epoxy[1]	54.85	0.8	0.61	0.316	221.98	62.04	5.50	26.56	11.0
Very high carbon/Epoxy[1]	66.25	0.87	0.72	0.316	324.0	58.20	3.93	26.20	9.72
High modulus graphite/Epoxy[2]	42	1	0.7	0.25	90	90	2	28	4
High strength Graphite/Epoxy[2]	18.5	1.6	0.65	0.25	169	162	6	25	7
Boron/Epoxy[2]	30	3	1	0.3	200	400	12	20	6
Kevlar® Epoxy[2]	11	0.8	0.3	0.34	200	40	4	40	6.4

TABLE F.2
Thermal Properties of Composites

Material	α_1 (1/°F)	α_2 (1/°F)
E-Glass epoxy[2]	3.5 (10^{-6})	11.4 (10^{-6})
High modulus Graphite/Epoxy[2]	0.58	16.5
High strength Graphite/Epoxy[2]	0.25	15.2
Graphite/Epoxy[3]	0.88	31.0
E-glass/Epoxy[3]	6.3	20.0

REFERENCES

1. *Composites Design Guide*, StorageTek Composites, Boulder, CO, 1999.
2. Jones, R. M., *Mechanics of Composite Materials*, 2nd ed., Taylor & Francis, Inc., Philadelphia, PA,1999.
3. Gibson, R., *Principles of Compsite Materials Mechanics*, McGraw-Hill, Inc., Washington, DC, 1994.

Appendix G: Thermal and Electrical Properties

G.1 THERMAL PROPERTIES

The thermal conductivity (k) of polymers is an important property in thermal-stress applications and also in thermal insulations. For the latter, polymer foams with air pockets or other gas (CCL_3F) pockets have k values as low as 0.15 BTU in./(°F ft^2 h). Values of thermal conductivity for several materials are listed in Table G.1. Use of foam insulations that contain formaldehyde are cautioned against because of recent evidence of possible adverse health effects. The professional engineer (P.E.), upon registration, pledges to protect the public health and safety. Therefore, such concerns should always be considered in design. Of course, this concern also applies to all engineering professionals.

The coefficient of thermal expansion (α) is important as well in thermal-stress problems, where the thermal strain is $\varepsilon = \alpha \Delta T$. Some typical room-temperature values of polymers and other materials are shown in Table G.2.

Another important thermal property is the heat-distortion temperature (HDT), which is a temperature limit at which the material can support a load for any significant time. To determine this property, a tensile load or bending load is applied to a specimen, and the strain or deflection is measured as the temperature is increased. For example, in the ASTM test D1637, the HDT temperature is determined when the elongation becomes 2%. As the temperature approaches the HDT from below, the strain will suddenly begin to increase near the HDT. For amorphous polymers, HDT is approximately

TABLE G.1
Thermal Conductivity Values for Various Materials

Material	k [BTU in./(°F ft^2 h)]
Copper	2700
Steel	350
Aluminum	1400
Marble	14–20
Glass	5.5
Typical solid plastic	1.0
Nylon 6,6 (PA 66)	0.243
Air	0.18
Rigid plastic foam with air	0.22
Rigid plastic foam with CCL_3F	0.15

TABLE G.2
Thermal Expansion Values for Various Materials

Material	α [(10^{-6} in./in.)/°F]
Copper	9
Steel	6
Aluminum	13
PMMA	80
PS	125
PVC	90–300
LDPE	230–360
HDPE	200–230
PA 6 (Nylon 6)	108
PA 66 (Nylon 6,6)	160
PTFE	180
EP	100
CR	430
ABS	125–230

equal to the glass-transition temperature T_g. For crystalline polymers, HDT is closer to the melting temperature T_m.

G.2 ELECTRICAL PROPERTIES

The engineer may be confronted with an application that requires knowledge of the electrical properties as well as the mechanical properties. A part may need to be an electrical insulator and as well as have some level of mechanical strength. The electrical resistance R is defined as

$$R = \frac{\rho L}{A},$$

where ρ = resistivity (ohms-length units), L = length or depth of the specimen, and A = cross-sectional area of electrical contact.

A conductor has a resistivity on the order of $\rho < 10^3$, whereas an insulator has a $\rho > 10^8$. Another electrical property is the dielectric strength, which is a measure of when the structure of an insulator breaks down. In the ASTM test D149, this property is measured by placement of a thin specimen between two steel plates and application of a voltage across the thickness of the specimen.

$$\text{Dielectric strength} = \frac{\text{Voltage}}{\text{Thickness}}$$

Typical electrical properties are listed in Table G.3.

TABLE G.3
Electrical Properties of Some Materials

Material	Resistivity ρ, Ω (cm)	Dielectric Strength (V/mil)
Aluminum	$2.66(10^{-6})$	—
Copper	$1.68(10^{-6})$	—
ABS	$1–5(10^{15})$	350–500
EP	$1(10^{15})$	300–500
PA6	$1(10^{15})$	400 dry
LDPE	$1(10^{16})$	450–1000
PVC	$1–2(10^{12})$	350–500
Rubber	—	125–225

Index

A

Abrasive friction, 289
Abrasive wear, 290
ABS. *See* Acrylonitrile butadiene styrene (ABS)
Acetal copolymer, 31
Acrylic. *See* Polymethyl methalcrylate (PMMA)
Acrylonitrile butadiene styrene (ABS), 116, 140
  ASTM standards, 224
  creep rupture curve, 82
  dielectric strength, 387
  drop-test impact strength values, 100
  electrical resistance, 387
  fracture toughness values, 110
  glass transition temperatures, 22
  melting temperatures, 22
  notched izod impact strength values, 97
  polymer ID, 140
  tensile impact strength values, 110
  thermal expansion values, 386
  vibration weld strengths, 270
  wear factors and PV limits, 296
  welding compatibility matrix, 263, 268, 275
Additive model development, 328
Adherends, 243. *See also* Adhesion; Adhesive
  cohesive failure, 246
  polymeric adhesives function, 253
Adhesion, 243. *See also* Adherends; Adhesive
  adhesive failure, 246
  adhesive joints, 244
  average shear stress, 247
  chemical, 253
  cohesive failure, 246
  enhancement in composites, 253
  exercises, 258
  failure modes, 246
  fiberglass silation, 253, 254
  lap shear specimens, 246–247
  master curve, 249, 250
  measurement, 246–249
  peel test, 115
  spontaneous, 292
  time–temperature transformation, 249
  viscoelasticity, 249–250
  wetting and work, 243–246
  Young's equation, 245
  Young–Dupré equation, 245
Adhesive, 243, 250. *See also* Adherends; Adhesion; Thermosets
  adhesive joints, 244
  anaerobic, 251–252
  common polymeric, 250–252
  curing of, 254–257
  exercises, 258
  failure, 246
  friction, 289
  mechanical properties, 252
  polymeric, 253
  shear, 247
  in situ, 252
  tensile properties, 252
  thermoset, 254
  uses, 246
  wear, 290, 292
Advanced viscoelastic models, 67
  eight-decade Prony model, 68
  four-parameter fractional solid model, 68
  fractional model fit parameters, 69
  Mittag–Leffler function, 68
  Prony series, 67
  with spring-pot, 67–68
Alloys, metal, 162
Aluminum
  dielectric strength, 387
  electrical resistance, 387
  wedge test use, 248
American Society for the Testing of Materials (ASTM), 42
  D-1784 class specification, 153
  rigid PVC, 154
  Standard D-2241, 155
  test bar, 228
Amorphous polymers, 20, 21
Amoton's Laws, 286
Angle ply laminate, 163
  configuration, 164
  nonsymmetric, 189
Anisotropic body, 162
Anisotropic lamina
  generalized Hooke's law, 181
  influence of end constraints, 192
Archard wear law, 290–291
Asperites, 284
Assembly techniques, 259. *See also* Polymer fusing
  boss, 278
  inserts, 278–279
  screws, 276–278

ASTM. *See* American Society for the Testing of Materials (ASTM)
Axial ratio, 230–231

B

B-2 stealth bomber, 6, 7
Bauer process, 346
Beam reactions, 363
  cantilever beam, 363–364
  coordinate system, 363
  fixed–fixed beam, 366
  simply supported beam, 365–366
Beam-bending tests, 95
Biopolymers, 2
Blends, 9, 312
  viscoelastic spectrum, 314, 315
Blister adhesion test, 248, 249
Blow molding, 229. *See also* Polymer processing
  axial ratio, 230–231
  blow ratio, 229, 230, 231
  blow-molded cylinder, 232
  blowup ratio BUR, 231
  cooling phase, 236–241
  cycle, 229
  hoop ratio, 230
  inflation, 232–236
  pressure response, 233
  process, 230
  theoretical vs. experimental data, 236
  thickness ratio, 231
Boltzmann superposition principle, 64. *See also* Viscoelastic correspondence principle; Time–temperature equivalence principle
  applications, 65, 66
  creep test, 64–65
  general stress history, 64, 65
  total creep strain, 66
Brittle fracture theory, 106
  crack, 107, 108
  external work done on bar, 106, 107
  fracture modes, 109
  Griffith factor, 108
  stress-intensity factor, 109
  tensile strength of glass filaments, 106
Brittle materials, 32–33
  Coulomb–Mohr failure theory, 34
  Mohr static failure theory, modified, 34
  tensile strength, 33
Buckling load
  Euler elastic, 83
  for Kelvin solid, 84, 85
  for Maxwell fluid, 85
Bulk melting, 301
Bulk surface–temperature calculations, 301
Butadiene chain units, 24
Butt joint, 262
  energy director, 265
  strength, 276
BUTYL. *See* Butyl rubber (BUTYL)
Butyl rubber (BUTYL), 8
  viscoelastic spectrum, 313

C

CAD. *See* Computer-aided design (CAD)
Cantilever beam, 70. *See also* Beam reactions
  concentrated moment, 364
  deflection, 104
  distributed load, 364
  impact loading, 104
  torsion, 36
  transverse end load, 363
  transverse intermediate span load, 363–364
Cedar stripper canoe, 5–6
Cellular piezoelectric polymers, 345. *See also* Piezoelectric polymers
Charpy impact specimens, 96
  notches uses, 95
  rate effects, 102
Charpy impact tests
  fracture mechanics analysis, 115, 116
  impact fatigue tests, 116
Chloroprene rubber (CR), 21
  glass transition temperatures, 22
  melting temperatures, 22
  thermal expansion values, 386
Circular cap joint, 265
Clear polycarbonate kayak, 6
Clusters of spherulites, 18
CMM. *See* Coordinate measurement machine (CMM)
CNC machines. *See* Computer Numerical Control machines (CNC machines)
Coatings, 302
Coefficient of friction, 285–286, 288
  composite, 299
  rolling object acceleration, 292
  static, 289
  temperature dependence, 303
Composite, 4, 295, 383, 384. *See also* Polymer
  adhesion, 243
  adhesives, 243
  development timeline, 5
  examples, 5–7
  fiber-reinforced, 5
  specific wear rate, 300
  step cure, 257
  wear of, 299
Composite lamina, 168
  deformation, 177–178
  notation, 166
  wear mode, 299, 300

Composite laminate failure, 197. *See also* Sandwich beam
- bending moment stress resultants, 209
- bending stiffness matrix of plate, 211
- bending strains in plate, 210
- distortional energy density theory of failure, 198
- exercises, 217–219
- extensional stiffness matrix of plate, 211
- fiber microbuckling, 199
- fiber pullout, 205–206
- fiber shear failure, 199
- fiber/matrix kink band formation, 199
- fiber/matrix splitting, 199
- force stress resultants, 209
- Kirchoff hypothesis, 210
- laminate analysis algorithm, 214
- macromechanical failure strength properties, 200
- maximum shear stress theory of failure, 197–198
- maximum stress theory of failure, 200–205
- microstructural failure mechanisms, 199
- principal stress and strain directions, 199
- strain distributions in plate, 210
- thermal stresses, 216
- Tsai–Hill theory of failure, 201
- Tsai–Wu theory of failure, 202
- von Mises theory of failure, 198
- World-Wide Failure Exercise, The, 203

Composite material mechanics, 161
- cross-ply and angle-ply configuration, 164
- engineering elastic constants determination, 190–194
- exercises, 194
- fiber volume ratio, 170
- lamina macromechanics, 178–190
- lamina micromechanics, 167–178
- lamina notation, 166
- macroscopic analysis, 166–167
- matrix volume ratio, 170
- micromechanical analysis, 166
- nomenclature and definitions, 161
- ply orientation designation for unidirectional plies, 163
- structures analysis, 165
- transverse strain, 171

Composite stiffness, laminated, 206
- laminated plates, 211
- orthotropic plate, 208–211
- sandwich beam, 206–208

Computer database design selection procedure, 140. *See also* Polymer selection
- beam impact problem, 142
- polymer names and descriptions, 140–141
- property names and descriptions, 141

Computer Numerical Control machines (CNC machines), 327–327

Computer solution, 135. *See also* Graphical solution
- feasible and infeasible regions, 136
- global minimum, 137
- gradient in design space, 136

Computer-aided design (CAD), 227

1-configuration, 23

Connections, mechanical fastener, 276

Constant-height test, 98

Constrained layer damping, 320
- earthquake bearings for buildings, 322
- oil pans for automobile engines, 322, 323
- use of spacers, 323

Contact
- angle analysis, 244–246
- geometries, 282

Conversion factors, 359

Coordinate measurement machine (CMM), 338

COPA. *See* Co-polyamides (COPA)

COPE. *See* Co-polyesters (COPE)

Co-polyamides (COPA), 9

Co-polyesters (COPE), 9

Copolymers, 9
- acetal, 31
- block, 24
- PC-ABS, 332, 334
- random, 22, 23
- types, 22, 23, 24
- X–Y, 8

Copper, 387

Corona poling method, 346

Coulomb's Law, 286

Coulomb–Mohr failure theory, 34

CR. *See* Chloroprene rubber (CR)

CR. *See* Polychloroprene (CR)

Crazing, 149

Creep, 2
- data, 51, 53
- isochronous creep curves, 51, 52
- Kelvin solid, 61
- Maxwell fluid, 59
- time hardening, 50–51
- ultimate strength in, 81

Creep buckling
- of columns, 84
- Euler elastic buckling load, 83
- of shells, 91–92

Creep failure. *See also* Fatigue failure; Polymer
- under compression, 83–85
- exercises, 92–93
- under tension, 81–83

Creep rupture
- curves, 81, 82
- of PVC, 81, 82–83
- strength, 155

Creep strain, 38, 149–150
  time-dependent, 30
Creep test, 38–39. *See also* Relaxation test
  Boltzmann superposition principle, 64–65
  Kelvin solid, 60, 61
  Maxwell fluid, 58, 59
Cross-ply laminate, 163
  configuration, 164
Crystalline polymer, 18, 20
  fringed-micelle structure, 20
  volume–temperature curve, 21
Curing, 254
  composite step cure, 257
  modes, 254
  thermoset adhesive, 254–255
  torque rheometry, 255–256
  undercure, 256
Cyanoacrylates, 251, 252. *See also* Adhesive

D

Dampers role, 319
Damping, 43, 46, 307. *See also* Isolation
  application of, 320–323
  constrained layer, 320
  dampers role, 319
  execises, 323
  free layer, 320
  fundamentals, 313
  layer, 320
  layers, 320
  loss factor, 309
  low, 311
  low, 311
  materials, 311
  materials for, 311
  methods, 309–310
  reduced frequency nomograph, 310
  ring down of system, 319
  rubber, 311, 313
  single-degree-of-freedom oscillator, 315
  spacers, 323
  thermomechanical spectrum in, 308–309
  two-degree-of-freedom oscillator, 317–319
  vibrating structure dynamics, 313
  viscoelastic spectrum, 311, 312, 313, 314
Damping analysis, 47. *See also* Polymer
  mechanical model, 48
  motion equation, 49
  relation to strain/stress, 50
d-configuration, 23
Deflection temperature under load (DTUL), 45
Delamination wear, 290
Design optimization, 129. *See also* Polymer selection
  computer solution, 135–137
  design variables, 129
  graphical solution, 130–135
  Microsoft Excel Solver Routine, 137–139
Dielectric constant, 344–345
Dielectric strength, 386
  ABS, 387
  aluminum, 387
  copper, 387
  EP, 387
  LDPE, 387
  PA, 387
  PVC, 387
  rubber, 387
Distortional energy density, 198
DMA. *See* Dynamic mechanical analysis (DMA)
Drop-weight impact test, 98
DTUL. *See* Deflection temperature under load (DTUL)
Ductile fracture theory, 109–110
Dupont's Teflon, 156
Dynamic mechanical analysis (DMA), 41
  complex modulus, 42
  flexure modulus, 46
  uses, 44
Dynamic tests, 40
  damping measure, 43–44, 46
  DMA test, 41, 42
  dynamic modulus, 42–43
  dynamic property data, 44–47
  torsional pendulum test, 41
  vibrating reed test, 41

E

Earthquake bearings, 322
Elastic modulus, 30, 167
  engineering orthotropic, 168
  magnitude, 168
  in Maxwell fluid, 58
  results from micromechanics analysis, 176
  in rubber, 71
Elastomers, 8–9, 320
Electric displacement, 344, 347
Electrical resistance, 386
  ABS, 387
  aluminum, 387
  copper, 387
  EP, 387
  LDPE, 387
  PA, 387
  PVC, 387
  rubber, 387
Electrical resistance of copper, 387
Electromagnetic dynamic mechanical analyzer, 41
Elongation
  polymers, 29–30
  single-fiber fragmentation test, 247

Energy directors, 262
crossed, 265
triangular, 262, 265
Engineering constants. *See* Poisson's ratio; Shear modulus; Young's modulus
Engineering materials, 2
EP. *See* Epoxy (EP)
Epoxy (EP), 8, 140, 252. *See also* Thermosets
adhesives, 250, 251
dielectric strength, 387
electrical resistance, 387
flexure modulus, 46
polymer ID, 140
thermal expansion values, 386
Ethane, 17
degree of polymerization, 18
rotation about C–C bond in, 17–18
Ethylene CH_2 molecule, 7
Eulerian strain, 47
Extrusion, 221. *See also* Polymer processing
die designs, 223
foam core pipe dimensions, 225
multimanifold die, 224
pressure drop, 223
product cooling, 222
PVC pipe manufacturing, 223–226
sandwich pipe, 224
shell bending stiffness, 225–226
single screw extruder, 221, 222
twin screw extruder, 221–223

F

Fan impeller blade, 147, 148. *See also* Polymer design applications
creep strain, 149–150
impact failure, 150–151
Fatigue failure. *See also* Creep failure; Polymer
causes, 85
cycles-to-failure vs. frequency, 86, 87
energy loss, 86
exercises, 92–93
fatigue testing, 87–89
hysteresis loop, 86
nonisothermal fatigue behavior, 85
phase angle, 86
strain cycle, 85
stress cycle, 85
Fatigue wear, 290
FDM. *See* Fused deposition modeling (FDM)
Ferroelectric materials, 341. *See also* Piezoelectric materials
Fiber
microbuckling, 199
Poisson ratio, 171
shear failure, 199
volume ratio, 170
Fiber/matrix
kink band formation, 199
splitting, 199
Fiberglass silation, 253
tensile test, 254
Fiber-reinforced composites, 5, 247
in macromechanics, 167
in macroscopic analysis, 166–167
mechanical analysis, 165, 166
principle directions for sliding, 295
reservation, 4
Fiber-reinforcing material, 162
Fixed–fixed beam, 366. *See also* Beam reactions
Flash temperatures, 300, 301
Fluorocarbon resins, 156. *See also* Polymer design applications
average pressure, 156
constrained bearing, 157–158
design with, 156
power loss, 157
radial wear, 156
running clearance, 157
Folded-chain structure
PE crystals, 20
Four-parameter model, 63–64, 68
Fractional model fit parameters, 69
Fracture instability theory, 110
adhesion-peel test, 114–115
in double-cantilever beam, 113–114
energy balance, 110
energy release rate, 111–113
fracture resistance, 111
Fracture toughness, 105. *See also* Impact strength; Polymer
brittle fracture theory, 106–109
ductile fracture theory, 109–110
exercises, 117–119
fracture instability theory, 110–115
values, 110, 111
Free layer, 320
Friction, 285
coefficient of friction, 285–286, 288–289, 294
environment on, 302–303
induced vibration, 303–304
polymer modification, 293–295
surface topography on, 302
Friction welding, 267
linear vibration welding, 267, 268, 269
orbital welding, 267, 269
spin welding, 269
Fringed micelle structure, 18, 20
Fused deposition modeling (FDM), 329–330
human skull with tumor, 339
properties, 332–333

G

General purpose (GP), 145
  grades, 153
General purpose phenolic (GP PF), 145
Glass fiber. *See* Fiberglass
Glass-reinforced resins, 231
GP. *See* General purpose (GP)
GP PF. *See* General purpose phenolic (GP PF)
Graphical solution, 130. *See also* Computer solution
  advantages, 132–133
  decrease of solution region, 135
  functions of time, 134
  geometric constraints, 132
  inequality constraints, 131
  Johnson buckling, 132
  minimum-weight nylon tubular column, 133–134
  normalized increases in functions, 135
  shell buckling, 123
  tubular column under compression, 130
Griffith theory. *See* Brittle fracture theory
Group technology, 227

H

HDPE. *See* High-density polyethylene (HDPE)
HDT. *See* Heat deflection temperature (HDT)
Heat calculation, total, 301
Heat deflection temperature (HDT), 333
Heated tool welding, 260–261
Hereditary integral
  formulation, 70
  for linear viscoelasticity, 64
Hertzian contact equations, 283
Heterogeneous body, 162
HI. *See* High impact (HI)
High impact (HI), 145
High-damping rubber, 311, 313
High-density polyethylene (HDPE), 9
  code number, 103
  creep rupture curves, 82
  identification codes, 10
  specific heat variation, 238
  thermal expansion values, 386
High-impact modified PS (HIPS), 98
  notched Izod impact strength values, 97
  rate effects, 102
HIPS. *See* High-impact modified PS (HIPS)
Homogeneous body, 161
Hoop ratio, 230
Hot gas welding, 272
Hydrophilic material, 243
Hydrophobic material, 243
Hysteresis, 346. *See also* Piezoelectric polymers
  applied electric field, 347
  complex piezoelectric coefficient, 347
  electric displacement, 347
  energy dissipation per volume per cycle, 347
  linear model, 348
  loop and energy loss, 86
  loop for PVDF, 347
  nonlinear model, 348, 349
Hysteretic losses. *See* Damping

I

Impact specimen analysis, 116
Impact strength, 95. *See also* Fracture toughness; Polymer
  Charpy impact specimens, 96
  drop-weight impact test, 98, 99
  ductile–brittle transition in, 97–98
  exercises, 117–119
  Izod impact specimens, 96
  measure, 95
  rate effects, 101–102
  stiffness and impact properties, 102–105
  thickness effects, 100, 101
  types S and L tension-impact specimens, 99
  values of polymers, 96–97, 100
Induction welding, 274, 275
  butt joint strength, 276
  polymer compatibility matrix, 275
Inertia, area moments of, 361
  circular section, 362
  hollow rectangular section, 361
  rectangular coordinate system, 361
  solid rectangular section, 361
  tubular section, 362
Infrared (IR), 9
Injection molding, 226, 279. *See also* Polymer processing
  ASTM test bar, 228
  components, 226–227
  group technology, 227
  machine, 226
  mold component family tree, 228
Inserts, 278–279
Interphase, 243
IR. *See* Infrared (IR)
IR. *See* Synthetic polyisoprene (IR)
Isochronous creep curves, 51, 52
Isolation, 316, 317. *See also* Damping
  earthquake bearings, 322
  isolator, 311, 320
Isolator, 311, 320
Isotropic body, 161–162
Izod impact specimens, 96
  notches uses, 95
Izod impact tests
  fracture mechanics analysis, 115, 116
  impact fatigue tests, 116

K

Kelvin solid, 55, 56. *See also* Maxwell fluid
buckling load for, 84
creep test, 60, 61
Prony series, 67
relaxation test, 61–63
retardation time, 60, 61
viscosity representation, 71
Kelvin solid model, 55, 56
Kirchoff
hypothesis, 210
strain, 47

L

Lamina, 162
mechanical properties, 185
orthotropic, 181–184
principal material axes, 182
types, 162, 163
Laminate, 162
MATLAB®, 367–375
Laminate macromechanics, 178
analysis algorithm, 214
anisotropic materials, contracted notation, 180–181
macroscopic analysis, 166–167
orientation code, 164–165
orthotropic lamina, 181–184
stress–strain relationships, 179, 184–190
Laminate micromechanical analysis, 166, 168
lamina elastic moduli results, 176–177
Poisson's ratio determination, major, 171–172
shear modulus, apparent, 174–176
tensile strength prediction, 177–178
Young's modulus determination, 168–170, 172–174
Laminate program, 377
compdata. txt file, 379
composite, 377
compout. txt file imported into excel, 380–381
input data, 378
Laminated object manufacturing (LOM), 331
materials, 334, 337
strengths and weaknesses, 331
Laser welding, 269, 272
Layer. *See* Lamina
LDPE. *See* Low density polyethylene (LDPE)
LEFM. *See* Linear elastic fracture mechanics (LEFM)
Linear elastic fluid model, 55, 56
Linear elastic fracture mechanics (LEFM), 109
Linear elastic solid model, 56
creep solution, 64
stress-strain behavior, 57
Linear vibration welding, 267–269
Linear viscoelastic polymer
creep compliance master curve, 39
relaxation modulus, 40
Linear viscoelasticity hereditary integral, 64
Linear viscous fluid, 55
creep solution, 64
stress-strain behavior, 57
Liquid-state densities of straight-chain hydrocarbons, 19
LOM. *See* Laminated object manufacturing (LOM)
Longitudinal modulus, 176, 183
Loss factor, 309
Low damping, 311
Low density polyethylene (LDPE), 3, 10, 140
code number, 103
creep rupture curves, 82
dielectric strength, 387
electrical resistance, 387
fracture toughness values, 110
polymer ID, 140
thermal expansion values, 386

M

Macromechanical failure strength properties, 200
Major Poisson's ratio, 176, 183
Mating test configurations, 300
Matrix
Poisson ratio, 171
volume ratio, 170
Matrix material, 162
fiber pullout test, 248
polymers as, 252
single-fiber fragmentation test, 247
Maximum stress theory of failure, 35, 197–198, 200
failure stress comparison, 204, 205
shortcomings of, 201
uniaxial strength curves, 202
Maxwell fluid model, 55, 56, 57. *See also* Kelvin solid
creep rate, 59
creep test, 58, 59
elastic modulus, 58
Prony series, 67
relaxation test, 58, 59
relaxation time, 59
Mechanics of materials approach, 168
deformation assumption, 169
hoop stress in thin cylinder, 155
Medium impact (MI), 128, 145
Medium impact phenolic (MI PF), 145
Melamine-formaldehyde (MF), 8
polymer ID, 140

Melting temperatures of straight-chain hydrocarbons, 19
Metallic powders, 145
Metals, 34
  alloys, 162
  creep, 33
  in mechanical design, 2
  von Mises criterion, 35
MF. *See* Melamine-formaldehyde (MF)
MI. *See* Medium impact (MI)
MI PF. *See* Medium impact phenolic (MI PF)
Microsoft Excel Solver routine, 137–139
Miner's rule, 81
Mineral fillers, 145
Minor Poisson's ratio, 183
Mittag–Leffler function, 68
Modulus of toughness, 95
Mohr static failure theory, modified, 34
MQ. *See* Silicone rubber (MQ)
Murnaghan strain, 47

N

Natural rubber (NR), 2, 8, 311
  glass transition temperatures, 22
  melting temperatures, 22
  polyisoprene, 312
  viscoelastic spectrum, 311–312
NBR. *See* Nitrile butadiene rubber (NBR)
NDE. *See* Nondestructive evaluation (NDE)
Neoprene. *See* Polychloroprene (CR)
Nitrile. *See* Nitrile butadiene rubber (NBR)
Nitrile butadiene rubber (NBR), 9
  glass transition temperatures, 22
  melting temperatures, 22
Nomograph, reduced frequency, 310–311
Nondestructive evaluation (NDE), 108
Nonmechanical strain, 342
N-parameter fluid model, 55, 56
N-parameter solid model, 55, 56
NR. *See* Natural rubber (NR)
Nylon. *See* Polyamide (PA)

O

Octave Code, 367–375
Orbital welding, 267
Organic materials, natural, 341
Orthotropic body, 162
Orthotropic lamina, 181–184
  compliance matrix, 183–184
  deformation, 189
  Hooke's Law for, 182
  longitudinal Young's modulus, 182–183
  major Poisson's ratio, 171, 183
  strength and failure theories, 199–200
Oscillations, 304

P

P.E. *See* Professional engineer (P.E.)
PA. *See* Polyamide (PA)
Parison, 229, 230
PBD. *See* Polybutadiene (PBD)
PC. *See* Polycarbonate (PC)
PC-PEI. *See* PC-polyetherimide (PC-PEI)
PC-polyetherimide (PC-PEI), 261
PDMS. *See* Polydimethyl siloxane (PDMS)
PDMS. *See* Silicone rubber (PDMS)
PE. *See* Polyethylene (PE)
Peak surface, 284
  parameter comparison, 285
PEEK. *See* Polyertherether ketone (PEEK)
Peel test, 247
Perfluoro alkoxy alkane (PFA), 156
PET. *See* Polyethylene terephthalate (PET)
PF resins. *See* Phenolic resins (PF resins)
PFA. *See* Perfluoro alkoxy alkane (PFA)
Phenolic resins (PF resins), 8, 140, 145. *See also* Polymer design applications
  applications for molded, 146
  classification, 145
  code number, 103
  composition, 146
  with fillers, 145
  folded-chain structure, 20
  GP PF, 145
  longitudinal fracture, 31
  MI PF, 145
  polymer ID, 140
  relative molecular mass, 18
  SPI grade number, 146
  wear factors and PV limits, 296
PIB. *See* Polyisobutylene (PIB)
Piezoelectric ceramics, 341
Piezoelectric materials, 341, 345–346. *See also* Ferroelectric materials
Piezoelectric polymers, 341
  bending actuator, 354–356
  cellular, 345
  composites, 348–350
  compressive stress gauge, 350–351
  exercise, 356–357
  human-powered electronics, 354
  hysteresis, 346–348
  material properties, 345–346
  pressure in cylinder, 352–354
  shear properties, 346
  stress gauge, 350
  tensile stress gauge, 352
Piezoelectric shear properties of biomaterials, 346
Piezoelectric strain behavior, 342
  axial and lateral strains, 343
  dielectric constant, 344–345

electric displacement, 344
nonmechanical strain, 342
shear strain, 342, 343, 344
stress–strain relationships, 344
total strain, 342
PIP. *See* Polyisoprene (PIP)
Plasticity, 286
Plastics, 1
identification, 9–11
machining, 338
organization of plastics industry, 12, 13
sales report, 15
screw torque, 278
uses in industries, 13, 14
world consumption, 13
Plateau surface, 284
parameter comparison, 285
Ply. *See* Lamina
Ply orientation designation, 163
PMMA. *See* Polymethyl methalcrylate (PMMA)
Poisson's ratio
atmospheric, 38
major, 171, 176
minor, 183, 193
value, 32
Poly(n-octyl methacrylate) storage compliance, 72
Polyacetal (POM), 8, 140
melting temperature, 23
welding compatibility matrix, 263, 268, 275
Polyamide (PA), 8
code number, 103
dielectric strength, 387
electrical resistance, 387
polymer ID, 140
thermal conductivity values, 385
thermal expansion values, 386
vibration weld strengths, 270
welding compatibility matrix, 263–264, 268, 275
Polybutadiene (PBD), 8, 312
glass transition temperatures, 22
melting temperatures, 22
Polycarbonate (PC), 8, 37, 140, 147. *See also* Polymer design applications
butt joint strength, 276
code number, 103
comparison, 147
creep rupture curves, 82
drop-test impact strength values, 100
fracture toughness values, 110
glass transition temperatures, 22
isochronous stress–strain curves, 149, 158
melting temperatures, 22
notched Izod impact strength values, 97
polymer ID, 140
properties, 332, 333
snap/fit design, 152
stretch limits BUR, 231
tensile impact strength values, 100
thermoplastic, 147
vibration weld strengths, 270
wear factors and PV limits, 296
welding compatibility matrix, 263, 268, 275
Polychloroprene (CR), 9
discovery, 3
glass transition temperatures, 22
melting temperatures, 22
thermal expansion values, 386
Polydimethyl siloxane (PDMS), 251
Polyertherether ketone (PEEK), 46
friction coefficient, 294
properties, 46
in stainless-steel mesh, 274
wear factors and PV limits, 296
welding compatibility matrix, 263–264, 268, 275
Polyester, 8
adhesives, 251. *See also* Adhesive
composite coefficient of friction, 299
experimental agreement use, 298
Polyethylene (PE), 8, 11
elongation under tension, 31
folded-chain structure, 20
glass transition temperatures, 22
melting temperatures, 22
notched Izod impact strength values, 97
properties, 19
relative molecular mass, 18
temperature rise and increase, 88
tensile impact strength values, 100
welding compatibility matrix, 263–264, 268, 275
Polyethylene terephthalate (PET), 9
code number, 103
flexure modulus, 46
glass transition temperatures, 22
identification codes, 10
melting temperatures, 22
stretch limits BUR, 231
tensile impact strength values, 100
uses, 231
welding compatibility matrix, 263–264, 268, 275
Polyisobutylene (PIB), 21
Polyisoprene (PIP), 312
Polymer, 1, 2, 7–8. *See also* Composite
adhesion, 243
adhesives, 243
advantages, 161
chains, 1, 18, 24
chemical structures, 14, 15–21
code numbers for materials, 103
creep curves, 51, 52

Polymer (*Continued*)
crystalline and amorphous, 20–21
curing, 254
damping, 307, 308
development, 3–4
ductile–brittle transition in, 97–98
exercises, 25–26, 51, 52–54
glass transition temperatures, 21, 22–24
heat generation in sliding, 300
large strain definitions, 47
melting temperatures, 21, 22–24
Nobel Prizes, 4
notch sensitivity factor, 90–91
PC and PF, 147
polymer-based products, 12
polymer-on-polymer sliding, 302
properties, 27–33, 38–47
raw materials, 11
stretch limits BUR, 231
tensile strength, 33
thermal conductivity, 385
time hardening creep, 50–51
toughness, 103
types, 8–9
wear factors and PV limits for, 296
Polymer design applications, 145
creep strain, 149–150
design with fluorocarbon resins, 156–158
exercises, 158–159
fan impeller blade, 147–148
impact failure, 150–151
phenolic resins with fillers, 145–146
polycarbonate, 147
PVC pipe example design, 152–156
snap/fit design, 151–152
Polymer fusing, 259, 260. *See also* Adhesion; Energy directors; Friction welding; Induction welding; Ultrasonic welding
butt joint, 262
circular cap joint, 265
exercises, 279
heated tool welding, 260–261
hot gas welding, 272
laser welding, 269
linear vibration welding, 267–269
orbital welding, 267
resistance welding, 273–274
shear joints, 265
spin welding, 269
strength of, 259–260
tongue and groove joint, 265
weldability diagram, 266
Polymer mathematical models, 57
Polymer mechanical models, 55–56
Polymer processing, 221
blow molding, 229
exercises, 241–242
extrusion, 221–223
injection molding, 226–228
PVC pipe manufacturing, 223–226
thermoforming, 228–229
vacuum forming, 228, 229
Polymer selection
exercises, 142–143
geometrical configurations, 122
loading conditions, 122
material availability, 122
material properties, 121–122
performance parameters, 122
rating factors, 128–129
rectangular beam in bending, 123–125
thermal gradient through beam, 126–128
Polymeric adhesives, 253. *See also* Adhesive
anaerobic, 251, 252
cyanoacrylates, 251
epoxy, 250–251
mechanical properties, 252
polyester, 251
polyurethane, 251
silicone, 251
Polymerization degree, 18
Polymethyl methalcrylate (PMMA), 3, 8, 37, 140
code number, 103
glass transition temperatures, 22
heated tool welding use, 261
melting temperatures, 22
polymer ID, 140
thermal expansion values, 386
vibration weld strengths, 271
welding compatibility matrix, 263, 268, 275
Polyphenylene sulfide (PPS), 276
wear factors and PV limits, 296
welding compatibility matrix, 263, 268, 275
Polyphenylsulfone (PPSF), 329, 332–333, 334
Polypropylene (PP), 8, 11, 15, 18
code number, 103
creep rupture curves, 82
fracture toughness values, 110
glass transition temperatures, 22
identification codes, 10
lap shear specimens, 274
melting temperatures, 22, 23
notched Izod impact strength values, 97
optimal weld strength, 266
polymer ID, 140
stretch limits BUR, 231
tensile impact strength values, 100
wear factors and PV limits, 296
welding compatibility matrix, 263–264, 268
Polystyrene (PS), 8, 15, 140
code number, 103
fracture toughness values, 110
glass transition temperatures, 22
identification codes, 10

melting temperatures, 22
notched Izod impact strength values, 97
polymer ID, 140
stretch limits BUR, 231
thermal expansion values, 386
welding compatibility matrix, 263, 268, 275
Polytetrafluorethylene (PTFE), 2, 8, 293–295
coefficient of friction, 156, 297
friction coefficient, 294
impact strength, 97
melting temperature, 23
polymer ID, 140
thermal expansion values, 386
on wear rate, 294, 297
Polyurethane (PUR), 8, 20, 99, 312–313
adhesives, 251, 252
code number, 103
melting temperature, 23
polymer ID, 140
viscoelastic spectrum for, 315
Polyvinyl alcohol (PVAL), 21
Polyvinyl chloride (PVC), 8, 11, 15, 152
advantages, 153
ASTM standard, 153, 224
code number, 103
creep failure curves, 81, 82
creep rupture, 82–83, 155
dielectric strength, 387
drop-test impact strength value, 100
electrical resistance, 387
GP grades, 153
melting temperature, 23
monomer, 15
notched Izod impact strength value, 97
patent, 3
polymer ID, 140, 141
rigid, 153, 154
stretch limits, 231
tensile impact strength values, 100
thermal expansion values, 386
working pressure, 153
Polyvinyl chloride (PVC) pipe. *See also* Polymer design applications
ASTM standards, 224
creep rupture strength, 155
design, 152–156
manufacture, 223–226
physical dimension, 155
pressure-rated pipe, 155
Polyvinylidene fluoride (PVDF), 341
electric dipoles, 342
hysteresis loop for, 347
permittivity data, 346
properties, 345
uniaxial stretching, 345
POM. *See* Polyacetal (POM)
Powdered Teflon PF, 145
PP. *See* Polypropylene (PP)
PPS. *See* Polyphenylene sulfide (PPS)
PPSF. *See* Polyphenylsulfone (PPSF)
Pressure drop, 223
Principal stress and strain directions, 199
Product development, 325, 326
Professional engineer (P. E.), 385
Prony series, 67
Prototyping process, 326
PS. *See* Polystyrene (PS)
PTFE. *See* Polytetrafluorethylene (PTFE)
PUR. *See* Polyurethane (PUR)
PV (product of pressure and velocity), 291, 304
curve for polymeric materials, 291
limit, 291–292, 296
PVAL. *See* Polyvinyl alcohol (PVAL)
PVC. *See* Polyvinyl chloride (PVC)
PVDF. *See* Polyvinylidene fluoride (PVDF)
Pyroelectric materials, 341. *See also* Piezoelectric materials

R

Random
copolymers, 22, 23
reinforcement configuration, 163
Rapid prototyping (RP), 325. *See also* Fused deposition modeling (FDM); Laminated object manufacturing (LOM); Selective laser sintering (SLS); Stereolithography (SLA)
additive model development, 328
in bioengineering field, 338
exercises, 339–340
materials, 332–334
Medtronic polycarbonate surgical ratchet, 327
Polaris snowmobile rack, 327
product development, tooling, and manufacture, 325–326
in reverse engineering, 338
slice, 328
subtractive model development, 328
system selection, 334, 338
techniques, 326–332
Rectangular beam
bending, 123–125, 126
material property and comparison, 128
thermal gradient through, 126–127
Reinforcement configuration, 163
Relaxation test, 40. *See also* Creep test
Kelvin solid, 61–63
Maxwell fluid, 58, 59
Resistance welding, 273–274
Room-temperature vulcanizing rubber. *See* Silicone
RP. *See* Rapid prototyping (RP)

Rubber. *See also* Coefficient of friction; Natural rubber (NR); Polyamide; Silicone
consumption, 13
dielectric strength, 387
elastic modulus, 71
electrical resistance, 387
Epoxy compatibility, 251
low-damping, 311
neoprene discovery, 3
nitrile, 303
shear modulus, 309
as thermosets, 1
Rule of mixtures, 170
inverse, 173, 176
longitudinal modulus, 176
macroscopic ply stress, 177
major Poisson's ratio, 172, 176

S

Safety, factor of, 34, 293
Safety approach, factor of, 34, 36–37
Sandwich beam, 206
bending, 206
stiffness, 207, 208
total weight of, 205
SBR. *See* Styrene–butadiene rubber (SBR)
Screws, 276
self-tapping, 277, 278
thread forming/tapping, 277
twin, 221
Selective laser sintering (SLS), 331–332
materials, 334, 337–338
Shear joints, 265
Shear modulus, 32, 176
apparent, 174
for isotropic material, 179
Shearing deformation, total, 174
Sheet molding compounds (SMCs), 228
Shift factors, 71
storage compliance, 73
WLF, 74
Silicone, 251, 394. *See also* Adhesive
as internal lubricant, 294
linear chain of, 16
mold, 338
polymer ID, 140
properties, 252
sealants, 1
Silicone rubber (MQ), 21
Silicone rubber (PDMS), 9
as filler, 293
Simply supported beam. *See also* Beam reactions
distributed load, 365–366
transverse intermediate span load, 365
transverse mid-span load, 365
Single-degree-of-freedom oscillator, 315
frequency–response function for, 317, 319
Single-fiber fragmentation test, 247–248
SLA. *See* Stereolithography (SLA)
SLI file. *See* SLIce files (SLI file)
Slice, 328
SLIce files (SLI file), 328
Slice software, 328
SLS. *See* Selective laser sintering (SLS)
Smart materials, 341. *See also* Piezoelectric materials
SMCs. *See* Sheet molding compounds (SMCs)
Snap/fit design, 151–152
Society of Plastics Industry (SPI), 146
Specific wear rate, 300
Spherulites, clusters of, 18
SPI. *See* Society of Plastics Industry (SPI)
Spin welding, 269
Stacking sequence, 163, 164
Staking, 266
Static failure theories, 33. *See also* Polymer
for brittle material, 34
for ductile materials, 35
factor of safety approach, 34, 36–37
stress tensor, 33
von Mises criterion, 35, 38
STereoLithgography file format files (STL files), 328
Stereolithography (SLA), 330. *See also* Selective laser sintering (SLS)
materials, 334, 335–336
STL files. *See* STereoLithgography file format files (STL files)
Stratasys, 329, 334
Stress tensor, 33
Stress–strain curve, 28. *See also* Elastic modulus; Tensile test
elastic modulus, 30
PC isochronous, 149, 158
strain rate effect, 28–29
Styrene–butadiene rubber (SBR), 8
Subtractive model development, 328
Super ventricular tachycardia (SVT), 341
Surface attraction, 244
Surface roughness, 243. *See also* Surface topography
effect of, 287, 302
Surface topography, 284
SVT. *See* Super ventricular tachycardia (SVT)
Synthetic polyisoprene (IR), 9
Synthetic rubber. *See* Polychloroprene (CR)

T

Teflon. *See* Polytetrafluorethylene (PTFE)
Tensile strength, 221
of brittle material, 33
of FDM materials, 332